SEMA SIMAI Springer Series

Volume 16

More information about this series at http://www.springer.com/series/10532

Daniele Boffi • Luca F. Pavarino • Gianluigi Rozza •
Simone Scacchi • Christian Vergara
Editors

Mathematical and Numerical Modeling of the Cardiovascular System and Applications

Springer

Editors
Daniele Boffi
Department of Mathematics "F. Casorati"
University of Pavia
Pavia, Italy

Luca F. Pavarino
Department of Mathematics "F. Casorati"
University of Pavia
Pavia, Italy

Gianluigi Rozza
SISSA mathLab, Mathematics Area
International School for Advanced Studies
Trieste, Italy

Simone Scacchi
Department of Mathematics
University of Milan
Milano, Italy

Christian Vergara
Department of Mathematics
Politecnico di Milano
Milano, Italy

ISSN 2199-3041 ISSN 2199-305X (electronic)
SEMA SIMAI Springer Series
ISBN 978-3-319-96648-9 ISBN 978-3-319-96649-6 (eBook)
https://doi.org/10.1007/978-3-319-96649-6

Library of Congress Control Number: 2018952632

This Springer imprint is published by the registered company Springer Nature Switzerland AG
The registered company address is: Gewerbestrasse 11, 6330 Cham, Switzerland

Preface

This book is dedicated to the outstanding career of Piero Colli Franzone, one of the founders of the field of computational electrocardiology. Piero graduated in mathematics from the University of Pavia in 1969. He subsequently became a researcher at the Institute of Numerical Analysis of the National Research Council (CNR) in Pavia (1971–1973), Associated Professor of Mathematical Analysis at the University of Pavia (1973–1981), and Full Professor of Numerical Analysis, first at the University of Udine (1981–1983) and then at the University of Pavia (1983–2016), where he is now Professor Emeritus. During his career, Piero has made pioneering contributions in the field of the inverse problem of electrocardiology, in the development of eikonal models for the propagation of electrical signals in cardiac tissue, in the development of an anisotropic source model for the simulation of electrocardiograms, and more recently in the modeling of cardiac mechanics and electromechanical coupling.

Although Piero's works have mainly focused on computational cardiac electrophysiology and mechanics, this book also contains contributions on the modeling and simulation of the entire cardiovascular system. The increasing use of numerical models and methods in the field of diagnosis/prognosis/therapy of cardiovascular diseases (the main cause of death in Western countries) is considered very important and increasingly appreciated by the scientific and medical community. However, its penetration into clinical practice is slowed by various cultural factors and methodological challenges, in particular by the high computational cost of the complex mathematical, multiscale, and nonlinear models used in computational cardiology.

The quantitative description of cardiovascular activity is based on the construction, analysis, and simulation of models of both the heart and the circulatory system. Cardiac models are composed of nonlinear multiphysical submodels that describe the three main components (electrophysiology, mechanics, and hemodynamics) of cardiac activity, with spatial and temporal scales very different from each other and with specific computational difficulties. The bioelectric models are based on systems of partial and ordinary differential equations that include both the microscopic models of the membrane ionic channels and the macroscopic cardiac models (bidomain, monodomain, and eikonal models). The mechanical models are based on nonlinear orthotropic elastic models for the large deformations of the almost incompressible cardiac tissue and active tension models for the generation of the myocyte contraction. The hemodynamic models are based on the equations of incompressible fluid dynamics in moving domains for blood flow in the cardiac chambers and in the vascular system. Another important process which is described by the interaction between fluid and structures is given by the cardiac valve modeling.

The numerical simulation of these complex multiscale models coupled together by various reaction and feedback terms is very intense, and several aspects are still problematic. The integrated simulation of the three main cardiac phases (bioelectrical, mechanical, and fluid dynamical) is currently a very competitive goal: several leading research groups in the world are working intensively toward this goal, which, if reached in the next few years, could have a major impact on our basic scientific knowledge (regarding cardiac arrhythmogenesis, myocardial infarction, ischemia, heart failure, valve pathologies, etc.), on health technologies, and on the design and development of new biomedical devices.

This volume contains selected invited papers from the conference "Mathematical and Numerical Modeling of the Cardiovascular System and Applications," held at the Aula Volta of the University of Pavia, Italy, on February 21–22, 2017 (http://matematica.unipv.it/pieroconference/), which was also the final workshop of the PRIN Project (funded by the Italian government) 2012HBLYE4 "Metodologie Innovative nella Modellistica Differenziale Numerica" coordinated by Alfio Quarteroni. The conference comprised contributions from recognized international experts in diverse fields of cardiovascular modeling and simulation. The event was sponsored by the Department of Mathematics of the Universities of Pavia and Milan, SISSA, the Polytechnic of Milan, the MIUR Grants PRIN 201289A4LX, PRIN 2012HBLYE4, the INdAM Grant Project GNCS 2017, and the Progetto Cariplo-Regione Lombardia: Problemi variazionali di evoluzione e trasporto ottimo.

The first chapter, entitled "A Distributed Lagrange Formulation of the Finite Element Immersed Boundary Method for Fluids Interacting with Compressible Solids" and coauthored by Daniele Boffi, Lucia Gastaldi, and Luca Heltai, reports on recent advances in the modeling and the numerical approximation of fluids interacting with compressible solids. A version of the Finite Element Immersed Boundary Method is presented which is based on a new variational formulation, thanks to the introduction of a distributed Lagrange multiplier. Stability estimates and numerical validation are included.

The second chapter, entitled "High-Order Operator-Splitting Methods for the Bidomain and Monodomain Models" and coauthored by Jessica Cervi and Raymond J. Spiteri, focuses on high-order operator splitting methods for the reaction-diffusion systems of computational electrocardiology. The methods considered are third and fourth order accurate and are based on use of a combination of explicit and implicit Runge–Kutta schemes. Numerical tests in one and three spatial dimensions and employing both simplified and detailed ionic models are performed to assess the accuracy of the methods.

The third chapter, entitled "Electro-mechanical Modeling and Simulation of Reentry Phenomena in the Presence of Myocardial Infarction" and coauthored by Piero Colli Franzone, Luca F. Pavarino, and Simone Scacchi, presents a review of the cardiac electromechanical coupling models and of the numerical methods for their discretizations. It then reports the results of numerical simulations focusing on the genesis of reentrant arrhythmias in the presence of infarct scars.

The fourth chapter, entitled "Ergotropic Effect in Cardiac Tissue After Electromagnetic and β-Adrenergic Stimulus" and coauthored by Lorenzo Fassina, Marisa Cornacchione, Maria E. Mognaschi, Giovanni Magenes, and Fabio Naro, demonstrates ergotropic effects in murine ventricular cardiomyocytes after electromagnetic field and/or isoproterenol stimulation by means of an innovative image processing technique. The study adds important and significant findings with regard to the structure–function relationship in spontaneously beating primary cultures of murine cardiomyocytes.

The fifth chapter, entitled "Role of Electrotonic Current in Excitable Cells" and coauthored by Emilio Macchi, Ezio Musso, and Stefano Rossi, provides a thorough overview of our understanding of electrotonic current in excitable cells. After a short introduction, the authors go through cable models of conduction, electrotonic currents and potentials, subthreshold current injection, the stimulation threshold, and propagating action potentials. Finally, the authors discuss electrotonic modulation of repolarization by the activation sequence and present some experimental evidence of acute modulation of the activation-recovery interval and the effective refractory period at a test site during ventricular drive and sinus rhythm in the normal rat heart.

The sixth chapter, entitled "Reduced Order Modeling for Cardiac Electrophysiology and Mechanics: New Methodologies, Challenges & Perspectives" and coauthored by Andrea Manzoni, Diana Bonomi, and Alfio Quarteroni, presents an extensive and detailed description of POD-Galerkin methods developed to speed up the solution of parametric coupled electromechanical problems arising in cardiac electrophysiology. In the first part, the authors introduce the models used to describe the electrophysiology and the mechanics of living tissues. Then, after presenting the full-order discretization and reduced-order techniques, they detail the methods used to tackle the challenges of the system considered and the numerical experiments.

The seventh chapter, entitled "Aortic Endovascular Surgery" and coauthored by M. Conti, S. Morganti, A. Finotello, R.M. Romarowski, A. Reali, and F. Auricchio, reports on the use of computer simulations to support two of the most routinely performed endovascular procedures: thoracic endovascular aortic repair (*TEVAR*) and transcatheter aortic valve implantation (*TAVI*). The authors report a short review of simulations (structural for TEVAR and both structural and fluid dynamics for TAVI) and then present two examples of applications.

The eighth chapter, entitled "Combined Parameter and Model Reduction of Cardiovascular Problems by Means of Active Subspaces and POD-Galerkin Methods" and coauthored by Marco Tezzele, Francesco Ballarin, and Gianluigi Rozza, deals with a first example of combination of a priori geometrical parameter space reduction carried out by the active subspace approach combined with a classic computational reduction method for blood flows, based on the POD-Galerkin technique, and applied over a parametrized shape of a patient-specific carotid artery bifurcation. This model problem could be seen as a proof of concept for future, more involved, applications in blood flows.

The ninth chapter, entitled "Extended Finite Elements Method for Fluid-Structure Interaction with an Immersed Thick Non-Linear Structure" and coauthored by Christian Vergara and Stefano Zonca, first introduces an overview of extended finite element methods for general interfaces and heterogeneous coupled problems. Then, it addresses the fluid-structure interaction problem for a thick immersed structure whose thickness is, however, often smaller than the fluid mesh characteristic size. In particular, the authors proposed an inexact Newton method to handle the case of a nonlinear structure described by finite elasticity. Finally, the chapter presents some numerical 3D results.

Pavia, Italy Daniele Boffi
Pavia, Italy Luca F. Pavarino
Trieste, Italy Gianluigi Rozza
Milano, Italy Simone Scacchi
Milano, Italy Christian Vergara
Giugno 2018

Contents

About the Editors

Daniele Boffi has been Professor of Numerical Analysis at the Department of Mathematics of the University of Pavia since 2005. He received his PhD in mathematics from the Universities of Brescia-Milano-Pavia, Italy (1996). He has been a Visiting Professor at several institutions. His primary research interests are related to the numerical approximation of partial differential equations, including mixed finite elements, eigenvalue problems, computational electromagnetism, and the interaction of fluids and solids.

Luca F. Pavarino is Professor of Numerical Analysis at the Department of Mathematics of the University of Pavia. He holds a PhD (1992) and master's (1990) in mathematics from the Courant Institute, NYU, USA, and a B.S. degree (1987) in mathematics from the University of Pavia. He was a Postdoctoral Research Associate at Rice University, USA, and Professor at the University of Milan. His current research focuses on domain decomposition methods for PDEs, parallel computing, and biomathematics.

Gianluigi Rozza is Full Professor of Numerical Analysis and Scientific Computing at SISSA, International School for Advanced Studies. He holds a degree in aerospace engineering from Politecnico di Milano (2002) and a PhD in applied mathematics from Ecole Polytechnique Federale de Lausanne (2005). He was a Postdoctoral Research Associate at the Massachusetts Institute of Technology (MIT) Center for Computational Engineering (2006–2008) and then a Researcher and Lecturer at EPFL (2008–2012). He is the author of over 100 scientific publications and recipient of the 2014 ECCOMAS young investigator Jacques Louis Lions Award in Computational Mathematics for researchers under the age of 40. Professor Rozza has been an Associate Editor of the *SIAM/ASA Journal of Uncertainty Quantification* since 2013, of the *SIAM Journal of Numerical Analysis* since 2015, and of *Computing and Visualization in Science* since 2016.

Simone Scacchi is Associate Professor of Numerical Analysis at the Department of Mathematics of the University of Milan. He received a degree in mathematics from the University of Milan (2004) and a PhD in mathematics and statistics from the University of Pavia (2008). He was a Postdoctoral Research Associate at the University of Pavia (2008) and then Assistant Professor at the University of Milan (2008–2015).

Christian Vergara has been Associate Professor of Numerical Analysis at MOX, Dipartimento di Matematica, Politecnico di Milano, since 2015. He received a PhD in mathematical engineering from Politecnico di Milano in 2006. His primary research interests are the numerical approximation of fluid-structure interaction problems, the numerical modeling of the cardiovascular system, both in terms of blood dynamics and electrical activity of the heart, and the application to cases of clinical interests.

A Distributed Lagrange Formulation of the Finite Element Immersed Boundary Method for Fluids Interacting with Compressible Solids

Daniele Boffi, Lucia Gastaldi, and Luca Heltai

Abstract We present a distributed Lagrange multiplier formulation of the Finite Element Immersed Boundary Method to couple incompressible fluids with compressible solids. This is a generalization of the formulation presented in Heltai and Costanzo (Comput. Methods Appl. Mech. Eng. 229/232:110–127, 2012), that offers a cleaner variational formulation, thanks to the introduction of distributed Lagrange multipliers, that act as intermediary between the fluid and solid equations, keeping the two formulation mostly separated. Stability estimates and a brief numerical validation are presented.

1 Introduction

Fluid-structure interaction (FSI) problems are everywhere in engineering and biological applications, and are often too complex to be solved analytically.

Well established techniques, like the Arbitrary Lagrangian Eulerian (ALE) framework [9], enable the numerical simulation of FSI problems by coupling computational fluid dynamics (CFD) and computational structural dynamics (CSD), through the introduction of a deformable fluid grid, whose movement at the interface

D. Boffi (✉)
Dipartimento di Matematica "F. Casorati", Università di Pavia, Pavia, Italy

Department of Mathematics and System Analysis, Aalto University, Helsinki, Finland
e-mail: daniele.boffi@unipv.it

L. Gastaldi
DICATAM, Università di Brescia, Brescia, Italy
e-mail: lucia.gastaldi@unibs.it

L. Heltai
SISSA, Trieste, Italy
e-mail: heltai@sissa.it

D. Boffi et al. (eds.), *Mathematical and Numerical Modeling of the Cardiovascular System and Applications*, SEMA SIMAI Springer Series 16,
https://doi.org/10.1007/978-3-319-96649-6_1

is driven by the coupling with the CSD simulation, while the interior is deformed arbitrarily, according to some smooth deformation operator.

Although this technique has reached a great level of robustness, whenever changes of topologies are present in the physics of the problem, or when freely floating objects (possibly rotating) are considered, a deforming fluid grid that follows the solid may no longer be a feasible solution strategy.

The Immersed Boundary Method (IBM), introduced by Peskin in the seventies [10] to simulate the interaction of blood flow with heart valves, addressed this issue by reformulating the coupled FSI problem as a "reinforced fluid" problem, where the CFD system is solved everywhere (including in the regions occupied by the solid), and the presence of the solid is taken into account in the fluid as a (singular) source term (see [11] for a review).

In the original IBM, the body forces expressing the FSI are determined by modeling the solid body as a network of elastic fibers with a contractile element, where each point of the fiber acts as a singular force field (a Dirac delta distribution) on the fluid.

Finite element variants of the IBM were first proposed, almost simultaneously, by [2, 15], and [16]. However, only [2] exploited the variational definition of the Dirac delta distribution directly in the finite element (FE) approximations.

Such approximation was later generalized to thick hyper-elastic bodies (as opposed to fibers) [3], where the constitutive behavior of the immersed solid is assumed to be incompressible and visco-elastic with the viscous component of the solid stress response being identical to that of the fluid.

In [7], the authors present a formulation that is applicable to problems with immersed bodies of general topological and constitutive characteristics, without the use of Dirac delta distributions, and with interpolation operators between the fluid and the solid discrete spaces that guarantee semi-discrete stability estimates and strong consistency. Such formulation has been successfully used [13] to match standard benchmark tests [14].

The incompressible version of the FSI model presented in [7] can be seen as a special case of the Distributed Lagrange Multiplier method, introduced in [5]. In this paper we present a novel distributed Lagrange multiplier method that generalizes the compressible model introduced in [7].

We provide a general variational framework for Immersed Finite Element Methods (IFEM) based on the distributed Lagrange multiplier formulation that is suitable for general fluid structure interaction problems.

2 Setting of the Problem

Let $\Omega \subset \mathbb{R}^d$, with $d = 2, 3$, be a fixed open bounded polyhedral domain with Lipschitz boundary which is split into two time dependent subdomains Ω_t^f and Ω_t^s, representing the fluid and the solid regions, respectively. Hence Ω is the interior of

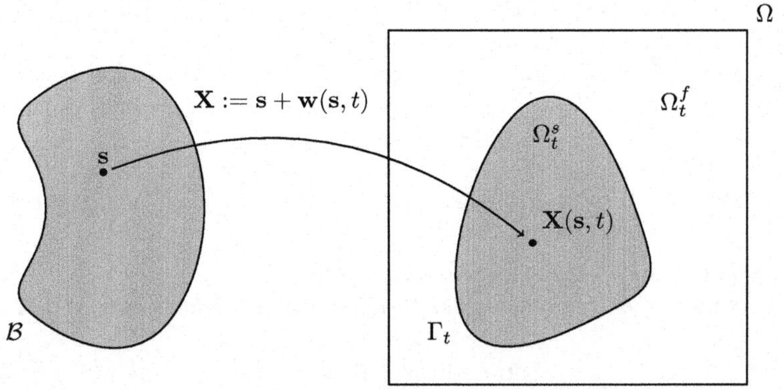

Fig. 1 Geometrical configuration of the FSI problem

$\overline{\Omega}_t^f \cup \overline{\Omega}_t^s$ and we denote by $\Gamma_t = \overline{\Omega}_t^f \cap \overline{\Omega}_t^s$ the moving interface between the fluid and the solid regions. For simplicity, we assume that $\Gamma_t \cap \partial\Omega = \emptyset$ (Fig. 1).

The current position of the solid Ω_t^s is the image of a reference domain $\mathcal{B}$ through a mapping $\mathbf{X} : \mathcal{B} \to \Omega_t^s$. The displacement of the solid is indicated by $\mathbf{w}$ and for any point $\mathbf{x} \in \Omega_t^s$ we have $\mathbf{x} = \mathbf{X}(\mathbf{s}, t) = \mathbf{X}_0(\mathbf{s}) + \mathbf{w}(\mathbf{s}, t)$ for some $\mathbf{s} \in \mathcal{B}$, where $\mathbf{X}_0 : \mathcal{B} \to \Omega_0^s$ denotes the mapping providing the initial configuration of Ω_t^s. For convenience, we assume that the reference domain $\mathcal{B}$ coincides with Ω_0^s so that $\mathbf{X}(s, t) = \mathbf{s} + \mathbf{w}(\mathbf{s}, t)$ and $\mathcal{B} \subset \Omega$. From the above definitions, $\mathbb{F} = \nabla_s \mathbf{X} = \mathbb{I} + \nabla_s \mathbf{w}$ stands for the deformation gradient and $J = \det(\mathbb{F})$ for its determinant.

We denote by $\mathbf{u}_f : \Omega \to \mathbb{R}^d$ and $p_f : \Omega \to \mathbb{R}$ the fluid velocity and pressure and assume that the solid velocity $\mathbf{u}_s$ is equal to the material velocity of the solid, that is

$$\mathbf{u}_s(\mathbf{x}, t) = \frac{\partial \mathbf{X}(\mathbf{s}, t)}{\partial t}\Big|_{\mathbf{x}=\mathbf{X}(\mathbf{s},t)} = \frac{\partial \mathbf{w}(\mathbf{s}, t)}{\partial t}\Big|_{\mathbf{x}=\mathbf{X}(\mathbf{s},t)} \tag{1}$$

We indicate with generic symbols $\mathbf{u}$, p, ρ the *Eulerian* fields, depending on $\mathbf{x}$ and t, which describe the velocity, pressure, and density, respectively, of a material particle (be it solid or fluid). By $\dot{\mathbf{u}}$ we denote the material derivative of $\mathbf{u}$, which in Eulerian coordinates is expressed by

$$\dot{\mathbf{u}}(\mathbf{x}, t) = \frac{\partial \mathbf{u}}{\partial t}(\mathbf{x}, t) + \mathbf{u}(\mathbf{x}, t) \cdot \nabla \mathbf{u}(\mathbf{x}, t). \tag{2}$$

In the Lagrangian framework, the material derivative coincides with the partial derivative with respect to time, so that

$$\dot{\mathbf{w}}(\mathbf{s}, t) = \partial \mathbf{w}(\mathbf{s}, t)/\partial t.$$

Continuum mechanics models are based on the conservation of three main properties: linear momentum, angular momentum, and mass.

When expressed in Lagrangian coordinates, mass conservation is guaranteed if the reference mass density ρ_0 is time independent. If expressed in Eulerian coordinates, however, mass conservation takes the form

$$\dot{\rho} + \rho \operatorname{div} \mathbf{u} = 0 \quad \text{in } \Omega, \tag{3}$$

and should be included in the system's equations.

Conservation of both momenta can be expressed, in *Eulerian* coordinates, as:

$$\rho \dot{\mathbf{u}} = \operatorname{div} \boldsymbol{\sigma} + \rho \mathbf{b} \quad \text{in } \Omega, \tag{4}$$

where ρ is the mass density distribution, $\mathbf{u}$ the velocity, $\boldsymbol{\sigma}$ is the Cauchy stress tensor (its symmetry implies that conservation of angular momenta is guaranteed by Eq. (4)), and $\mathbf{b}$ describes the external force density per unit mass acting on the system. Such a description is common to all continuum mechanics models (see, for example, [6]). The equations for fluids and solids are different according to their constitutive behavior, i.e., according to how $\boldsymbol{\sigma}$ relates to $\mathbf{u}$, $\mathbf{w}$, or p.

If the material is incompressible, it can be shown that $\operatorname{div} \mathbf{u} = 0$ everywhere, and the material derivative of the density is constantly equal to zero (from Eq. (3)). Notice that this does not imply that ρ is constant (neither in time nor in space), and it is still in general necessary to include Eq. (3) in the system.

For incompressible materials, however, the volumetric part of the stress tensor can be interpreted as a Lagrange multiplier associated with the incompressibility constraints. For incompressible fluids, the stress is decomposed into $\boldsymbol{\sigma}_f = -p_f \mathbb{I} + \nu_f \mathbf{D}\mathbf{u}_f$ where $\mathbf{u}_f$ is the fluid velocity, p_f the pressure, $\nu_f > 0$ is the viscosity coefficient and $\mathbf{D}\mathbf{u}_f = (1/2)\left(\boldsymbol{\nabla}\,\mathbf{u}_f + (\boldsymbol{\nabla}\,\mathbf{u}_f)^\top\right)$.

Hence the equations describing the fluid motion are the well-known Navier–Stokes equations, that is:

$$\begin{aligned} &\rho_f \dot{\mathbf{u}}_f - \operatorname{div}(\nu_f \mathbf{D}\mathbf{u}_f) + \nabla\, p_f = \rho_f \mathbf{b} && \text{in } \Omega_t^f \\ &\operatorname{div} \mathbf{u}_f = 0 && \text{in } \Omega_t^f, \end{aligned} \tag{5}$$

where we assumed that ρ_f is constant throughout Ω_t^f.

As far as the solid is concerned, we assume that it is composed by a viscous elastic material so that the Cauchy stress tensor can be decomposed into the sum of two contributions: a viscous part and a pure elastic part, as follows

$$\boldsymbol{\sigma}_s = \boldsymbol{\sigma}_s^v + \boldsymbol{\sigma}_s^e = \nu_s \mathbf{D}\mathbf{u}_s + \boldsymbol{\sigma}_s^e. \tag{6}$$

Here $\mathbf{u}_s$ is the solid velocity, $\nu_s \geq 0$ is the solid viscosity coefficient and $\boldsymbol{\sigma}_s^e$ denotes the elastic part of the solid Cauchy stress tensor. We assume this elastic part to behave hyper-elastically, i.e., we assume that there exists an elastic potential energy density $W(\mathbb{F})$ such that $W(\mathbf{R}\mathbb{F}) = W(\mathbb{F})$ for any rotation $\mathbf{R}$, that represents the amount of elastic energy stored in the current solid configuration, and that depends only on its deformation gradient $\mathbb{F} = \nabla_s \mathbf{X}$.

When expressed in Lagrangian coordinates, a possible measure for the elastic part of the stress is the so called first Piola–Kirchhoff stress tensor, defined as the Fréchet derivative of W w.r.t. to $\mathbb{F}$, i.e.:

$$\mathbb{P}_s^e := \frac{\partial W}{\partial \mathbb{F}}. \tag{7}$$

The first Piola–Kirchhoff stress tensor allows one to express the conservation of linear momentum in Lagrangian coordinates as

$$\rho_{s_0} \ddot{\mathbf{w}} = \mathbf{Div}(\mathbb{P}) + \rho_{s_0} \mathbf{B} \quad \text{in } \mathcal{B}, \tag{8}$$

where, similarly to its Eulerian counterpart, we assume that $\mathbb{P}$ is decomposed in an *additive* way into its viscous part $\mathbb{P}_s^v$ and into its elastic part $\mathbb{P}_s^e$, defined in Eq. (7).

For any portion $\mathcal{P} \subset \mathcal{B}$ of the solid (with outer normal $\mathbf{N}$) deformed to $\mathcal{P}_t$ (with outer normal $\mathbf{n}$), the following relation between $\mathbb{P}$ and $\boldsymbol{\sigma}$ holds:

$$\int_{\mathcal{P}} \mathbb{P}\mathbf{N}\, \mathbf{d}\Gamma_s = \int_{\mathcal{P}_t} \boldsymbol{\sigma}\mathbf{n}\, \mathbf{d}\Gamma_x, \qquad \forall \mathcal{P} \subset \mathcal{B}, \qquad \mathcal{P}_t := \mathbf{X}(\mathcal{P}, t), \tag{9}$$

that is, we can express pointwise the *viscous* part of the solid stress in Lagrangian coordinates by rewriting the first Piola–Kirchhoff stress $\mathbb{P}_s^v$ in terms of $\boldsymbol{\sigma}_s^v := \nu_s \mathbf{D}\mathbf{u}_s$, and the *hyper-elastic* part of the solid stress in Eulerian coordinates by expressing the Cauchy stress $\boldsymbol{\sigma}_s^e$ in terms of $\mathbb{P}_s^e := \partial W / \partial \mathbb{F}$:

$$\begin{aligned} \mathbb{P}_s^v(\mathbf{s}, t) &:= J \quad \boldsymbol{\sigma}_s^v(\mathbf{x}, t)\mathbb{F}^{-\top}(\mathbf{s}, t) \qquad && \text{for } \mathbf{x} = \mathbf{X}(\mathbf{s}, t) \\ \boldsymbol{\sigma}_s^e(\mathbf{x}, t) &:= J^{-1}\mathbb{P}_s^e(\mathbf{s}, t)\, \mathbb{F}^{\top}(\mathbf{s}, t) \qquad && \text{for } \mathbf{x} = \mathbf{X}(\mathbf{s}, t). \end{aligned} \tag{10}$$

With these definitions, the conservation of linear momentum for the solid equation can be expressed either in Lagrangian coordinates as

$$\rho_{s_0} \ddot{\mathbf{w}} = \mathbf{Div}\, \mathbb{P}_s^v + \mathbf{Div}\, \frac{\partial W}{\partial \mathbb{F}} + \rho_{s_0} \mathbf{B} \quad \text{in } \mathcal{B} \tag{11}$$

or in Eulerian coordinates as

$$\rho_s \dot{\mathbf{u}}_s - \operatorname{div}(\nu_s \mathbf{D}\mathbf{u}_s) - \operatorname{div} \boldsymbol{\sigma}_s^e = \rho_s \mathbf{b} \quad \text{in } \Omega_t^s. \tag{12}$$

Notice that the conservation of mass for the solid equation is a simple kinematic identity that derives from the fact that ρ_{s_0} does not depend on time, i.e., we have

$$\frac{\dot{\rho}_s}{\rho_s} + \operatorname{div} \mathbf{u}_s = 0 \qquad \text{in } \Omega_t^s, \tag{13}$$

or, equivalently,

$$\rho_s(\mathbf{x}, t) = \rho_{s_0}(\mathbf{s})/J(\mathbf{s}, t) \quad \text{for } \mathbf{x} = \mathbf{X}(\mathbf{s}, t), \tag{14}$$

that is:

$$\operatorname{div} \mathbf{u}_s(\mathbf{x}, t) = \frac{\dot{J}}{J}(\mathbf{s}, t) \qquad \text{for } \mathbf{x} = \mathbf{X}(\mathbf{s}, t). \tag{15}$$

The equations in the solid and in the fluid are coupled through interface conditions along Γ_t, which enforce the continuity of the velocity, corresponding to the no-slip condition between solid and fluid, and the balance of the normal stress:

$$\begin{aligned} \mathbf{u}_f &= \mathbf{u}_s && \text{on } \Gamma_t \\ \boldsymbol{\sigma}_f \mathbf{n}_f + \boldsymbol{\sigma}_s \mathbf{n}_s &= 0 && \text{on } \Gamma_t, \end{aligned} \tag{16}$$

where $\mathbf{n}_f$ and $\mathbf{n}_s$ denote the outward unit normal vector to Ω_t^f and Ω_t^s, respectively.

The system is complemented with initial and boundary conditions. The boundary $\partial\Omega$ is split into two parts $\partial\Omega_D$ and $\partial\Omega_N$, where Dirichlet and Neumann conditions are imposed, respectively, with $\partial\Omega_D \cap \partial\Omega_N = \emptyset$. Since we assumed that $\partial\Omega \cap \Gamma_t = \emptyset$, the initial and boundary conditions are given by:

$$\begin{aligned} \mathbf{u}_f(0) &= \mathbf{u}_{f0} && \text{in } \Omega_0^f \\ \mathbf{u}_s(0) &= \mathbf{u}_{s0} && \text{in } \Omega_0^s \\ \mathbf{X}(0) &= \mathbf{X}_0 && \text{in } \mathcal{B} \\ \mathbf{u}_f &= \mathbf{u}_g && \text{on } \partial\Omega_D \\ (\nu_f \mathbf{D}\mathbf{u}_f - p_f \mathbb{I})\mathbf{n}_f &= \boldsymbol{\tau}_g && \text{on } \partial\Omega_N. \end{aligned} \tag{17}$$

In the following, we shall consider $\mathbf{u}_g = 0$ on $\partial\Omega_D$, for simplicity, and we define the space $H^1_{0,D}(\Omega)^d$ as the space of functions in $H^1(\Omega)^d$ such that their trace on $\partial\Omega_D$ is zero.

By multiplying by $\mathbf{v} \in H^1_{0,D}(\Omega)^d$ the first equation in (5) and (12), integrating by part and using the second interface condition in (16), we arrive at the following weak form of the fluid-structure interaction problem: find $\mathbf{u}_f$, p_f, $\mathbf{u}_s$, and $\mathbf{w}$ such that (1), $(16)_1$ and (17) are satisfied and it holds:

$$
\begin{aligned}
&\int_{\Omega_t^f} \rho_f(\dot{\mathbf{u}}_f - \mathbf{b})\mathbf{v}\,\mathbf{dx} + \int_{\Omega_t^f} \nu_f \mathbf{D}\mathbf{u}_f : \mathbf{D}\mathbf{v}\,\mathbf{dx} - \int_{\Omega_t^f} p_f \operatorname{div}\mathbf{v}\,\mathbf{dx} \\
&\quad + \int_{\Omega_t^s} \rho_s(\dot{\mathbf{u}}_s - \mathbf{b})\mathbf{v}\,\mathbf{dx} + \int_{\Omega_t^s} \nu_s \mathbf{D}\mathbf{u}_s : \mathbf{D}\mathbf{v}\,\mathbf{dx} + \int_{\Omega_t^s} \boldsymbol{\sigma}_s^e : \mathbf{D}\mathbf{v}\,\mathbf{dx} \\
&\quad = \int_{\partial\Omega_N} \boldsymbol{\tau}_g \cdot \mathbf{v}\,\mathrm{da} \qquad \forall \mathbf{v} \in H^1_{0,D}(\Omega)^d \\
&\int_{\Omega_t^f} \operatorname{div}\mathbf{u}_f q_f\,\mathbf{dx} = 0 \qquad \forall q_f \in L^2(\Omega_t^f)
\end{aligned}
\tag{18}
$$

where the notation $\dot{\mathbf{u}}$ indicates the material derivative with respect to time. Using a *fictitious domain* approach and extending the arguments of [5], we transform problem (18) by introducing the following new unknowns. Thanks to the continuity condition for fluid and solid velocity, we define $\mathbf{u} \in H^1_{0,D}(\Omega)^d$ and $p \in L^2(\Omega)$ as

$$
\mathbf{u} = \begin{cases} \mathbf{u}_f & \text{in } \Omega_t^f \\ \mathbf{u}_s & \text{in } \Omega_t^s. \end{cases}, \qquad p = \begin{cases} p_f & \text{in } \Omega_t^f \\ p_s = 0 & \text{in } \Omega_t^s. \end{cases}. \tag{19}
$$

Notice that the pressure field does not have any physical meaning in the solid, and it is (weakly) imposed to be zero.

Then we can write:

$$
\begin{aligned}
&\int_{\Omega} \rho_f \dot{\mathbf{u}}\mathbf{v}\,\mathbf{dx} + \int_{\Omega} \nu_f \mathbf{D}(\mathbf{u}) : \mathbf{D}(\mathbf{v})\,\mathbf{dx} - \int_{\Omega} p \operatorname{div}\mathbf{v}\,\mathbf{dx} + \int_{\Omega_t^s} (\rho_s - \rho_f)\dot{\mathbf{u}}\mathbf{v}\,\mathbf{dx} \\
&\quad + \int_{\Omega_t^s} (\nu_s - \nu_f)\mathbf{D}\mathbf{u} : \mathbf{D}\mathbf{v}\,\mathbf{dx} + \int_{\Omega_t^s} \boldsymbol{\sigma}_s^e : \mathbf{D}\mathbf{v}\,\mathbf{dx} + \int_{\Omega_t^s} p \operatorname{div}\mathbf{v}\,\mathbf{dx} \\
&\quad = \int_{\Omega} \mathbf{f}\mathbf{v}\,\mathbf{dx} + \int_{\partial\Omega_N} \boldsymbol{\tau}_g \cdot \mathbf{v}\,\mathrm{da} \qquad \forall \mathbf{v} \in H^1_{0,D}(\Omega)^d \\
&\int_{\Omega} \operatorname{div}\mathbf{u}\, q\,\mathbf{dx} - \int_{\Omega_t^s} \operatorname{div}\mathbf{u}\, q\,\mathbf{dx} + \int_{\Omega_t^s} \frac{1}{\kappa} p q\,\mathbf{dx} = 0 \qquad \forall q \in L^2(\Omega)
\end{aligned}
\tag{20}
$$

where

$$
\mathbf{f} = \begin{cases} \rho_f \mathbf{b} & \text{in } \Omega_t^f \\ \rho_s \mathbf{b} & \text{in } \Omega_t^s \end{cases},
$$

and κ plays the role of a bulk modulus constant.

In the solid the Lagrangian framework should be preferred, hence we transform the integrals over Ω_t^s into integrals on the reference domain $\mathcal{B}$. Recalling (1) and (14) the equations in (20) are rewritten in the following form:

$$\begin{aligned}
&\int_\Omega \rho_f \dot{\mathbf{u}}(t)\mathbf{v}\,\mathrm{d}\mathbf{x} + \int_\Omega \nu_f \mathbf{D}\mathbf{u}(t) : \mathbf{D}\mathbf{v}\,\mathrm{d}\mathbf{x} - \int_\Omega p\,\mathrm{div}\,\mathbf{v}\,\mathrm{d}\mathbf{x} \\
&\quad + \int_\mathcal{B} (\rho_{s_0} - \rho_f J)\ddot{\mathbf{w}}(t)\mathbf{v}(\mathbf{X}(\mathbf{s},t))\,\mathrm{d}\mathbf{s} + V\Big(\dot{\mathbf{w}}(t), \mathbf{v}(\mathbf{X}(\mathbf{s},t))\Big) \\
&\quad + \int_\mathcal{B} \mathbb{P}^e_s(t) : \nabla_s \mathbf{v}(\mathbf{X}(\mathbf{s},t))\,\mathrm{d}\mathbf{s} + \int_\mathcal{B} Jp(\mathbf{X}(\mathbf{s},t),t)\mathbb{F}^{-\top} : \nabla_s \mathbf{v}(\mathbf{X}(\mathbf{s},t))\,\mathrm{d}\mathbf{s} \\
&\qquad = \int_\Omega \mathbf{f}(t)\mathbf{v}\,\mathrm{d}\mathbf{x} + \int_{\partial\Omega_N} \boldsymbol{\tau}_g \cdot \mathbf{v}\,\mathrm{d}\mathrm{a} \qquad \forall \mathbf{v} \in H^1_{0,D}(\Omega)^d \\
&\int_\Omega \mathrm{div}\,\mathbf{u}(t)\,q\,\mathrm{d}\mathbf{x} - \int_\mathcal{B} J\,q(\mathbf{X}(\mathbf{s},t))\mathbb{F}^{-\top} : \nabla_s \dot{\mathbf{w}}(t)\,\mathrm{d}\mathbf{s} \\
&\qquad + \int_\mathcal{B} \frac{1}{\kappa} p(t) q J\,\mathrm{d}\mathbf{x} = 0 \qquad \forall q \in L^2(\Omega)
\end{aligned} \tag{21}$$

where, for all $\mathbf{X}, \mathbf{z} \in H^1(\mathcal{B})^d$

$$V(\mathbf{X},\mathbf{z}) = \frac{\nu_s - \nu_f}{4} \int_\mathcal{B} (\nabla_s \mathbf{X}\mathbb{F}^{-1} + \mathbb{F}^{-\top}\nabla_s \mathbf{X}^\top) : (\nabla_s \mathbf{z}\mathbb{F}^{-1} + \mathbb{F}^{-\top}\nabla_s \mathbf{z}^\top) J\,\mathrm{d}\mathbf{s}. \tag{22}$$

Since $\mathbf{X}(t) : \mathcal{B} \to \Omega^s_t$ is one to one and belongs to $W^{1,\infty}(\mathcal{B})^d$, $\mathbf{z} = \mathbf{v}(\mathbf{X}(t))$ is an arbitrary element of $H^1(\mathcal{B})^d$ when $\mathbf{v}$ varies in $H^1_{0,D}(\Omega)^d$.

Let $\boldsymbol{\Lambda}$ be a functional space to be defined later on and $\mathbf{c} : \boldsymbol{\Lambda} \times H^1(\mathcal{B})^d \to \mathbb{R}$ a bilinear form such that

$$\begin{aligned}
&\mathbf{c} \text{ is continuous on } \boldsymbol{\Lambda} \times H^1(\mathcal{B})^d \\
&\mathbf{c}(\boldsymbol{\mu}, \mathbf{z}) = 0 \text{ for all } \boldsymbol{\mu} \in \boldsymbol{\Lambda} \text{ implies } \mathbf{z} = 0.
\end{aligned} \tag{23}$$

For example, we can take $\boldsymbol{\Lambda}$ as the dual space of $H^1(\mathcal{B})^d$ and define $\mathbf{c}$ as the duality pairing between $H^1(\mathcal{B})^d$ and $(H^1(\mathcal{B})^d)'$, that is:

$$\mathbf{c}(\boldsymbol{\mu}, \mathbf{z}) = \langle \boldsymbol{\mu}, \mathbf{z} \rangle \quad \forall \boldsymbol{\mu} \in (H^1(\mathcal{B})^d)',\ \mathbf{z} \in H^1(\mathcal{B})^d. \tag{24}$$

Alternatively, one can set $\boldsymbol{\Lambda} = H^1(\mathcal{B})^d$ and define

$$\mathbf{c}(\boldsymbol{\mu}, \mathbf{z}) = (\nabla_s \boldsymbol{\mu}, \nabla_s \mathbf{z})_\mathcal{B} + (\boldsymbol{\mu}, \mathbf{z})_\mathcal{B} \quad \forall \boldsymbol{\mu},\ \mathbf{z} \in H^1(\mathcal{B})^d. \tag{25}$$

With the above definition for $\mathbf{c}$, we introduce an unknown $\boldsymbol{\lambda} \in \boldsymbol{\Lambda}$ such that

$$
\begin{aligned}
\mathbf{c}(\boldsymbol{\lambda}, \mathbf{z}) = &\int_{\mathcal{B}} (\rho_{s0} - \rho_f J)\ddot{\mathbf{w}}(t)\mathbf{z}\,\mathrm{ds} + V(\dot{\mathbf{w}}(t), \mathbf{z}) \\
&+ \int_{\mathcal{B}} \mathbb{P}(\mathbb{F}(t)) : \nabla_s \mathbf{z}\,\mathrm{ds} + \int_{\mathcal{B}} Jp(\mathbf{X}(\mathbf{s}, t), t)\mathbb{F}^{-\top} : \nabla_s \mathbf{z}\,\mathrm{ds} \qquad \forall \mathbf{z} \in H^1(\mathcal{B})^d.
\end{aligned} \tag{26}
$$

Hence we can write the following problem.

Problem 1 Let us assume that $\mathbf{u}_0 \in H^1_{0,D}(\Omega)^d$, $\mathbf{X}_0 \in W^{1,\infty}(\mathcal{B})^d$, and that for all $t \in [0, T]$ $\boldsymbol{\tau}_g(t) \in H^{-1/2}(\partial\Omega_N)$, and $\mathbf{f}(t) \in L^2(\Omega)$. For almost every $t \in]0, T]$, find $(\mathbf{u}(t), p(t)) \in H^1_{0,D}(\Omega)^d \times L^2(\Omega)$, $\mathbf{w}(t) \in H^1(\mathcal{B})^d$, and $\boldsymbol{\lambda}(t) \in \boldsymbol{\Lambda}$ such that it holds

$$
\begin{aligned}
&\rho_f(\dot{\mathbf{u}}(t), \mathbf{v}) + a(\mathbf{u}(t), \mathbf{v}) - (\operatorname{div}\mathbf{v}, p(t)) \\
&\quad + \mathbf{c}(\boldsymbol{\lambda}(t), \mathbf{v}(\mathbf{X}(t))) = (\mathbf{f}(t), \mathbf{v}) + (\boldsymbol{\tau}_g(t), \mathbf{v})_{\partial\Omega_N} \quad \forall \mathbf{v} \in H^1_{0,D}(\Omega)^d && \text{(27a)} \\
&-(\operatorname{div}\mathbf{u}(t), q) + (J\, q(\mathbf{X}(\mathbf{s}, t))\mathbb{F}^{-\top}, \nabla_s \dot{\mathbf{w}}(t))_{\mathcal{B}} \\
&\quad - \frac{1}{\kappa}(Jp(t), q)_{\mathcal{B}} = 0 \qquad \forall q \in L^2(\Omega) && \text{(27b)} \\
&\left(\delta_\rho \ddot{\mathbf{w}}(t), \mathbf{z}\right)_{\mathcal{B}} + (\mathbb{P}^e_s(t), \nabla_s \mathbf{z})_{\mathcal{B}} + V(\dot{\mathbf{w}}(t), \mathbf{z}) \\
&\quad + (Jp(\mathbf{X}(\mathbf{s}, t), t)\mathbb{F}^{-\top}, \nabla_s \mathbf{z})_{\mathcal{B}} - \mathbf{c}(\boldsymbol{\lambda}(t), \mathbf{z}) = 0 \ \forall \mathbf{z} \in H^1(\mathcal{B})^d && \text{(27c)} \\
&\mathbf{c}\left(\boldsymbol{\mu}, \mathbf{u}(\mathbf{X}(\cdot, t), t) - \dot{\mathbf{w}}(t)\right) = 0 \qquad \forall \boldsymbol{\mu} \in \boldsymbol{\Lambda} && \text{(27d)} \\
&\mathbf{X}(\mathbf{s}, t) = \mathbf{s} + \mathbf{w}(\mathbf{s}, t) \quad \text{for } \mathbf{s} \in \mathcal{B} && \text{(27e)} \\
&\mathbf{u}(0) = \mathbf{u}_0 \quad \text{in } \Omega, \qquad \mathbf{X}(0) = \mathbf{X}_0 \quad \text{in } \mathcal{B}. && \text{(27f)}
\end{aligned}
$$

Here $\delta_\rho = \rho_{s0} - \rho_f J$, $(\cdot, \cdot)$ and $(\cdot, \cdot)_{\mathcal{B}}$ stand for the scalar product in $L^2(\Omega)$ and $L^2(\mathcal{B})$, respectively, and

$$
\begin{aligned}
a(\mathbf{u}, \mathbf{v}) &= (\nu_f \mathbf{D}\mathbf{u}, \mathbf{D}\mathbf{v}) \quad \forall \mathbf{u}, \mathbf{v} \in H^1_{0,D}(\Omega)^d \\
(\boldsymbol{\tau}_g, \mathbf{v})_{\partial\Omega_N} &= \int_{\partial\Omega_N} \boldsymbol{\tau}_g \cdot \mathbf{v}\,\mathrm{da} \\
(\mathbb{P}, \mathbb{Q})_{\mathcal{B}} &= \int_{\mathcal{B}} \mathbb{P} : \mathbb{Q}\,\mathrm{ds} \quad \text{for } \mathbb{P}, \mathbb{Q} \text{ tensors in } L^2(\mathcal{B}).
\end{aligned}
$$

Proposition 1 *Let* $(\mathbf{u}, p, \mathbf{w}, \boldsymbol{\lambda})$ *be a solution of Problem 1. We have that* $p(t) = 0$ *in* Ω^s_t *for* $t \in]0, T]$ *and* $(\operatorname{div}\mathbf{u}, q)_{\Omega^f_t} = 0$ *for all* $q \in L^2(\Omega^f_t)$.

Proof The constraint in (27d) together with (23) implies that $\mathbf{u}(t) = \dot{\mathbf{w}}(t)$ in Ω_t^s. Using this fact and changing variable in the last two integrals in (27b), we arrive at

$$-(\operatorname{div}\mathbf{u}(t), q) + \int_{\Omega_t^s} \operatorname{div}\mathbf{u}(t) q \,\mathrm{d}\mathbf{x} - \int_{\Omega_t^s} \frac{1}{\kappa} p(t) q \,\mathrm{d}\mathbf{x} = 0 \,\forall q \in L^2(\Omega).$$

Taking $q = p(t)$ in Ω_t^s and vanishing in Ω_t^f, we end up with

$$\int_{\Omega_t^s} \frac{1}{\kappa} p^2(t) \,\mathrm{d}\mathbf{x} = 0,$$

which implies that $p(t) = 0$ in Ω_t^s. Taking $q = 0$ in Ω_t^f in (27b) Ω_t^s we obtain that the velocity is divergence free in the fluid domain.

Since the pressure variable does not carry any physical meaning, the choice of the constant κ is arbitrary. In this work we choose κ so that the entries of the mass matrix of the pressure have the same scale w.r.t. the other entries of the problem, i.e., we select κ to be equal to the bulk modulus of the solid problem.

The following theorem gives the estimate of the energy.

Theorem 1 *Let* $(\mathbf{u}, p, \mathbf{w}, \boldsymbol{\lambda})$ *be the solution of Problem 1, then the following estimate holds true*

$$\begin{aligned} &\frac{1}{2}\frac{d}{dt}\|\rho^{1/2}\mathbf{u}(t)\|_{0,\Omega}^2 + \|\nu^{1/2}\mathbf{D}\mathbf{u}(t)\|_{0,\Omega}^2 + \frac{d}{dt}\int_B W(\mathbb{F}(t))\,\mathrm{d}\mathbf{s} \\ &\qquad \le C\left(\|\mathbf{f}(t)\|_{0,\Omega}^2 + \|\boldsymbol{\tau}_g(t)\|_{H^{-1/2}(\partial\Omega_N)}\right). \end{aligned} \tag{28}$$

Here

$$\rho = \begin{cases} \rho_f & \text{in } \Omega_t^f \\ \rho_s(\mathbf{x}, t) & \text{in } \Omega_t^s \end{cases}, \qquad \nu = \begin{cases} \nu_f & \text{in } \Omega_t^f \\ \nu_s & \text{in } \Omega_t^s \end{cases}.$$

Proof We take the following test functions in the equations listed in Problem 1: $\mathbf{v} = \mathbf{u}(t)$, $q = -p(t)$, $\mathbf{z} = \partial\mathbf{w}(t)/\partial t$, and $\boldsymbol{\mu} = -\boldsymbol{\lambda}(t)$. Summing up all the equations, and taking into account Proposition 1 and the constraint in (27d) together with (23), we have

$$\begin{aligned} &\rho_f \int_\Omega \dot{\mathbf{u}}(t)\cdot\mathbf{u}(t)\,\mathrm{d}\mathbf{x} + \nu_f\|\mathbf{D}\mathbf{u}(t)\|_{0,\Omega}^2 + (\nu_s - \nu_f)\|\mathbf{D}\mathbf{u}\|_{0,\Omega_t^s}^2 \\ &\qquad + \frac{1}{2}\int_B \delta_\rho \frac{\partial}{\partial t}(\dot{\mathbf{w}}(t))^2\,\mathrm{d}\mathbf{s} + \left(\mathbb{P}_s^e(t), \nabla_s\,\dot{\mathbf{w}}(t)\right)_B \\ &\qquad = (\mathbf{f}(t), \mathbf{u}(t)) + (\boldsymbol{\tau}_g(t), \mathbf{u}(t))_{\partial\Omega_N}. \end{aligned} \tag{29}$$

Using again the constraint (27d), we can deal with the first integral on the second line as follows:

$$
\begin{aligned}
\frac{1}{2}\int_B \delta_\rho \frac{\partial}{\partial t}(\dot{\mathbf{w}})^2\,\mathrm{ds} &= \frac{1}{2}\int_B \rho_{s0}\frac{\partial}{\partial t}(\dot{\mathbf{w}})^2\,\mathrm{ds} - \frac{1}{2}\int_B \rho_f J\frac{\partial}{\partial t}(\dot{\mathbf{w}})^2\,\mathrm{ds}\\
&= \frac{1}{2}\frac{d}{dt}\int_B \rho_{s0}(\dot{\mathbf{w}})^2\,\mathrm{ds} - \frac{1}{2}\frac{d}{dt}\int_{\mathcal{B}} \rho_f J\dot{\mathbf{w}}\,\mathrm{ds} + \frac{1}{2}\int_B \rho_f\frac{\partial J}{\partial t}(\dot{\mathbf{w}})^2\,\mathrm{ds}\\
&= \frac{1}{2}\frac{d}{dt}\int_{\Omega_t^s}(\rho_s(\mathbf{x},t)-\rho_f)\mathbf{u}^2(\mathbf{x},t)\,\mathrm{d}\mathbf{x} + \frac{1}{2}\int_{\Omega_t^s}\rho_f(\operatorname{div}\mathbf{u})\mathbf{u}^2(\mathbf{x},t)\,\mathrm{d}\mathbf{x}.
\end{aligned}
$$

We add this relation to the first integral in (29), and we take into account the definition of the material derivative; hence we obtain after integration by parts

$$
\begin{aligned}
&\rho_f\int_\Omega \dot{\mathbf{u}}(t)\cdot\mathbf{u}(t)\,\mathrm{d}\mathbf{x} + \frac{1}{2}\int_B \delta_\rho\frac{\partial}{\partial t}(\dot{\mathbf{w}}(t))^2\,\mathrm{ds}\\
&= \frac{\rho_f}{2}\int_\Omega\left(\frac{\partial\mathbf{u}^2(t)}{\partial t} + \mathbf{u}(t)\cdot\nabla\,\mathbf{u}(t)^2\right)\mathrm{d}\mathbf{x} + \frac{1}{2}\frac{d}{dt}\int_{\Omega_t^s}(\rho_s(\mathbf{x},t)-\rho_f)\mathbf{u}^2(t)\,\mathrm{d}\mathbf{x}\\
&\qquad + \frac{1}{2}\int_{\Omega_t^s}\rho_f(\operatorname{div}\mathbf{u}(t))\mathbf{u}^2(t)\,\mathrm{d}\mathbf{x}\\
&= \frac{\rho_f}{2}\frac{d}{dt}\int_\Omega \mathbf{u}^2(t)\,\mathrm{d}\mathbf{x} - \frac{\rho_f}{2}\int_\Omega(\operatorname{div}\mathbf{u}(t))\mathbf{u}^2(t)\,\mathrm{d}\mathbf{x}\\
&\quad + \frac{1}{2}\frac{d}{dt}\int_{\Omega_t^s}(\rho_s(\mathbf{x},t)-\rho_f)\mathbf{u}^2(t)\,\mathrm{d}\mathbf{x} + \frac{1}{2}\int_{\Omega_t^s}\rho_f(\operatorname{div}\mathbf{u}(t))\mathbf{u}^2(t)\,\mathrm{d}\mathbf{x}\\
&= \frac{1}{2}\frac{d}{dt}\int_\Omega \rho\mathbf{u}^2(t)\,\mathrm{d}\mathbf{x}
\end{aligned}
\tag{30}
$$

Notice that the quantity $(1/2)\rho\mathbf{u}^2$ represents the kinetic energy density per unit volume.

From the definition of the deformation gradient, we deduce that

$$
\begin{aligned}
\left(\mathbb{P}_s^e(t),\nabla_s\,\dot{\mathbf{w}}(t)\right)_{\mathcal{B}} &= \int_{\mathcal{B}}\mathbb{P}_s^e(t):\frac{\partial\mathbb{F}}{\partial t}(t)\,\mathrm{ds} = \int_{\mathcal{B}}\frac{\partial W(\mathbb{F}(t))}{\partial\mathbb{F}}:\frac{\partial\mathbb{F}(t)}{\partial t}\,\mathrm{ds}\\
&= \int_B\frac{\partial W(\mathbb{F}(t))}{\partial t}\,\mathrm{ds} = \frac{d}{dt}\int_{\mathcal{B}}W(\mathbb{F}(t))\,\mathrm{ds}.
\end{aligned}
$$

The integral on the right hand side represents the elastic energy of the solid.

Putting together these expressions with (29) we obtain the desired stability estimate.

3 Time Discretization

For an integer N, let $\Delta t = T/N$ be the time step, and $t_n = n\Delta t$ for $n = 0, \ldots, N$. We discretize the time derivatives with backward finite differences and use the following notation:

$$\partial_t \mathbf{u}^{n+1} = \frac{\mathbf{u}^{n+1} - \mathbf{u}^n}{\Delta t}, \qquad \partial_{tt} \mathbf{u}^{n+1} = \frac{\mathbf{u}^{n+1} - 2\mathbf{u}^n + \mathbf{u}^{n-1}}{\Delta t^2}.$$

By linearization of the nonlinear terms, we arrive at the following semi-discrete problem:

Problem 2 For $n = 1, \ldots, N$, find $(\mathbf{u}^n, p^n) \in H^1_{0,D}(\Omega)^d \times L^2(\Omega)$,$\mathbf{w}^n \in H^1(\mathcal{B})^d$, and $\boldsymbol{\lambda}^n \in \boldsymbol{\Lambda}$ such that it holds

$$\begin{aligned}
&\rho_f(\partial_t \mathbf{u}^{n+1}, \mathbf{v}) + b(\mathbf{u}^n, \mathbf{u}^{n+1}, \mathbf{v}) + a(\mathbf{u}^{n+1}, \mathbf{v}) \\
&\quad - (\operatorname{div} \mathbf{v}, p^{n+1}) + \mathbf{c}(\boldsymbol{\lambda}^{n+1}, \mathbf{v}(\mathbf{X}^n)) \\
&\quad = (\mathbf{f}^{n+1}, \mathbf{v}) + (\boldsymbol{\tau}_g^{n+1}, \mathbf{v})_{\partial\Omega_N} \qquad \forall \mathbf{v} \in H^1_{0,D}(\Omega)^d \qquad (31a)
\end{aligned}$$

$$\begin{aligned}
&- (\operatorname{div} \mathbf{u}^{n+1}, q) + (J^n q(\mathbf{X}^n)(\mathbb{F}^n)^{-\top}, \nabla_s \partial_t \mathbf{w}^{n+1})_{\mathcal{B}} \\
&\quad - \frac{1}{\kappa}(J^n p^{n+1}, q)_{\mathcal{B}} = 0 \qquad \forall q \in L^2(\Omega) \qquad (31b)
\end{aligned}$$

$$\begin{aligned}
&\left(\delta_\rho^n \partial_{tt} \mathbf{w}^{n+1}, \mathbf{z}\right)_{\mathcal{B}} + (\mathbb{P}_s^{en+1}, \nabla_s \mathbf{z})_{\mathcal{B}} + V_n(\partial_t \mathbf{w}^{n+1}, \mathbf{z}) \\
&\quad + (J^n p^{n+1}(\mathbf{X}^n)(\mathbb{F}^n)^{-\top}, \nabla_s \mathbf{z})_{\mathcal{B}} - \mathbf{c}(\boldsymbol{\lambda}^{n+1}, \mathbf{z}) = 0 \quad \forall \mathbf{z} \in H^1(\mathcal{B})^d \qquad (31c)
\end{aligned}$$

$$\mathbf{c}\left(\boldsymbol{\mu}, \mathbf{u}^{n+1}(\mathbf{X}^n) - \partial_t \mathbf{w}^{n+1}\right) = 0 \qquad \forall \boldsymbol{\mu} \in \boldsymbol{\Lambda} \qquad (31d)$$

$$\mathbf{X}^{n+1} = \mathbf{s} + \mathbf{w}^{n+1} \quad \text{in } \mathcal{B} \qquad (31e)$$

$$\mathbf{u}^0 = \mathbf{u}_0 \quad \text{in } \Omega, \qquad \mathbf{X}^0 = \mathbf{X}_0 \quad \text{in } \mathcal{B}. \qquad (31f)$$

In (31c), $V_n(\mathbf{X}, \mathbf{z})$ indicates that the Jacobian J and the deformation gradient $\mathbb{F}$ appearing in the definition of V (see (22)) are computed at time t_n, moreover, using the definition of the first Piola–Kirchhoff stress tensor (7) we set $\mathbb{P}_s^{en+1} = \frac{\partial W}{\partial \mathbb{F}}(\mathbb{F}^{n+1})$ where different linearizations can be obtained according to the specific hyper-elastic model in use.

In (31a) we have $b(\mathbf{u}, \mathbf{v}, \mathbf{w}) = \rho_f(\mathbf{u} \cdot \nabla \mathbf{v}, \mathbf{w})$.

Problem 2 can be written in operator matrix form as follows:

$$
\left[\begin{array}{cc|c|c}
M_f/\Delta t + A_f^n & B_f^\top & 0 & C_f^{n\top} \\
B_f & M_p^n & B_s^n & 0 \\
\hline
0 & B_s^{n\top}/\Delta t & M_s^n/\Delta t^2 + A_s^{v,n}/\Delta t + A_s^{e,n} & -C_s^\top \\
\hline
C_f^n & 0 & -C_s/\Delta t & 0
\end{array}\right]
\left[\begin{array}{c}
\mathbf{u}^{n+1} \\
p^{n+1} \\
\hline
\mathbf{w}^{n+1} \\
\hline
\boldsymbol{\lambda}^{n+1}
\end{array}\right]
$$

$$
= \left[\begin{array}{c}
M_f\mathbf{u}^n/\Delta t + \mathbf{f}^{n+1} + \boldsymbol{\tau}_g^{n+1} \\
B_s^{n\top}\mathbf{w}^n/\Delta t \\
\hline
M_s^n(2\mathbf{w}^n - \mathbf{w}^{n-1})/\Delta t^2 + A_s^v\mathbf{w}^n/\Delta t \\
\hline
C_s\mathbf{w}^n/\Delta t
\end{array}\right]
$$

where

$$
\langle M_f\mathbf{u}, \mathbf{v}\rangle = (\mathbf{u}, \mathbf{v}), \qquad \langle M_s^n\mathbf{w}, \mathbf{z}\rangle = (\delta_\rho^n\mathbf{w}, \mathbf{z})_\mathcal{B}, \qquad \langle M_p^n p, q\rangle = \frac{1}{\kappa}(J^n p, q)_\mathcal{B},
$$

$$
\langle A_f^n\mathbf{u}, \mathbf{v}\rangle = b(\mathbf{u}^n, \mathbf{u}, \mathbf{v}) + a(\mathbf{u}, \mathbf{v}),
$$

$$
\langle A_s^{v,n}\mathbf{w}, \mathbf{z}\rangle = V_n(\mathbf{w}, \mathbf{z}), \qquad \langle A_s^{e,n}\mathbf{w}, \mathbf{z}\rangle = (2\,\nabla_s\mathbf{w}\cdot\frac{\partial W}{\partial \mathbb{C}}(\mathbb{F}^n), \mathbf{z})_\mathcal{B},
$$

$$
\langle B_f\mathbf{v}, q\rangle = -(\operatorname{div}\mathbf{v}, q), \qquad \langle B_s^n\mathbf{z}, q\rangle = (J^n q(\mathbf{X}^n)(\mathbb{F}^n)^{-\top}, \nabla_s\mathbf{z})_\mathcal{B},
$$

$$
\langle C_f^n\mathbf{v}, \boldsymbol{\mu}\rangle = \mathbf{c}(\boldsymbol{\mu}, \mathbf{v}(\mathbf{X}^n)), \qquad \langle C_s\mathbf{w}, \boldsymbol{\mu}\rangle = \mathbf{c}(\boldsymbol{\mu}, \mathbf{w}).
$$

We have used the second Piola-Kirchhoff stress tensor to define the operator associated to the elastic stress tensor in the solid and $\mathbb{C} = \mathbb{F}^\top\mathbb{F}$.

By choosing the bilinear operator $\mathbf{c}$ as in Eq. (24), and solving explicitly for the Lagrange multiplier $\boldsymbol{\lambda}$ (inverting Eq. (27c) and substituting its result in the conservation equation), it is possible to recover exactly the system described in [7]. In this case, the discretisation of the two different formulations provide exactly the same numerical results.

On the other hand, the formulation we present here is more general, and allows greater flexibility in the choice of both the operator $\mathbf{c}$ and the numerical resolution scheme. The numerical results in [13] provide so far only a qualitative comparison of the two formulations.

4 Space-Time Discretization

In this section we introduce the finite element spaces needed for the space-time discretization of Problem 1. For this we consider two independent meshes in Ω and in $\mathcal{B}$. We use a stable pair $\mathbf{V}_h \times Q_h$ of finite elements to discretize fluid velocity and pressure. We denote by h the maximum edge size. For example, we can take a mesh made of simplexes and use the Hood–Taylor element of lowest degree or we can subdivide the domain Ω in parallelepipeds and apply the $Q_2 - P_1$ element. The main difference among the above finite elements consists in the fact that the pressure for the Hood–Taylor element is continuous while it is discontinuous in the $Q_2 - P_1$ case. It is well known that discontinuous pressure approximation enjoys better local mass conservation; possible strategies to improve mass conservation are presented in [4]. In the solid, we take a regular mesh of simplexes, where h_s stands for the maximum edge size and denote by $\mathbf{S}_h$ the finite element space containing piecewise polynomial continuous functions. Finally, the finite element space $\boldsymbol{\Lambda}_h$ for the Lagrange multiplier $\boldsymbol{\lambda}$ coincides with $\mathbf{S}_h$.

Problem 3 Given $\mathbf{u}_{0,h} \in \mathbf{V}_h$ and $\mathbf{X}_{0,h} \in \mathbf{S}_h$, for $n = 1, \ldots, N$ find $(\mathbf{u}_h^n, p_h^n) \in \mathbf{V}_h \times Q_h$, $\mathbf{w}_h^n \in \mathbf{S}_h$, and $\boldsymbol{\lambda}_h^n \in \boldsymbol{\Lambda}_h$ such that it holds

$$\begin{aligned}
&\rho_f(\partial_t \mathbf{u}_h^{n+1}, \mathbf{v}) + b(\mathbf{u}_h^n, \mathbf{u}_h^{n+1}, \mathbf{v}) + a(\mathbf{u}_h^{n+1}, \mathbf{v}) \\
&\quad - (\operatorname{div} \mathbf{v}, p^{n+1}) + \mathbf{c}(\boldsymbol{\lambda}_h^{n+1}, \mathbf{v}(\mathbf{X}^n)) \\
&\quad = (\mathbf{f}^{n+1}, \mathbf{v}) + (\boldsymbol{\tau}_g^{n+1}, \mathbf{v})_{\partial\Omega_N} && \forall \mathbf{v} \in \mathbf{V}_h && \text{(32a)}\\
&- (\operatorname{div} \mathbf{u}_h^{n+1}, q) + (J^n\, q(\mathbf{X}^n)(\mathbb{F}^n)^{-\top}, \nabla_s\, \partial_t \mathbf{w}_h^{n+1})_{\mathcal{B}} \\
&\quad - \frac{1}{\kappa}(J^n p^{n+1}(\mathbf{X}^n), q)_{\mathcal{B}} = 0 && \forall q \in Q_h && \text{(32b)}\\
&(\delta_\rho^n \partial_{tt} \mathbf{w}^{n+1}, \mathbf{z})_{\mathcal{B}} + (\mathbb{P}_{sh}^{en+1}, \nabla_s \mathbf{z})_{\mathcal{B}} + V_{h,n}(\partial_t \mathbf{w}_h^{n+1}, \mathbf{z}) \\
&\quad + (J^n p^{n+1}(\mathbf{X}^n)(\mathbb{F}^n)^{-\top}, \nabla_s \mathbf{z})_{\mathcal{B}} - \mathbf{c}(\boldsymbol{\lambda}_h^{n+1}, \mathbf{z}) = 0 && \forall \mathbf{z} \in \mathbf{S}_h && \text{(32c)}\\
&\mathbf{c}\left(\boldsymbol{\mu}, \mathbf{u}^{n+1}(\mathbf{X}^n) - \partial_t \mathbf{w}_h^{n+1}\right) = 0 && \forall \boldsymbol{\mu} \in \boldsymbol{\Lambda}_h && \text{(32d)}\\
&\mathbf{X}_h^{n+1} = \mathbf{s} + \mathbf{w}_h^{n+1} \quad \text{in } \mathcal{B} && && \text{(32e)}\\
&\mathbf{u}^0 = \mathbf{u}_{0,h} \quad \text{in } \Omega, \qquad \mathbf{X}^0 = \mathbf{X}_{0,h} \quad \text{in } \mathcal{B}. && && \text{(32f)}
\end{aligned}$$

5 Numerical Validation

We have implemented the model described in the previous sections in a custom C++ code, based on the open source finite element software library `deal.II` [1], and on a modification of the code presented in [8].

We use the classic inf-sup stable pair of finite elements $Q_2 - P_1$, for the velocity and pressure in the fluid part, and standard Q_2 elements for the solid part.

In the following test cases the solid is modeled as a compressible neo-Hookean material, and the constitutive response function for the first Piola–Kirchhoff stress of the solid is given as

$$\mathbb{P}_s^e = \mu^e \left(\mathbb{F} - J^{-2\nu/(1-2\nu)} \mathbb{F}^{-\mathrm{T}} \right), \tag{33}$$

where μ^e is the shear modulus and ν is the Poisson's ratio for the solid.

The tests are designed to validate the correct handling of the coupling between incompressible fluids and compressible solids.

This paper is not focused on the behavior of the numerical solution with respect to the involved discretization parameters (for instance, fluid and solid mesh size). We expect the results to be not too different from what we obtained in the case of incompressible solids (see, for instance, [5]).

5.1 *Recovery and Rise of an Initially Compressed Disk in a Stationary Fluid*

This test case has been presented initially in Heltai and Costanzo [7] and simulates the motion of a compressible, viscoelastic disk having an undeformed radius R. The disk is initially squeezed (i.e. initial dimension of disk is $\lambda_0 R$ with $\lambda_0 < 1$), and left to recover in a square control volume of edge length L that is filled with an initially stationary, viscous fluid (see Fig. 2). The referential mass density of the disk is less than that of the surrounding fluid, i.e. $\rho_{\mathrm{s}_0} < \rho_{\mathrm{f}}$. The bottom and the sides of the control volume have homogeneous Dirichlet boundary condition, while the top side has homogeneous Neumann boundary conditions. This ensures that fluid can freely enter and exit the control volume along the top edge. As the disk tries to recover its undeformed state, it expands and causes the flux to exit the control volume. Thus the change in the area of the disk from its initial state, over a certain interval of time,

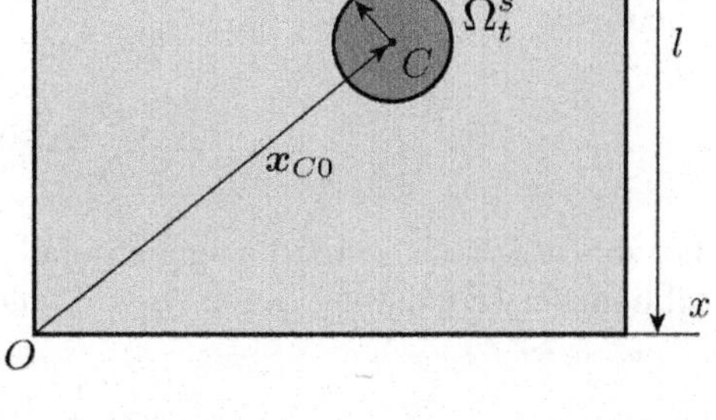

Fig. 2 Initial configuration of the system with the compressible disk

matches the amount of fluid efflux from the control volume over the same interval of time. We use this idea to estimate the error in our numerical method.

In this test, the following parameters have been used: $R = 0.125\,\mathrm{m}$, $l = 1.0\,\mathrm{m}$, $\rho_{s_0} = 0.8\,\mathrm{kg/m^3}$, $\rho_f = 1.0\,\mathrm{kg/m^3}$, $\mu^e = 20\,\mathrm{Pa}$, $\mu_s = 2.0\,\mathrm{Pa{\cdot}s}$, $\mu_f = 0.01\,\mathrm{Pa{\cdot}s}$, $\lambda_0 = 0.7$. The body force on the system is $\mathbf{b} = (0, -10)\,\mathrm{m/s^2}$. The initial location of the center of the disk is $\mathbf{x}_{C0} = (0.6, 0.4)\,\mathrm{m}$. We have used $Q_2 - P_1$ elements for the control volume and the mesh comprises 1024 cells and 11522 DoFs. Q_2 elements have been used for disk and the mesh comprises 224 cells with 1894 DoFs.

In Fig. 3a–f, we can see the velocity field over the entire control volume due to the motion of the disk as well as the pressure in the fluid for several instants of time spanning the duration of the simulation. The initial deformation of the disk is such that the density of the disk is greater than that of the fluid. But as soon as the disk is released, it starts to expand while remaining at almost the same vertical height, instead of descending, as can be seen from Fig. 4a. This causes a surge of fluid outflow that as shown in Fig. 4b. The expansion of the disk results in an increased buoyancy on the disk that begins to rise through the fluid. As the solid rises the hydrostatic pressure from the fluid decreases and hence it grows further (see Fig. 5a) till it reaches the top of the domain. The amount of fluid ejected from the control volume due to expansion of the disk does not quite keep pace with this expansion (see Fig. 5a). The difference in these two amounts is a measure of the error in our numerical method (see Fig. 5b).

5.2 Deformation of a Compressible Annulus Under the Action of Point Source of Fluid

This test was first proposed by Roy [12], as a toy model to describe the behavior of hydrocephalus in the brain. In this test we observe the deformation of a hollow cylinder, submerged in a fluid contained in a rigid prismatic box, due to the influx of fluid along the axis of the cylinder. In the two-dimensional context the test comprises an annular solid with inner radius R and thickness w that is concentric with a fluid-filled square box of edge length l (see, Fig. 6). A point mass source of fluid of strength Q is located at $\mathbf{x}_C$ which corresponds to the center C of the control volume. The radially symmetric nature of point source ensures that momentum balance law remains unaltered. However, we need to modify the balance of mass to account for the mass influx from the point source by adding the following term to the pressure equation:

$$\frac{Q}{\rho_f}\delta\left(\mathbf{x} - \mathbf{x}_C\right) \tag{34}$$

For this test we have used a single point source whose strength is a constant, and all boundary conditions on the control volume are of homogeneous Dirichlet type.

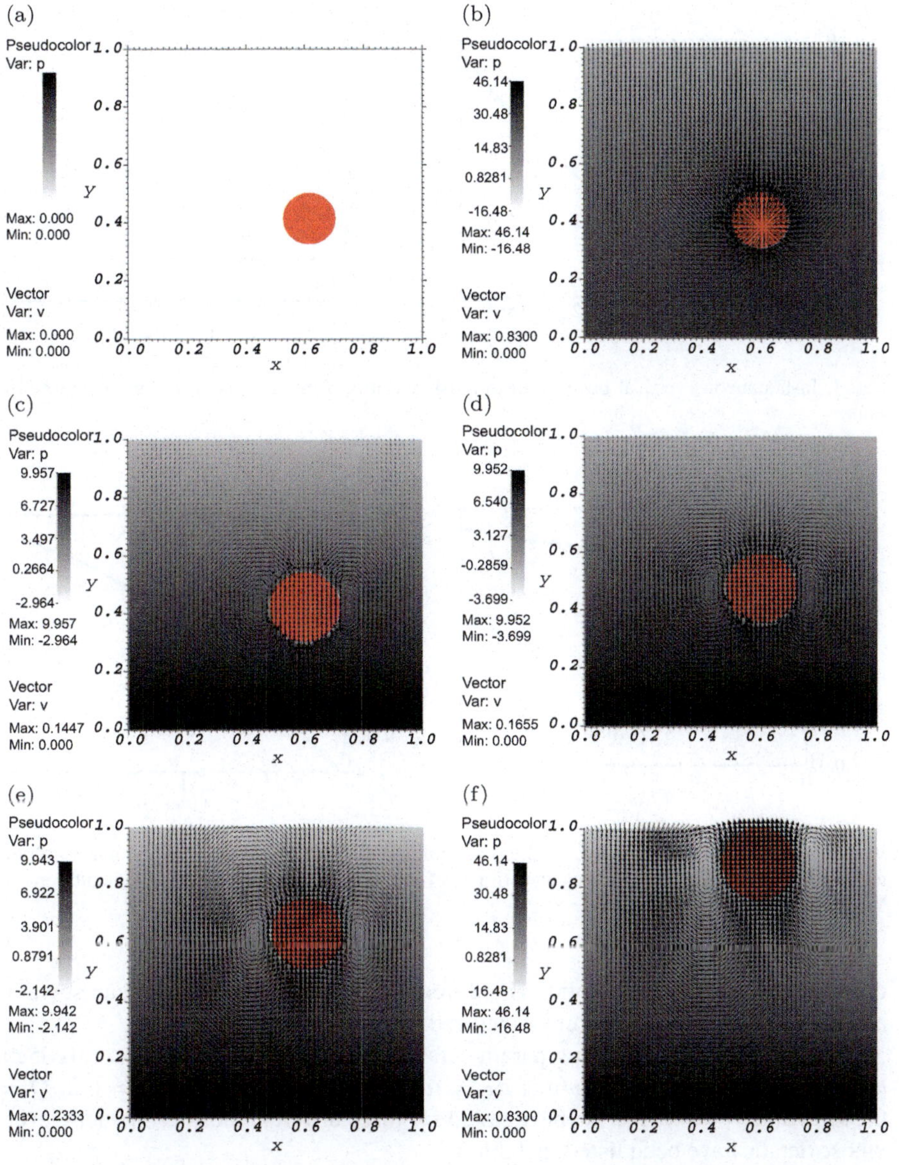

Fig. 3 The velocity field over the entire control volume, the pressure in the fluid and the mesh of the disk. **(a)** $t = 0$ s. **(b)** $t = 0.01$ s. **(c)** $t = 0.5$ s. **(d)** $t = 1.0$ s. **(e)** $t = 2.0$ s. **(f)** $t = 3.0$ s

The solid is compressible and hence volume of solid and thereby the volume of fluid in the control volume can change. The homogeneous Dirichlet boundary condition implies that the fluid cannot leave the control volume and hence the amount of fluid that accumulates in the control volume due to the point source must equate the

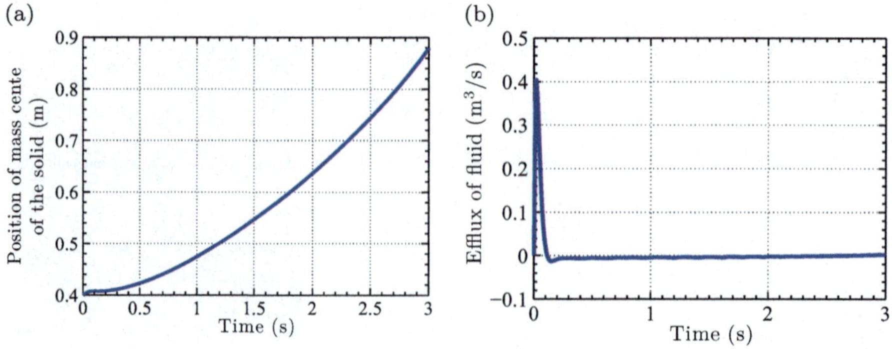

Fig. 4 Instantaneous vertical position of the disk and flux of the fluid. (**a**) Vertical position. (**b**) Flux

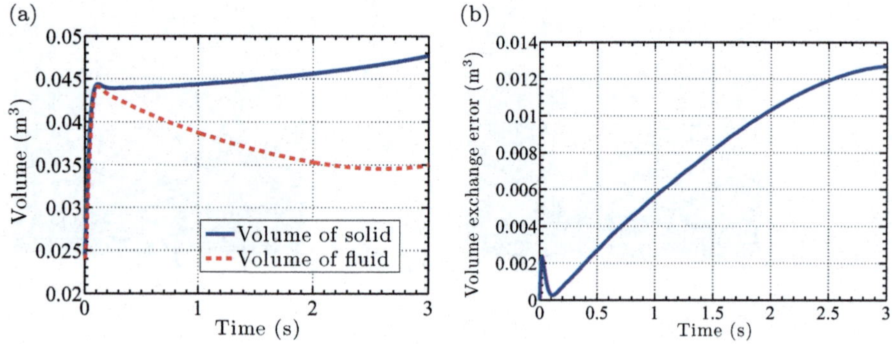

Fig. 5 Instantaneous area change of the disk, amount of fluid ejected from the control volume and the difference between them as an estimate of the error in our numerical implementation. (**a**) Volume. (**b**) Error

decrease in volume of the solid. The difference in these two volumes can serve as an estimate of the numerical error incurred.

We have used the following parameters for this test: $R = 0.25$ m, $w = 0.05$ m, $l = 1.0$ m, $\rho_\mathbf{f} = \rho_{\mathbf{s}_0} = 1\,\mathrm{kg/m^3}$, $\mu_\mathbf{f} = \mu_\mathbf{s} = 1$ Pa·s, $\mu^e = 1$ Pa, $\nu = 0.3$, $Q = 0.1$ kg/s and $dt = 0.01$ s. We have tested for three different mesh refinement levels whose details have been listed in Table 1.

The initial state of the system is shown in Fig. 7a. As time progresses, the fluid entering the control volume deforms and compresses the annulus as shown in Fig. 7b for $t = 1$ s. When we look at the difference in the instantaneous amount of fluid entering the control volume and the decrease in the volume of the solid, we see that the difference increases over time (see Fig. 8). This is not surprising since the mesh of the solid becomes progressively distorted as the fluid emanating from the point source push the inner boundary of the annulus. The error significantly reduces with the increase in the refinements of the fluid and the solid meshes.

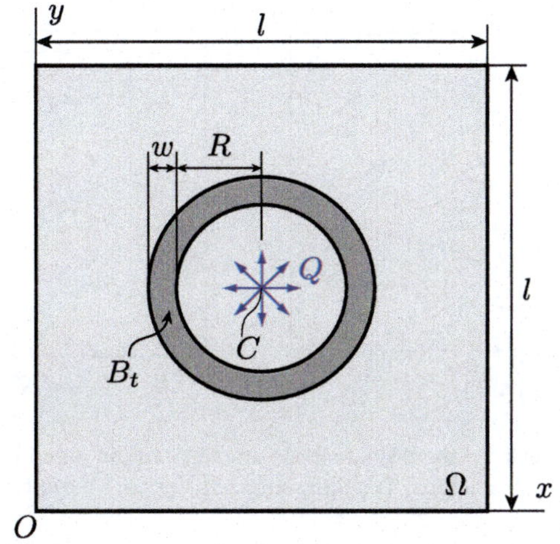

Fig. 6 Initial configuration of an annulus immersed in a square box filled with fluid. At the center C of the box there is a point source of strength Q

Table 1 Number of cells and DoFs used in the different simulations involving the deformation of a compressible annulus under the action of a point source

	Solid		Control volume	
	Cells	DoFs	Cells	DoFs
Level 1	6240	50960	1024	9539
Level 2	24960	201760	4096	37507
Level 3	99840	802880	16384	148739

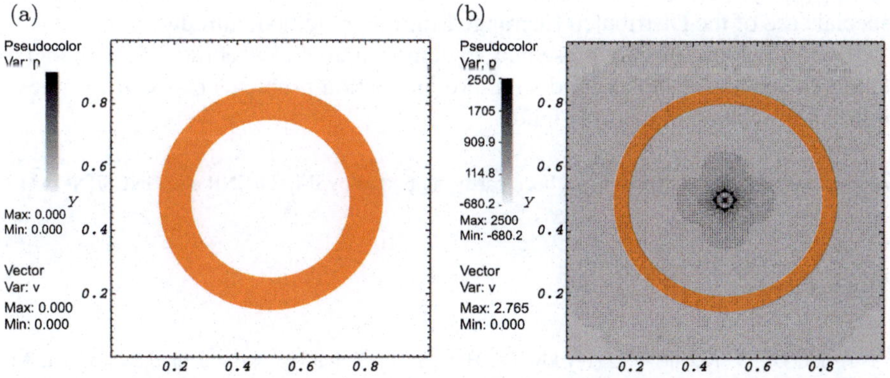

Fig. 7 The velocity and the mean normal stress field over the control volume. Also shown is the annulus mesh. (**a**) $t = 0$ s. (**b**) $t = 1.0$ s

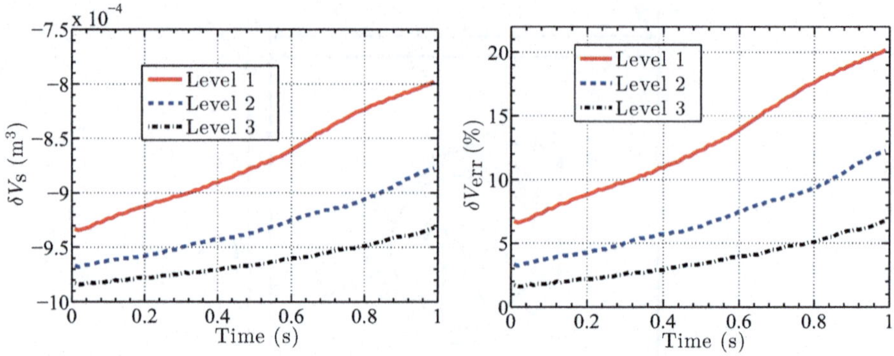

Fig. 8 The difference between the instantaneous amount of fluid entering due to the source and the change in the area of the annulus. The difference reduces with mesh refinement

6 Conclusions

The Finite Element Immersed Boundary Method (FEIBM) is a well established formulation for fluid structure interaction problems of general types. In most implementations (see, for example, [3]), the solid constitutive behavior is constrained to be viscous (with the same viscosity of the surrounding fluid) and incompressible.

The first attempt to allow for solids of arbitrary constitutive type was presented in [7], whose formulation is applicable to problems with immersed bodies of general topological and constitutive characteristics. Such formulation, however, did not expose the intrinsic structure of the underlying problem.

In this work we presented a novel distributed Lagrange multiplier method that generalizes the compressible model introduced in [7] and which can be seen as a special case of the Distributed Lagrange Multiplier method, introduced in [5].

Two validation tests are presented, to demonstrate the capability of the model to take into account complex fluid structure interaction problems between compressible solids and incompressible fluids.

Acknowledgements This work has been partly supported by IMATI/CNR and GNCS/INDAM.

References

1. Arndt, D., Bangerth, W., Davydov, D., Heister, T., Heltai, L., Kronbichler, M., Maier, M., Pelteret, J.-P., Turcksin, B., Wells, D.: The deal.II library, version 8.5. J. Numer. Math. **25**(3), 137–145 (2017)
2. Boffi, D., Gastaldi, L.: A finite element approach for the immersed boundary method. Comput. Struct. **81**(8–11), 491–501 (2003)
3. Boffi, D., Gastaldi, L., Heltai, L., Peskin, C.S.: On the hyper-elastic formulation of the immersed boundary method. Comput. Methods Appl. Math. Eng. **197**(25–28), 2210–2231 (2008)

4. Boffi, D., Cavallini, N., Gardini, F., Gastaldi, L.: Local mass conservation of Stokes finite elements. J. Sci. Comput. **52**(2), 383–400 (2012)
5. Boffi, D., Cavallini, N., Gastaldi, L.: The finite element immersed boundary method with distributed Lagrange multiplier. SIAM J. Numer. Anal. **53**(6), 2584–2604 (2015)
6. Gurtin, M.E.: An Introduction to Continuum Mechanics, vol. 158. Academic, New York (1982)
7. Heltai, L., Costanzo, F.: Variational implementation of immersed finite element methods. Comput. Methods Appl. Mech. Eng. **229/232**, 110–127 (2012)
8. Heltai, L., Roy, S., Costanzo, F.: A fully coupled immersed finite element method for fluid structure interaction via the Deal.II library. Arch. Numer. Softw. **2**(1), 1–27 (2014)
9. Hughes, T.J.R., Liu, W.K., Zimmerman, T.K.: Lagrangian-Eulerian finite element formulations for incompressible viscous flows. Comput. Methods Appl. Mech. Eng. **29**, 329–349 (1981)
10. Peskin, C.S.: Numerical analysis of blood flow in the heart. J. Comput. Phys. **25**(3), 220–252 (1977)
11. Peskin, C.S.: The immersed boundary method. Acta Numer. **11**, 479–517 (2002)
12. Roy, S.: Numerical simulation using the generalized immersed finite element method: an application to hydrocephalus. PhD thesis, The Pennsylvania State University (2012)
13. Roy, S., Heltai, L., Costanzo, F.: Benchmarking the immersed finite element method for fluid-structure interaction problems. Comput. Math. Appl. **69**, 1167–1188 (2015)
14. Turek, S., Hron, J.: Proposal for numerical benchmarking of fluid-structure interaction between an elastic object and laminar incompressible flow. In: Bungartz, H.-J., Schäfer, M. (eds.) Fluid-Structure Interaction. Lecture Notes in Computational Science and Engineering, vol. 53, pp. 371–385. Springer, Berlin, Heidelberg (2006). https://doi.org/10.1007/3--540-34596--5_15
15. Wang, X., Liu, W.K.: Extended immersed boundary method using FEM and RKPM. Comput. Methods Appl. Mech. Eng. **193**(12–14), 1305–1321 (2004)
16. Zhang, L., Gerstenberger, A., Wang, X., Liu, W.K.: Immersed finite element method. Comput. Methods Appl. Mech. Eng. **193**(21–22), 2051–2067 (2004)

Daniele Boffi has been Professor of Numerical Analysis at the Department of Mathematics of the University of Pavia since 2005. He received his PhD in mathematics from the Universities of Brescia-Milano-Pavia, Italy (1996). He has been a Visiting Professor at several institutions. His primary research interests are related to the numerical approximation of partial differential equations, including mixed finite elements, eigenvalue problems, computational electromagnetism, and the interaction of fluids and solids.

Lucia Gastaldi is Professor of Numerical Analysis at the University "La Sapienza" of Rome from November 1, 1996 to October 31, 1997 and, successively, at the University of Brescia until now. She served as director of the Dipartimento di Matematica, University of Brescia since November 1, 2007 to October 31, 2012. Her scientific interests are mainly focused on the numerical approximation of partial differential equations and, in particular, on the approximation of eigenproblems in mixed form, evolution problem in moving domains, fluid-structure interaction systems and immersed boundary method.

Luca Heltai is an Associate professor of Numerical Analysis at the International School of Advanced Studies in Trieste. He received a PhD in Mathematics from the University of Pavia, Italy (2007), and he is one of the main developers of the open source Differential Equations Analysis Library—deal.II (www.dealii.org). His main research themes focus on innovative numerical methods and algorithms to tackle physiological, biological and industrial fluid-structure interaction problems. He contributed in setting up the "Master in High Performance Computing" (www.mhpc.it), and the "Master in Data Science and Scientific Computing" (https://dssc.units.it), and he is currently the Director of the Master in High Performance Computing.

High-Order Operator-Splitting Methods for the Bidomain and Monodomain Models

Jessica Cervi and Raymond J. Spiteri

Abstract The bidomain and monodomain models are among the most widely used mathematical models to describe cardiac electrophysiology. They take the form of multi-scale reaction-diffusion partial differential equations that couple the dynamic behaviour on the cellular scale with that on the tissue scale. The systems of differential equations associated with these models are large and strongly non-linear, but they also have a distinct structure due to their multi-scale nature. For these reasons, numerical solutions to these systems are often found via operator-splitting methods. In this chapter, we provide a survey of operator-splitting methods for the numerical solution of differential equations. In particular, we focus on splitting methods with order higher than two that, according to the Sheng–Suzuki theorem, require backward time integration and historically have been considered unstable for solving deterministic parabolic systems. We demonstrate the stability of operator-splitting methods of up to order four to solve the bidomain and monodomain models on several examples arising in the field of cardiovascular modeling.

1 Introduction

The cardiac pumping function is a process that is controlled by a complex pattern of electrical activation of approximately 10 billion muscle cells. Because the myocardial electrical activity is fundamental for the function of the heart, extensive research has been done to understand the different mechanisms occurring at a cellular, tissue, and organ level [30]. It follows that the organ-level electrical activity in the heart is the result of billions of small-scale processes occurring in the cells,

J. Cervi (✉)
Department of Mathematics and Statistics, University of Saskatchewan, SK, Canada
e-mail: jessica.cervi@usask.ca

R. J. Spiteri
Department of Computer Science, University of Saskatchewan, SK, Canada
e-mail: spiteri@cs.usask.ca

D. Boffi et al. (eds.), *Mathematical and Numerical Modeling of the Cardiovascular System and Applications*, SEMA SIMAI Springer Series 16,
https://doi.org/10.1007/978-3-319-96649-6_2

yet the knowledge about how these processes interact is limited. For this reason, the use of mathematical models and computer simulation to reproduce the electrical activity of the heart has been a promising technique to study the heart and heart diseases.

Mathematically, the electrophysiological behaviour of the heart can be modelled using the bidomain model [35] or the monodomain model [30]. Both are multi-scale models that were developed to overcome the fact that the distance scales used to measure the electrical potentials are large compared to the length of a single cell [35]. For this reason, the bidomain and monodomain models use a continuum approach in order to handle the large number of cells. The bidomain model is a generalization of the monodomain model under the assumption of unequal anisotropy rates [30, 35]. The bidomain model is generally considered to be the most accurate and general model of cardiac tissue electrophysiology [11]; nonetheless, the monodomain model is still widely used by the research community, e.g., [20].

Solving the bidomain or monodomain models is challenging because of the resolutions in time and space required to produce clinically relevant data. A classic strategy to deal with complicated problems is the well-known "divide and conquer" strategy. In the context of evolutionary partial differential equations (PDEs), of which the bidomain and monodomain models are examples, a rather successful approach in this spirit has been operator splitting (OS). OS constitutes an appropriate choice when dealing with a PDE whose vector field can be decomposed into different pieces, each of which is integrable or whose solution is more amenable. OS methods have been successfully employed to find numerical solutions to the bidomain and monodomain models, e.g., [23, 31]. Although fully coupled methods exist for the numerical solution of the bidomain and monodomain models, e.g., [8], OS methods represent more the rule rather than the exception and in fact are the only option in the Cancer, Heart, and Soft Tissue Environment (Chaste) software environment [19].

The remainder of this chapter is structured as follows. In Sect. 2, the bidomain and monodomain models are described mathematically. In Sect. 3, the theoretical background behind OS methods is given, including example methods and algorithms for how the methods can be applied to the bidomain and monodomain models. In Sect. 4, the convergence of high-order OS methods is demonstrated on a diverse set of benchmark problems. Finally, in Sect. 5, conclusions from this study are given.

2 Mathematical Models

The bidomain model is widely accepted as a useful model for describing the electrical activity in myocardial tissue. It is a multi-scale reaction-diffusion model that couples PDEs describing the chemical reactions and the local flow of ions at the cell membrane level with a system of diffusion PDEs that describes the propagation of the electrical activation through the cardiac muscle. To account for both of these effects, the tissue is divided into two co-located domains, the intracellular and the extracellular. These domains are separated by the cell membrane, which acts as an insulator between them.

On a spatial domain $\Omega \subset \mathbb{R}^d$ of dimension d and time interval $[t_0, t_f]$, the bidomain model can be written as

$$\frac{\partial \mathbf{s}}{\partial t} = \mathbf{f}(t, \mathbf{s}, v), \tag{1a}$$

$$\chi C_m \frac{\partial v}{\partial t} + \chi I_{\text{ion}}(\mathbf{s}, v) = \nabla \cdot (\sigma_i \nabla v) + \nabla \cdot (\sigma_i \nabla u_e), \tag{1b}$$

$$0 = \nabla \cdot (\sigma_i \nabla v) + \nabla \cdot ((\sigma_i + \sigma_e) \nabla u_e), \tag{1c}$$

subject to boundary conditions on $\partial\Omega \times [t_0, t_f]$ given by

$$\hat{\mathbf{n}} \cdot (\sigma_i \nabla v + \sigma_i \nabla u_e) = 0, \tag{1d}$$

$$\hat{\mathbf{n}} \cdot (\sigma_e \nabla u_e) = 0, \tag{1e}$$

where $\mathbf{s} = \mathbf{s}(\mathbf{x}, t)$ is a vector describing the cellular state at location $\mathbf{x} \in \Omega$ and time $t \in [t_0, t_f]$, $v = v(\mathbf{x}, t)$ is the transmembrane potential, and $u_e = u_e(\mathbf{x}, t)$ is the extracellular potential. The ionic current, represented by $I_{\text{ion}}(v, \mathbf{s})$, and $\mathbf{f}(t, \mathbf{s}, v)$ are non-linear terms related to the cell model, and σ_i and σ_e are the intracellular and extracellular conductivities, respectively. Finally, χ is the area of the cell membrane per unit volume, C_m is the capacitance of the cell membrane per unit area, and $\hat{\mathbf{n}}$ is the outward unit normal to $\partial\Omega$.

By making the assumption of equal anisotropy rates throughout the domain, i.e., $\sigma_i = \lambda\sigma_e$, where λ is a constant scalar, it is possible to write a simplified version of the bidomain model describing only the dynamics of the transmembrane potential v in the intracellular tissue. Such a model is known as monodomain model and can be written as

$$\frac{\partial \mathbf{s}}{\partial t} = \mathbf{f}(t, \mathbf{s}, v), \tag{2a}$$

$$\chi C_m \frac{\partial v}{\partial t} + \chi I_{\text{ion}}(\mathbf{s}, v, t) = \frac{\lambda}{1+\lambda} \nabla \cdot (\sigma_i \nabla v), \tag{2b}$$

subject to

$$\hat{\mathbf{n}} \cdot (\sigma_i \nabla v) = 0. \tag{2c}$$

3 Operator-Splitting Methods

Operator-splitting methods represent powerful divide-and-conquer strategies to solve mathematical models, including the bidomain and monodomain models, where a system can be decomposed into smaller and simpler problems that can be solved with the aid of targeted and specialized techniques. Before giving details about the specific splitting methods presented in this chapter and ways to derive them, it is useful to present the general theoretical framework behind OS methods.

Consider the following Cauchy problem

$$\frac{d\mathbf{y}}{dt} = \mathscr{A}(\mathbf{y}) = \mathscr{A}^{[1]}(\mathbf{y}) + \mathscr{A}^{[2]}(\mathbf{y}), \qquad \mathbf{y}(0) = \mathbf{y}_0, \tag{3}$$

with associated exact flow ϕ_t, where the (generally non-linear) operator $\mathscr{A}(\cdot)$ has been additively split into two terms $\mathscr{A}^{[1]}(\cdot)$ and $\mathscr{A}^{[2]}(\cdot)$ and $\mathbf{y}_0$ is the given initial condition. We now view the original problem as split into two separate subproblems

$$\frac{d\mathbf{y}^{[1]}}{dt} = \mathscr{A}^{[1]}(\mathbf{y}^{[1]}), \qquad \frac{d\mathbf{y}^{[2]}}{dt} = \mathscr{A}^{[2]}(\mathbf{y}^{[2]}), \tag{4}$$

with associated flows $\phi_t^{[1]}$ and $\phi_t^{[2]}$, respectively, that in principle are somehow more desirable to solve.

As shown in the next sections, it is possible to generate numerical methods for such systems with a high order of accuracy. Although the theory throughout this chapter is illustrated in terms of a basic 2-additive splitting, it is generalizable to N-additively split or partitioned systems.

In this chapter, we focus our attention on the so-called *fractional-step methods*, e.g., [15]. To understand how we can derive methods belonging to this class, we consider an evolution equation in the form (3) and define the s-stage method $\mathscr{S}_{\Delta t}(\mathbf{y})$ with "abscissae" $\boldsymbol{\alpha} = \{(\alpha_j^{[1]}, \alpha_j^{[2]})\}_{j=1}^{s}$ in a recursive manner as

$$\mathscr{S}_{\Delta t}(\mathbf{y}) = \mathscr{S}^{(s)}_{\boldsymbol{\alpha}_s \Delta t} \circ \mathscr{S}^{(s-1)}_{\boldsymbol{\alpha}_{s-1} \Delta t} \circ \cdots \circ \mathscr{S}^{(1)}_{\boldsymbol{\alpha}_1 \Delta t}(\mathbf{y}) \approx \phi_{\Delta t}(\mathbf{y}), \tag{5}$$

with building block

$$\mathscr{S}^{(j)}_{\boldsymbol{\alpha}_j \Delta t}(\mathbf{y}) = \phi^{[2]}_{\alpha_j^{[2]} \Delta t} \circ \phi^{[1]}_{\alpha_j^{[1]} \Delta t}(\mathbf{y}). \tag{6}$$

The solution information from the end point of a given fractional step (or *sub-flow*) is used as the initial condition for the start of the subsequent sub-flow. An example of a 2-additive, two-stage OS method with positive sub-flows $\alpha_j^{[i]} \Delta t$, $i, j = 1, 2$, is depicted in Fig. 1 to advance the numerical solution from time t_n to time $t_{n+1} = t_n + \Delta t$. The solid lines denote sub-flows and dashed lines denote the transfer of solution information between sub-flows.

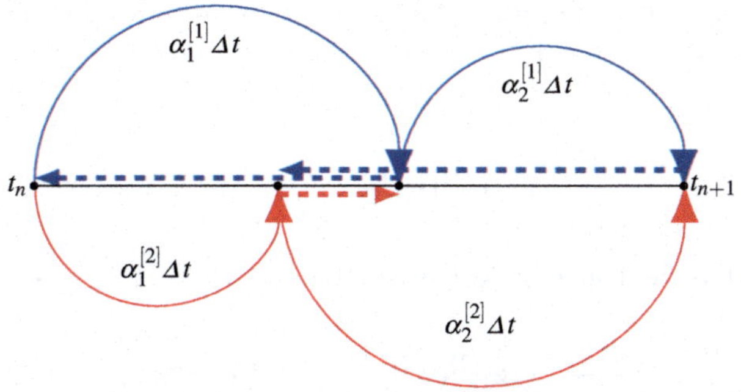

Fig. 1 Schematic of a two-stage OS method. The solid lines denote the sub-flows and the dashed lines denote the transfer of solution information between sub-flows

The local splitting error of an OS method can be written as

$$\mathscr{L}_{\Delta t}(\mathbf{y}) = \phi_{\Delta t} - \mathscr{S}_{\Delta t}(\mathbf{y}). \tag{7}$$

A fractional-step OS method has order of accuracy p if and only if the Taylor expansion of (7) is $\mathscr{O}((\Delta t)^{p+1})$. However, in practice, the sub-flows must be approximated numerically. Accordingly, for the overall error of a time integration to have order p, each of the individual sub-flows must also be approximated to order p.

Using the notation of Lie derivatives, the building block can be expressed as

$$\mathscr{S}^{(j)}_{\boldsymbol{\alpha}_j \Delta t}(\mathbf{y}) = e^{\Delta t \alpha_j^{[2]} \mathscr{D}^{[2]}} e^{\Delta t \alpha_j^{[1]} \mathscr{D}^{[1]}}, \quad j = 1, 2, \ldots, s.$$

For notational convenience, we further compact the notation for the building block to

$$\mathscr{S}^{(j)}_{\boldsymbol{\alpha}_j \Delta t}(\mathbf{y}) \doteq (\boldsymbol{\alpha}_j \Delta t). \tag{8}$$

We further denote the reversal of the default order (6) in which the differential operators $\mathscr{A}^{[j]}$, $j = 1, 2$, are applied by

$$(\breve{\boldsymbol{\alpha}}_j \Delta t) \doteq e^{\Delta t \alpha_j^{[1]} \mathscr{D}^{[1]}} e^{\Delta t \alpha_j^{[2]} \mathscr{D}^{[2]}}, \tag{9}$$

where $\breve{\boldsymbol{\alpha}}_j = (\alpha_j^{[2]}, \alpha_j^{[1]})$.

3.1 Basic Operator-Splitting Methods

We start our discussion about specific operator-splitting methods by considering the celebrated Godunov [12] and Strang [29] methods.

Consider the split system (4). The Godunov (also known as Lie–Trotter [34]) OS method advances an approximate solution by time step Δt by first solving the problem

$$\frac{d\mathbf{y}^{[1]}}{dt} = \mathscr{A}^{[1]}(\mathbf{y}^{[1]}), \qquad \mathbf{y}^{[1]}(0) = \mathbf{y}_0,$$

for $t \in [0, \Delta t]$ to yield an approximate solution $\mathbf{y}^{[1]}_{\Delta t}$. Next, it solves

$$\frac{d\mathbf{y}^{[2]}}{dt} = \mathscr{A}^{[2]}(\mathbf{y}^{[2]}), \qquad \mathbf{y}^{[2]}(0) = \mathbf{y}^{[1]}_{\Delta t},$$

for $t \in [0, \Delta t]$ to obtain $\mathbf{y}^{[2]}_{\Delta t}$ such that

$$\mathbf{y}(\Delta t) \approx \mathbf{y}^{[2]}_{\Delta t} = e^{\Delta t \mathscr{D}^{[2]}} e^{\Delta t \mathscr{D}^{[1]}} \mathbf{y}_0$$

The above method is perhaps the most famous and simplest OS method. This method computes an approximation to the exact solution of (3) that is first-order accurate in the sense that

$$\|\mathbf{y}(\Delta t) - \mathbf{y}^{[2]}_{\Delta t}\| = \mathscr{O}((\Delta t)^2).$$

This result can be derived by computing the Taylor series for the exact solution $\mathbf{y}(t)$ and using that of the approximate solution $\mathbf{y}^{[2]}_{\Delta t}$ obtained by operator spitting. We have

$$\mathbf{y}(\Delta t) = \mathbf{y}_0 + \Delta t \left.\frac{d\mathbf{y}}{dt}\right|_0 + \left.\frac{(\Delta t)^2}{2}\frac{d^2\mathbf{y}}{dt^2}\right|_0 + \mathscr{O}((\Delta t)^3).$$

From (4), by applying direct differentiation k times, we have

$$\frac{d^k\mathbf{y}}{dt^k} = (\mathscr{A}^{[1]} + \mathscr{A}^{[2]})^k \mathbf{y},$$

which, when substituted into the above Taylor series, gives

$$\mathbf{y}(\Delta t) = \mathbf{y}_0 + \Delta t(\mathscr{A}^{[1]} + \mathscr{A}^{[2]})\mathbf{y}_0 + \frac{(\Delta t)^2}{2}(\mathscr{A}^{[1]} + \mathscr{A}^{[2]})^2\mathbf{y}_0 + \mathscr{O}((\Delta t)^3).$$

The expansions of the approximate solutions $\mathbf{y}^{[1]}(\Delta t)$ and $\mathbf{y}^{[2]}(\Delta t)$ are

$$\mathbf{y}^{[1]}(\Delta t) = \mathbf{y}_0 + \Delta t \mathscr{A}^{[1]}\mathbf{y}_0 + \frac{(\Delta t)^2}{2}(\mathscr{A}^{[1]})^2\mathbf{y}_0 + \mathscr{O}((\Delta t)^3)$$

$$\mathbf{y}^{[2]}(\Delta t) = \mathbf{y}^{[1]}_{\Delta t} + \Delta t \mathscr{A}^{[2]}\mathbf{y}^{[1]}_{\Delta t} + \frac{(\Delta t)^2}{2}(\mathscr{A}^{[2]})^2\mathbf{y}^{[1]}_{\Delta t} + \mathscr{O}((\Delta t)^3)$$

Inserting the series expansion for $\mathbf{y}^{[1]}(\Delta t)$ into the expansion of $\mathbf{y}^{[2]}(\Delta t)$, the local splitting error is

$$\mathbf{y}(\Delta t) - \mathbf{y}^{[2]}(\Delta t) = \frac{(\Delta t)^2}{2}[\mathscr{A}^{[1]}, \mathscr{A}^{[2]}]\mathbf{y}_0 + \mathscr{O}((\Delta t)^3),$$

where $[\mathscr{A}^{[1]}, \mathscr{A}^{[2]}] = \mathscr{A}^{[1]}\mathscr{A}^{[2]} - \mathscr{A}^{[2]}\mathscr{A}^{[1]}$ is a commutator. We see that the error after one time step is proportional to $(\Delta t)^2$. Thus, the Godunov–Lie–Trotter OS method is first order.

It is possible to increase the accuracy of an OS method by adding stages or interchanging the order of applying $\mathscr{A}^{[1]}$ and $\mathscr{A}^{[2]}$. For example, carrying out two symmetric half steps gives the method

$$\mathbf{y}_{n+1} = \left(e^{\frac{1}{2}\Delta t \mathscr{D}^{[1]}} e^{\frac{1}{2}\Delta t \mathscr{D}^{[2]}}\right)\left(e^{\frac{1}{2}\Delta t \mathscr{D}^{[2]}} e^{\frac{1}{2}\Delta t \mathscr{D}^{[1]}}\right)\mathbf{y}_n.$$

Following the notation in (8), the Strang–Marchuk method is represented by

$$(\boldsymbol{\alpha}_1 \Delta t)(\check{\boldsymbol{\alpha}}_2 \Delta t),$$

for $\alpha_j^{[i]} = 1/2$ for $i, j = 1, 2$.

We refer to this method, independently proposed by Strang [29] and Marchuk [17], as SM2. After a series expansion, the local splitting error is found to be

$$\mathbf{y}(\Delta t) - \mathbf{y}^{[1]}(\Delta t) = \frac{1}{24}(\Delta t)^3\left([\mathscr{D}^{[1]}, [\mathscr{D}^{[1]}, \mathscr{D}^{[2]}]] + 2([\mathscr{D}^{[2]}, [\mathscr{D}^{[2]}, \mathscr{D}^{[1]}]]\right)\mathbf{y}\left(\frac{\Delta t}{2}\right) + \mathscr{O}((\Delta t)^5).$$

Therefore, the SM2 method is a second-order method.

3.2 *High-Order Operator-Splitting Methods*

Extensive research has been done in recent years to find high-order OS methods. In 1989, Sheng showed that OS methods with order higher than two require sub-flows that integrate backwards in time [25]. This result was proved separately by Suzuki in 1991 [32], and it is now known as the Sheng–Suzuki theorem. In particular, every OS method of at least third order applied to (3) with at least two additive operators requires at least one backward-in-time sub-flow for each solution operator. This result was first proved in [13]; a simpler proof appears in [6]. Furthermore, a fourth-order method for which $\alpha^{[1]} = \alpha^{[2]}$ must contain at least two backwards-in-time sub-flows [27]. Although numerous high-order OS methods have been proposed (see, e.g., [3, 7, 27]), there have been claims in the literature that these methods are not suitable for parabolic equations because such equations are well-known to be ill-posed in the negative time direction. Nonetheless, in this section, we present a few high-order OS methods that are ultimately shown to be stable and converge at their expected orders in simulations involving the bidomain and monodomain models.

The first high-order OS method that we consider was proposed in [26, 27]. In these works, a family of high-order OS methods is derived by using the Campbell–Baker–Hausdorff (CBH) formula. The idea is to approximate the exact N-additive flow $\exp(\sum_{i=1}^{N} \mathscr{A}^{[i]})$ by the product of single exponentials $\exp(\mathscr{A}^{[i]})$. A well-known strategy to achieve this result is to recursively use the CBH formula, which reads

$$\exp\left(t\mathscr{D}^{[1]}\right)\exp\left(t\mathscr{D}^{[2]}\right) = \exp\left[t(\mathscr{D}^{[1]}+\mathscr{D}^{[2]}) + \frac{1}{2}t^2\mathscr{D}_{12} + \frac{1}{12}t^3(\mathscr{D}_{112}+\mathscr{D}_{221}) + \dots\right],$$

where

$$\mathscr{D}_{kl\dots mn} \doteq [\mathscr{D}^{[k]}, [\mathscr{D}^{[l]}, \cdots [\mathscr{D}^{[m]}, \mathscr{D}^{[n]}]\cdots]]$$

and

$$[\mathscr{D}^{[m]}, \mathscr{D}^{[n]}] \doteq \mathscr{D}^{[m]}\mathscr{D}^{[n]} - \mathscr{D}^{[n]}\mathscr{D}^{[m]}.$$

In this way, order conditions are derived, and integer solutions to the order conditions that also minimize 2-norm of the leading local error coefficients to produce a number of third- and fourth-order methods [27]. In [26], a specific third-order method is presented and shown to be stable for a simple convection-diffusion PDE. We denote this method by S3. It is represented by

$$(\boldsymbol{\alpha}_1\Delta t)(\check{\boldsymbol{\alpha}}_2\Delta t)(\check{\boldsymbol{\alpha}}_3\Delta t)(\check{\boldsymbol{\alpha}}_4\Delta t)(\check{\boldsymbol{\alpha}}_5\Delta t)(\boldsymbol{\alpha}_6\Delta t)(\boldsymbol{\alpha}_7\Delta t)(\check{\boldsymbol{\alpha}}_8\Delta t)(\check{\boldsymbol{\alpha}}_9\Delta t). \quad (10)$$

The coefficients $\alpha_j^{[i]}$ for the method (10) are displayed in Table 1.

Another family of high-order OS methods was proposed by Koch, Neuhauser, and Thalhammer in [14] and further analyzed by Auzinger and co-workers in [3, 4]. The derivation of the order conditions was based on the literal formulation that an

Table 1 Coefficients $\alpha_j^{[i]}$ for the S3 method

j	$\alpha_j^{[1]}$	$\alpha_j^{[2]}$
1	1/6	1/6
2	1/6	1/6
3	1/6	1/6
4	1/6	1/6
5	1/6	1/6
6	−1/3	−1/3
7	1/6	1/6
8	1/6	1/6
9	1/6	1/6

OS method is of order p if and only if the local error $\mathscr{L}(\Delta t) = \mathscr{O}((\Delta t)^{p+1})$, and hence

$$\frac{d}{dt}\mathscr{L}(0) = \frac{d^2}{dt^2}\mathscr{L}(0) = \cdots = \frac{d^p}{dt^p}\mathscr{L}(0) = 0.$$

To select optimal methods among those that satisfy the order conditions, the authors in [5] minimize the local error measure (LEM)

$$\text{LEM} := \left(\sum_{k=1}^{l_{p+1}} |\lambda_{p+1,k}|^2 \right)^{1/2},$$

where the $\lambda_{p+1,k}$ are the coefficients of the Lyndon (or Lyndon–Shirshov) monomials used as a basis to express the order conditions. This measure has the advantage of being convenient to calculate because it uses precisely the same framework used to set up the order conditions.

The first high-order OS method considered here is one proposed by Auzinger and co-workers [5]. This method, denoted in [5] as "Emb 3/2 AKS" and here simply as "AKS3", is third-order accurate and has three stages. It is optimal in the sense that it has minimal LEM among similar methods with comparable effort. Following the notation (8), it can be written as

$$(\boldsymbol{\alpha}_1 \Delta t)(\boldsymbol{\alpha}_2 \Delta t)(\boldsymbol{\alpha}_3 \Delta t), \tag{11}$$

where the coefficients $\alpha_j^{[i]}$ for $i = 1, 2$, $j = 1, 2, 3$ for the method (11) are displayed in Table 2. The AKS3 method belongs to the class of *embedded exponential operator-splitting methods*; it is in fact the main integrator of a 3(2) embedded pair. Such methods were presented for the first time in [14]. It is also a *palindromic* method [3]. Palindromic methods are characterized by the property

$$\alpha_j^{[2]} = \alpha_{s+1-j}^{[1]}, \quad j = 1, 2, \ldots, s.$$

We also consider a third-order OS method originally proposed by Ruth in 1983 [24]. This method, denoted here by R3, has three stages, and it was considered in [4]

Table 2 Coefficients $\alpha_j^{[i]}$ for the AKS3 method

j	$\alpha_j^{[1]}$	$\alpha_j^{[2]}$
1	0.268330095781759925	$\alpha_3^{[1]}$
2	−0.187991618799159782	$\alpha_2^{[1]}$
3	0.919661523017399857	$\alpha_1^{[1]}$

Table 3 Coefficients $\alpha_j^{[i]}$ for the R3 method

j	$\alpha_j^{[1]}$	$\alpha_j^{[2]}$
1	1	−1/24
2	−2/3	3/4
3	2/3	7/24

Table 4 Coefficients $\alpha_j^{[i]}$ for the Y4 method

j	$\alpha_j^{[1]}$	$\alpha_j^{[2]}$
1	0.675603595979828817	1.351207191959657634
2	−0.175603595979828817	−1.702414383919315268
3	−0.175603595979828817	1.351207191959657634
4	0.675603595979828817	0.0

as part of a pair of embedded methods. It was derived in the context of Hamilton's conditions and Hamiltonian systems; we refer to [24] for details. Following the notation used for the previous methods, the R3 method can be written as

$$(\boldsymbol{\alpha}_1 \Delta t)(\boldsymbol{\alpha}_2 \Delta t)(\boldsymbol{\alpha}_3 \Delta t), \tag{12}$$

where the coefficients $\alpha_j^{[i]}$ for $i = 1, 2, \ j = 1, 2, 3$, for the method (12) are displayed in Table 3.

As an example of fourth-order method, we consider a method proposed by Yoshida [36]. This method, denoted here by Y4, has four stages, and it was derived using a three-fold composition of the Strang–Marchuk method and repeated application of the CBH formula; see [36] for details. Following the notation used for the previous methods, the Y4 scheme can be written as

$$(\boldsymbol{\alpha}_1 \Delta t)(\boldsymbol{\alpha}_2 \Delta t)(\boldsymbol{\alpha}_3 \Delta t)(\boldsymbol{\alpha}_4 \Delta t), \tag{13}$$

where the coefficients $\alpha_j^{[i]}$ for $i = 1, 2, \ j = 1, 2, \ldots, 5$, for the method (13) are displayed in Table 4. The method proposed by Yoshida is remarkable in the sense that is a fourth-order OS method with four stages and real coefficients $\alpha_j^{[i]}$.

3.3 Operator Splitting for the Bidomain Model

To apply operator splitting to the bidomain model, we write (1) as

$$\mathbf{M}\dot{\mathbf{Y}} = \mathscr{A}(\mathbf{Y}) = \mathscr{A}^{[1]}(\mathbf{Y}) + \mathscr{A}^{[2]}(\mathbf{Y}), \tag{14}$$

Algorithm 1 OS for the bidomain model

1: Define $\alpha_0^{[i]} = 0, i = 1, 2, t_{n\alpha_j^{[i]}} = t_n + \alpha_j^{[i]} \Delta t, \mathbf{Y}_{n\alpha_j^{[i]}} \approx \mathbf{Y}(t_n \alpha_j^{[i]})$.
2: **for** $j = 1$ to s **do**
3: Solve the ODE sub-system
4: $\mathbf{M}\dot{\mathbf{Y}} = \mathscr{A}^{[1]}(\mathbf{Y})$
5: for $t_{n\alpha_{j-1}^{[1]}} \le t \le t_{n\alpha_j^{[1]}}$ subject to the IC $\mathbf{Y}_{n\alpha_{j-1}^{[1]}} = \mathbf{Y}_{n\alpha_{j-1}^{[2]}}$ to obtain $\mathbf{Y}_{n\alpha_j^{[1]}}$
6: Solve the DAE sub-system
7: $\mathbf{M}\dot{\mathbf{Y}} = \mathscr{A}^{[2]}(\mathbf{Y})$
8: from $t_{n\alpha_{j-1}^{[2]}} \le t \le t_{n\alpha_j^{[2]}}$ subject to the IC $\mathbf{Y}_{n\alpha_{j-1}^{[2]}} = \mathbf{Y}_{n\alpha_j^{[1]}}$ and the BC (1d), (1e) to obtain $\mathbf{Y}_{n\alpha_j^{[2]}}$.
9: **end for**
10: Set $\mathbf{Y}_{n+1} = \mathbf{Y}_{n\alpha_s^{[2]}}$ as the solution at time $t_{n+1} = t_n + \Delta t$.

where $\mathbf{Y} \doteq (\mathbf{s}^T, v, u_e)^T$,

$$\mathbf{M} = \begin{pmatrix} \mathbf{I} & 0 & 0 \\ 0 & 1 & 0 \\ 0 & 0 & 0 \end{pmatrix},$$

$$\mathscr{A}^{[1]}(\mathbf{Y}) = \begin{pmatrix} \mathbf{f}(t, \mathbf{Y}) \\ -\frac{1}{C_m} I_{\text{ion}}(\mathbf{Y}) \\ 0 \end{pmatrix}, \text{ and } \mathscr{A}^{[2]}(\mathbf{Y}) = \begin{pmatrix} 0 \\ \frac{1}{\chi C_m} (\nabla \cdot (\sigma_i \nabla v) + \nabla \cdot (\sigma_i \nabla u_e)) \\ \nabla \cdot (\sigma_i \nabla v) + \nabla \cdot ((\sigma_i + \sigma_e) \nabla u_e) \end{pmatrix}.$$

Doing so, we split (1b) into a non-linear PDE, involving only the I_{ion} term and no spatial derivatives, and a linear PDE. Upon spatial discretization, the non-linear PDE forms an ODE sub-system at each mesh point, and hence we loosely refer to this sub-system as such. Similarly, the discretized linear PDE system forms a differential-algebraic equation (DAE) sub-system. For the appropriate choice of $\alpha_j^{[i]}$, an algorithm for one integration step of an OS method applied to the bidomain model is described in Algorithm 1, where it is assumed the ODE sub-system is solved first. For the sub-steps in (10) and (11) that require solving the DAE sub-system first, the order of the solves and the associated indices are reversed.

To ensure the overall accuracy of a third-order OS method is third-order accurate in time, we need to approximate the flows of both the ODE and DAE sub-systems to at least third order in time. To approximate the flow of the ODE sub-system in line (4), we use the three-stage, third-order explicit Runge–Kutta method (ERK3) defined by the Butcher tableau

$$\begin{array}{c|ccc} 0 & 0 & 0 & 0 \\ 1/2 & 1/2 & 0 & 0 \\ 1 & -1 & 2 & 0 \\ \hline & 1/6 & 2/3 & 1/6 \end{array}$$

To approximate the flow of the PDE sub-system in line (7), we use the A-stable, two-stage, third-order SDIRK method defined by the Butcher tableau

$$\begin{array}{c|cc} \gamma & \gamma & 0 \\ 1-\gamma & 1-2\gamma & \gamma \\ \hline & 1/2 & 1/2 \end{array}$$

with $\gamma = (3+\sqrt{3})/6$ [1].

Similar considerations must be made at order four. The numerical method used to approximate the flow of the ODE system is the classic four-stage, fourth-order explicit Runge–Kutta (ERK4) method

$$\begin{array}{c|cccc} 0 & 0 & 0 & 0 & 0 \\ 1/2 & 1/2 & 0 & 0 & 0 \\ 1/2 & 0 & 1/2 & 0 & 0 \\ 1 & 0 & 0 & 1 & 0 \\ \hline & 1/6 & 1/3 & 1/3 & 1/6 \end{array}$$

The numerical method used to approximate the flow the DAE system is the A-stable, three-stage, fourth-order SDIRK method defined by the Butcher tableau

$$\begin{array}{c|ccc} (1+\gamma)/2 & (1+\gamma)/2 & 0 & 0 \\ 1/2 & -\gamma/2 & (1+\gamma)/2 & 0 \\ (1-\gamma)/2 & 1+\gamma & -(1-2\gamma) & (1+\gamma)/2 \\ \hline & 1/(6\gamma^2) & 1-1/(3\gamma^2) & 1/(6\gamma^2) \end{array}$$

with $\gamma = \frac{2}{\sqrt{3}}\cos\left(\frac{\pi}{18}\right)$ [9, 21].

3.4 *Operator Splitting for the Monodomain Model*

In a similar fashion to the bidomain model, we write the monodomain model (2) as

$$\frac{\partial \mathbf{Y}}{\partial t} = \mathscr{A}^{[1]}(\mathbf{Y}) + \mathscr{A}^{[2]}(\mathbf{Y}),$$

where $\mathbf{Y} \doteq (\mathbf{s}^T, v)^T$,

$$\mathscr{A}^{[1]}(\mathbf{Y}) = \begin{pmatrix} \mathbf{f}(t,\mathbf{Y}) \\ -\frac{1}{C_m} I_{\text{ion}}(\mathbf{Y}) \end{pmatrix}, \text{ and } \mathscr{A}^{[2]}(\mathbf{Y}) = \begin{pmatrix} \mathbf{0} \\ \frac{\lambda}{(1+\lambda)} \frac{1}{\chi C_m} \nabla \cdot (\sigma_i \nabla v) \end{pmatrix}.$$

Algorithm 2 OS for the monodomain model

Define $\alpha_0^{[i]} = 0$, $t_{n\alpha_j^{[i]}} = t_n + \alpha_j^{[i]}\Delta t$, $\mathbf{Y}_{n\alpha_j^{[i]}} \approx \mathbf{Y}(t_{n\alpha_j^{[1]}})$.
for $j = 1$ to s **do**
  Solve the ODE sub-system
    $\dot{\mathbf{Y}} = \mathscr{A}^{[1]}(\mathbf{Y})$
  for $t_{n\alpha_{j-1}^{[1]}} \le t \le t_{n\alpha_j^{[1]}}$ subject to the IC $\mathbf{Y}_{n\alpha_{j-1}^{[1]}} = \mathbf{Y}_{n\alpha_{j-1}^{[2]}}$ to obtain $\mathbf{Y}_{n\alpha_j^{[1]}}$.
  Solve the PDE sub-system
    $\dot{\mathbf{Y}} = \mathscr{A}^{[2]}(\mathbf{Y})$
  from $t_{n\alpha_{j-1}^{[2]}} \le t \le t_{n\alpha_j^{[2]}}$ subject to the IC $\mathbf{Y}_{n\alpha_{j-1}^{[2]}} = \mathbf{Y}_{n\alpha_j^{[1]}}$ and the BC in (2c) to obtain $\mathbf{Y}_{n\alpha_j^{[2]}}$.
end for
Set $\mathbf{Y}_{n+1} = \mathbf{Y}_{n\alpha_s^{[2]}}$ as the solution at time $t_{n+1} = t_n + \Delta t$.

After spatial discretization, this splitting reduces the non-linear PDE to a set of non-linear ODEs at each spatial mesh point and a linear PDE. Again to ensure the overall method is accurate up to desired order, we approximate the flows of the ODE and PDE sub-systems using the appropriate integrators for the ODE and the PDE systems to reach third- and fourth-order accuracy as we did in Sect. 3.3.

For the appropriate choice of $\alpha_j^{[i]}$, one integration step of an OS method applied to the monodomain can be performed as described in Algorithm 2.

As previously mentioned, for the sub-steps in (10) that require solving the PDE sub-system first, the order of the solves and the associated indices are reversed.

4 Numerical Experiments

In this section, we present the numerical experiments that were performed in order to assess the order of convergence of the high-order OS methods described in Sect. 3.2 when applied to the bidomain and monodomain models. The experiments consist of problems set in one and three spatial dimensions with a variety of different initial conditions and cardiac ionic cell models with a range of stiffness characteristics.

The Chaste software environment [19] is used for all the numerical experiments. Chaste uses linear finite elements for the spatial discretization and integrates in time as per the method of lines. The (Chaste default) method for handling the I_{ion} term in the finite element method uses ionic current interpolation [22]. The (Chaste default) pair of Krylov subspace solver and preconditioner in Chaste, i.e., the conjugate gradient solver with block Jacobi preconditioner, are used as the linear system solver.

Timings were recorded in Chaste 3.4 running in serial on a dual Quad-Core Intel Xeon 2.26GHz with 16GB of RAM running OS X El Capitan 10.11.6.

4.1 Accuracy

To assess the accuracy of the third- and fourth-order OS methods described in Sect. 3.2, we use the Mixed Root Mean Square (MRMS) error [18] defined by

$$[\text{MRMS}]_w = \sqrt{\frac{1}{N}\sum_{i=1}^{N}\left(\frac{\tilde{w}_i - w_i}{1 + |\tilde{w}_i|}\right)^2}, \tag{15}$$

where $\tilde{w}_i$ and w_i denote the reference and numerical solutions, respectively, for the quantity w at the space-time point i. The order of convergence of the numerical solution p was then computed in the standard way as

$$p = \frac{\log([\text{MRMS}]_{w1}/[\text{MRMS}]_{w2})}{\log(\Delta t_1/\Delta t_2)},$$

where the subscripts 1 and 2 refer to computations performed using time steps Δt_1 and Δt_2.

Reference solutions for the experiments were generated as follows. We compared solutions computed in Chaste by halving the time step and doubling the number of mesh points until there were 4 to 6 matching digits between successive approximations. Solutions were then stored from the finest mesh at $N_s = 101$ equally spaced points in the spatial interval and $N_t = 21$ equally spaced points in the time interval of the simulations for a total of $N = N_s N_t$ space-time points.

4.1.1 1D Experiments

The monodomain model was simulated in one spatial dimension with two cell models, those of FitzHugh–Nagumo (FHN) [10] and Ten Tusscher–Panfilov (TTP) for epicardial tissue [33].

The FNH model has two cellular state variables and is considered to be non-stiff [28], whereas the TTP model has 19 cellular variables and is considered to be highly stiff [28]. We use the following initial conditions for the variables in each model:

$$\mathbf{s}(x, 0) = \mathbf{s}_0,$$
$$v(x, 0) = v_0 + 100(1 - \sin(x)),$$

where $\mathbf{s}_0$ and v_0 are the default values for $\mathbf{s}$ and v according to the definition in CellML [2] for the cell model in use. The Chaste default parameters $\chi = 1400$/cm, $C_m = 1\ \mu\text{F/cm}^2$, and $\sigma_i = 1.75$ mS/cm were used. The spatial domain was [0, 1] cm and the time interval was [0, 40] ms for all the reported cases. We used a regular grid with $N = 51$ points giving a spatial resolution $\Delta x = 0.02$ mm. Convergence of the third- and fourth-order methods was successfully assessed using different

Table 5 Convergence of the S3 method applied to the monodomain model using the FHN and the TTP epicardial cell models

	FHN		TTP	
Δt (ms)	$[\text{MRMS}]_v$	p	$[\text{MRMS}]_v$	p
1.5e−5	0.03896	–	0.04058	–
1.0e−5	0.01245	2.84	0.01358	2.90
8.0e−6	0.00653	2.89	0.00635	3.04

Table 6 Convergence of the AKS3 method applied to the monodomain model using the FHN and the TTP epicardial cell models

	FHN		TTP	
Δt (ms)	$[\text{MRMS}]_v$	p	$[\text{MRMS}]_v$	p
1.5e−5	0.03683	–	0.03946	–
1.0e−5	0.01304	3.20	0.01058	3.24
8.0e−6	0.00787	2.92	0.00528	3.11

Table 7 Convergence for the R3 method applied to the monodomain model using the LR1 and the TTP epicardial cell models

	FHN		TTP	
Δt (ms)	$[\text{MRMS}]_v$	p	$[\text{MRMS}]_v$	p
1.5e−5	0.03748	–	0.03699	–
1.0e−5	0.01102	3.01	0.0101	3.20
8.0e−6	0.00571	2.95	0.00512	3.04

Table 8 Convergence for the Y4 method applied to the monodomain model using the LR1 and the TTP epicardial cell models

	FHN		TTP	
Δt (ms)	$[\text{MRMS}]_v$	p	$[\text{MRMS}]_v$	p
1.5e−5	0.03051	–	0.02985	–
1.0e−5	0.00621	3.92	0.00571	4.07
8.0e−6	0.00321	3.96	0.00237	3.94

combinations of time steps for both cell models considered. Results for all the high-order methods described are reported in Tables 5, 6, 7 and 8. confirming the expected order of convergence.

4.1.2 3D Experiments

The convergence of the high-order OS methods when applied to solve the bidomain model was verified by considering the cell models of Luo–Rudy Phase I (LR1) [16] and TTP for epicardial tissue [33]. The model of LR1 has 9 cellular state variables and is considered mildly stiff [28]. We use the following initial condition for the variables in our model:

$$\begin{aligned}
\mathbf{s}(x, y, z, 0) &= \mathbf{s}_0, \\
v(x, y, z, 0) &= v_0 + 100(1 - \sin(xyz)), \\
u_e(x, y, z, 0) &= 0,
\end{aligned}$$

Table 9 Convergence for the S3 method applied to the bidomain model using the LR1 and the TTP epicardial cell models

	LR1		TTP	
Δt (ms)	$[\text{MRMS}]_v$	p	$[\text{MRMS}]_v$	p
1.5e−5	0.03975	–	0.03948	–
1.0e−5	0.01199	2.90	0.01219	2.89
8.0e−6	0.00597	3.12	0.00611	3.09

Table 10 Convergence for the AKS3 method applied to the bidomain model using the LR1 and the TTP epicardial cell models

	LR1		TTP	
Δt (ms)	$[\text{MRMS}]_v$	p	$[\text{MRMS}]_v$	p
1.5e−5	0.03751	–	0.03846	–
1.0e−5	0.00995	3.27	0.01145	2.98
8.0e−6	0.00491	2.98	0.00583	3.02

Table 11 Convergence for the R3 method applied to the bidomain model using the LR1 and the TTP epicardial cell models

	LR1		TTP	
Δt (ms)	$[\text{MRMS}]_v$	p	$[\text{MRMS}]_v$	p
1.5e−5	0.03799	–	0.03699	–
1.0e−5	0.01012	3.21	0.01184	2.90
8.0e−6	0.00495	3.19	0.00578	3.20

Table 12 Convergence for the Y4 method applied to the bidomain model using the LR1 and the TTP epicardial cell models

	LR1		TTP	
Δt (ms)	$[\text{MRMS}]_v$	p	$[\text{MRMS}]_v$	p
1.5e−5	0.03147	–	0.03004	–
1.0e−5	0.00571	4.20	0.00583	4.04
8.0e−6	0.00234	3.99	0.00242	3.97

where, as before $\mathbf{s}_0$ and v_0 are the CellML default values for $\mathbf{s}$ and v according to the cell model in use [2]. We use the (Chaste default) parameter values $\chi = 1400/\text{cm}$, $C_m = 1\ \mu\text{F/cm}^2$, $\sigma_i = \text{diag}(\sigma_i^f, \sigma_i^n, \sigma_i^t)$, $\sigma_e = \text{diag}(\sigma_e^f, \sigma_e^n, \sigma_e^t)$, with $\sigma_i^f = \sigma_i^n = \sigma_i^t = 1.75$ mS/cm and $\sigma_e^f = \sigma_e^n = \sigma_e^t = 7$ mS/cm. The spatial domain was $[0, 1]\text{cm} \times [0, 1]\text{cm} \times [0, 1]\text{cm}$, and the time interval was $[0, 40]$ ms. We used a grid with $N = 51$ equally spaced points per direction resulting in a $\Delta x = 0.02$ mm. The accuracy of the third- and fourth-order methods when applied to the bidomain model was assessed for all the cell models considered in this study, and the results are reported in Tables 9, 10, 11 and 12, confirming the expected order of convergence for all the schemes.

5 Conclusions

The generation of clinically relevant numerical solutions to the bidomain and monodomain models for the simulation of cardiac electrophysiology is computationally demanding. Due to typical limitations of readily available computer hardware, the bidomain and monodomain models are typically solved with the use of OS methods. High-order OS methods hold the promise of greater efficiency for accurate calculations, and hence finding high-order numerical methods is of great interest.

However, OS methods of order greater than two require sub-flows in the direction of negative time. Because such backward time integration is ill-posed for diffusion problems, it is not well appreciated that it is possible to use high-order OS methods in practice for the bidomain and monodomain models. In this chapter, we demonstrate that third- and fourth-order OS methods can successfully simulate the bidomain and monodomain models and converge at their expected orders despite the necessary backward time integration.

References

1. Ascher, U.M., Ruuth, S.J., Spiteri, R.J.: Implicit-explicit Runge–Kutta methods for time-dependent partial differential equations. Appl. Numer. Math. **25**(2–3), 151–167 (1997)
2. Autumn, A.C., et al.: An overview of CellML 1.1, a biological model description language. Simulation **79**(12), 740–747 (2003)
3. Auzinger, W., Herfort, W.: Local error structures and order conditions in terms of lie elements for exponential splitting schemes. Opuscula Mathematica **34**(2) (2014)
4. Auzinger, W., Herfort, W., Hofstätter, H., Koch, O.: Setup of order conditions for splitting methods. In: Computer Algebra in Scientific Computing: 18th International Workshop, CASC 2016, Bucharest, Romania, September 19–23, 2016, Proceedings, pp. 30–42. Springer International Publishing (2016)
5. Auzinger, W., Hofstätter, H., Ketcheson, D., Koch, O.: Practical splitting methods for the adaptive integration of nonlinear evolution equations. part i: Construction of optimized schemes and pairs of schemes. BIT Numerical Mathematics 1–20 (2016)
6. Blanes, S., Casas, F.: On the necessity of negative coefficients for operator splitting schemes of order higher than two. Appl. Numer. Math. **54**(1), 23–37 (2005)
7. Blanes, S., Casas, F., Farres, A., Laskar, J., Makazaga, J., Murua, A.: New families of symplectic splitting methods for numerical integration in dynamical astronomy. Appl. Numer. Math. **68**, 58–72 (2013)
8. Cherry, E.M., Greenside, H.S., Henriquez, C.S.: A space-time adaptive method for simulating complex cardiac dynamics. Phys. Rev. Lett. **84**(6), 1343 (2000)
9. Crouzeix, M.: Sur lapproximation des équations différentielles opérationelles linéaires par des méthodes de Runge–Kutta. PhD thesis, Université Paris (1978)
10. FitzHugh, R.: Mathematical models of excitation and propagation in nerve. In: Schwan, H.P. (ed.) Biological Engineering, pp. 1–85. McGraw-Hill (1966)
11. Geraldo-Giorda, L.: Nonlinear Dynamics in Biological Systems, vol. 7. Springer International Publishing (2016)
12. Godunov, S.K.: A difference method for numerical calculation of discontinuous solutions of the equations of hydrodynamics. Matematicheskii Sbornik **89**(3), 271–306 (1959)
13. Goldman, G., Kaper, T.J.: Nth-order operator splitting schemes and nonreversible systems. SIAM J. Numer. Anal. **33**(1), 349–367 (1996)
14. Koch, O., Neuhauser, Ch., Thalhammer, M.: Embedded exponential operator splitting methods for the time integration of nonlinear evolution equations. Appl. Numer. Math. **63**, 14–24 (2013)
15. LeVeque, R.J.: Numerical Methods for Conservation Laws. Birkhauser, Basel (1992)
16. Luo, C.H., Rudy, Y.: A model of the ventricular cardiac action potential. depolarization, repolarization, and their interaction. Circulation Research **68**(6), 1501–1526 (1991)
17. Marchuk, G.I.: Splitting Methods. Nauka, Moscow (1988)

18. Marsh, M.E., Ziaratgahi, S.T., Spiteri, R.J.: The secrets to the success of the rush–larsen method and its generalizations. IEEE Trans. Biomed. Eng. **59**(9), 2506–2515 (2012)
19. Mirams, G.R., et al.: Chaste: An open source C++ library for computational physiology and biology. PLoS Comput. Biol. **9**(3) (2013)
20. Niederer, S.A., et al.: Verification of cardiac tissue electrophysiology simulators using an n-version benchmark. Philos. Trans. R. Soc. Lond. A Math. Phys. Eng. Sci. **369**(1954), 4331–4351 (2011)
21. Nørsett, S.P.: Semi explicit Runge–Kutta methods. Mathematics and Computation, no. 6. University of Trondheim (1974)
22. Pathmanathan, P., Mirams, G.R., Southern, J., Whiteley, J.P.: The significant effect of the choice of ionic current integration method in cardiac electro-physiological simulations. Int. J. Numer. Methods Biomed. Eng. **27**(11), 1751–1770 (2011)
23. Qu, Z., Garfinkel, A.: An advanced algorithm for solving partial differential equation in cardiac conduction. IEEE Trans. Biomed. Eng. **46**(9), 1166–1168 (1999)
24. Ruth, R.D.: A canonical integration technique. IEEE Trans. Nucl. Sci. **30** (1983)
25. Sheng, Q.: Solving linear partial differential equations by exponential splitting. IMA J. Numer. Anal. **9**, 199–212 (1989)
26. Sornborger, A.T.: Higher-order operator splitting methods for deterministic parabolic equations. Int. J. Comput. Math. **84**(6), 887–893 (2007)
27. Sornborger, A.T., Stewart, E.D.: Higher-order methods for simulations on quantum computers. Phys. Rev. A **60**(3), 765–789 (1999)
28. Spiteri, R.J., Dean, R.C.: Stiffness analysis of cardiac electrophysiological models. Ann. Biomed. Eng. **38**(12), 3592–3604 (2010)
29. Strang, G.: On the construction and comparison of difference schemes. SIAM J. Numer. Anal. **5**(3), 506–517 (1968)
30. Sundnes, J., Lines, G.T., Cai, X.: Computing the Electrical Activity in the Heart. Springer (2006)
31. Sundnes, J., Lines, G.T., Tveito, A.: An operator splitting method for solving the bidomain equations coupled to a volume conductor model for the torso. Mathematical Biosciences **194**(2), 233–248 (2005)
32. Suzuki, M.: Solving linear partial differential equa tions by exponential splitting. IMA J. Numer. Anal. **9**, 400–407 (1991)
33. Ten Tusscher, K.H.W.J., Panfilov, A.V.: Alternans and spiral breakup in a human ventricular tissue model. Am. J. Physiol. Heart Circ. Physiol. **291**(3), H1088–H1100 (2006)
34. Trotter, H.F.: On the product of semi-groups of operators. Proc. Am. Math. Soc. **10**(4), 545–551 (1959)
35. Tung, L.: A bi-domain model for describing ischemic myocardial d-c potentials. PhD thesis, Massachusetts Institute of Technology (1978)
36. Yoshida, H.: Construction of higher order symplectic integrators. Phys. Lett. A **150**(5), 262–268 (1990)

Jessica Cervi is a Ph.D. student in the Department of Mathematics and Statistics at the University of Saskatchewan. Her research interests are numerical analysis and scientific computing with particular emphasis to higher-order operator splitting methods and cardiac simulations.

Raymond J. Spiteri is a Professor in the Department of Computer Science. He obtained a Ph.D. in Mathematics (Institute of Applied Mathematics) from the University of British Columbia in 1997. His areas of research expertise are numerical analysis, scientific computing, and high-performance computing, with special emphasis on time-stepping methods, problem-solving software environments, and application to problems with industrial relevance.

Electro-Mechanical Modeling and Simulation of Reentry Phenomena in the Presence of Myocardial Infarction

Piero Colli Franzone, Luca F. Pavarino, and Simone Scacchi

Abstract In this work we present a parallel solver for the numerical simulation of the cardiac electro-mechanical activity. We first review the most complete mathematical model of cardiac electro-mechanics, the so-called electro-mechanical coupling (EMC) model, which consists of the following four sub-models, strongly coupled together: the Bidomain model for the electrical activity at tissue scale, constituted by a parabolic system of two reaction-diffusion partial differential equations (PDEs); the finite elasticity system for the mechanical behavior at tissue scale; the membrane model for the bioelectrical activity at cellular scale, consisting of a stiff system of ordinary differential equations (ODEs); the active tension model for the mechanical activity at cellular scale, consisting of a system of ODEs. The discretization of the EMC model is performed by finite elements in space and an operator splitting strategy in time, based on semi-implicit finite differences. As a result of the discretization techniques adopted, the most computational demanding part at each time step is the solution of the non-linear algebraic system, deriving from the discretization of the finite elasticity equations, and of the linear system deriving from the discretization of the Bidomain equations. The former is solved by a Newton-GMRES-BDDC solver, i.e. the Jacobian system at each Newton iteration is solved by GMRES accelerated by the Balancing Domain Decomposition by Constraints (BDDC) preconditioner. The latter is solved by the Conjugate Gradient method, preconditioned by the Multilevel Additive Schwarz preconditioner. The performance of the resulting parallel solver is studied on the simulation of the induction of ventricular tachycardia in an idealized left ventricle affected by an infarct scar. The simulations are run on the Marconi-KNL cluster of the Cineca laboratory.

P. Colli Franzone · L. F. Pavarino
Department of Mathematics, University of Pavio, Pavia, Italy
e-mail: colli@imati.cnr.it; luca.pavarino@unipv.it

S. Scacchi (✉)
Department of Mathematics, University of Milano, Milano, Italy
e-mail: simone.scacchi@unimi.it

D. Boffi et al. (eds.), *Mathematical and Numerical Modeling of the Cardiovascular System and Applications*, SEMA SIMAI Springer Series 16,
https://doi.org/10.1007/978-3-319-96649-6_3

1 Introduction

The heartbeat represents a complex multiphysical phenomenon, consisting of three main aspects: bioelectrical, mechanical and fluid dynamical. The propagation of the electrical impulse through the cardiac tissue is mathematically described by non-linear reaction-diffusion equations. The subsequent mechanical contraction and relaxation process is described by the large deformation elasticity equations. The ejection of blood from atria to ventricles and then from ventricles to the circulatory system, induced by the mechanical contraction of the cardiac muscle, is described by the Navier-Stokes equations. For integrated models of the whole heart function, see the recent reviews [10, 46, 78, 79]. In this work, we will focus only on the coupling of the bioelectrical and mechanical phenomena.

The most complete mathematical model of the cardiac bioelectrical activity at macroscopic level is the so-called Bidomain model, see [27, 95]. In this model, the cardiac tissue is represented as the superposition of two anisotropic continuous media, the intra- and extra- cellular media, coexisting at every point of the tissue and separated by a distributed continuous cellular membrane; see [72] for a rigorous derivation of the macroscopic Bidomain model from homogenization of cellular models and [16] for an introduction to its mathematical analysis and numerical discretization. The anisotropy of the intra- and extra-cellular media is related to the macroscopic arrangement of the cardiac myocytes in a fiber structure. The model consists of a degenerate system of two parabolic reaction-diffusion partial differential equations (PDEs), that describe the spatio-temporal evolution of the intra- and extra-cellular electric potentials. The diffusion terms are linear, with diffusion tensors modeling the anisotropic structure of the cardiac tissue. Through the non-linear reaction term, the two PDEs are coupled with a system of ordinary differential equations (ODEs), the so-called membrane model, that describes the ionic exchanges across the cellular membrane. The unknowns of the membrane model are the gating variables, representing the percentage of channels open for a specific ion, and the intra- and extra-cellular concentrations of ions, among which the most important are sodium, potassium and calcium. The first membrane model was proposed by Hodgkin and Huxley in [32]. Although the Hodgkin-Huxley model was developed specifically for nerve cells, the ideas and the mathematical formalism have been used later in a huge number of models describing the ionic fluxes across the membrane of cardiac cells, see the reviews [64, 84] and the monographs [16, 77].

The numerical solution of the Bidomain system is computationally very expensive, because of the interaction of the different scales in space and time, the degenerate nature of the PDEs involved and the very severe ill-conditioning of the discrete systems arising at each time step. Fully implicit methods in time have been considered in few studies, see e.g. [54–56, 87] and require the solution of non-linear systems at each time step. Instead, a large body of numerical studies employ semi-implicit methods in time and/or operator splitting schemes, where the diffusion and reaction terms are treated separately, see e.g [15, 101]. In order to devise efficient iterative solvers for the linear systems arising at each time step, many different preconditioners have been proposed: block type [14, 26, 53, 69, 73], multigrid [4, 5, 74, 76, 89], Additive Schwarz [68, 69], Neumann-Neumann and BDDC [104, 105] preconditioners.

The intracellular calcium released during electrical activation couples the electrical and mechanical phenomena of the heartbeat, being an input of the active tension model, which describes the generation of active force within each myocyte, see [61, 81]. Active tension models consist of systems of ordinary differential—algebraic equations describing the calcium binding to troponin C, cooperativity between regulatory proteins, and cross-bride cycling.

The deformation of the cardiac muscle is modeled by the equations of finite elasticity, with the myocardium being an orthotropic, hyperelastic, and nearly incompressible material. The orthotropic passive properties of the cardiac tissue are described by strain energy functions of exponential or pole-zero type laws, see e.g. [29, 35, 58], while the active properties depend on the cellular active tension. During contraction, the stretch ratio (ratio of myocyte length before and after deformation) and its time derivative influence the active tension generation, including length-dependent calcium sensitivity. As a result, the active tension model and the finite elasticity system are strongly coupled. Finally, to take into account the pressure of cavitary blood on the interior surfaces of the ventricles, the non-linear elasticity PDEs are coupled through the boundary conditions with lumped-parameter models of the systemic and pulmonary circulatory systems.

The coupling of the two electrophysiological models (Bidomain and membrane model) with the two mechanical models (finite elasticity and active tension model) yields the so-called cardiac electro-mechanical coupling (EMC) model. This complex non-linear model poses great theoretical and numerical challenges. At the numerical level, the approximation and simulation of the cardiac EMC model is a very demanding and expensive task, because of the very different space and time scales associated with the electrical and mechanical models, as well as their non-linear and multiphysics interactions. Nevertheless, in recent years several research groups have developed electro-mechanical models by assembling the main four components with different levels of coupling and description details: see e.g. [24, 28, 31, 41, 59, 67, 86, 96] for electro-mechanical simulations based on the active-stress approach; [12, 57, 63, 82] for electro-mechanical simulations based on the active-strain approach; [19, 39, 40, 102] for electro-mechanical simulations of cardiac arrhythmias, including the mechano-electric feedback; [4] for a parallel highly scalable algebraic multigrid solver for cardiac electro-mechanics; [11] for a recent stress-driven electro-mechanical model.

The aim of this work is to review the main mathematical models that constitute the EMC model and to present our parallel solver for the solution of EMC model. The solver is based on a finite element discretization in space of the PDEs of the model and on operator splitting methods in time, employing semi-implicit finite difference schemes. The non-linear system arising from the discretization of finite elasticity is solved by a Newton-GMRES-BDDC solver, where the Jacobian system at each Newton iteration is solved by GMRES accelerated by the Balancing Domain Decomposition by Constraints (BDDC) preconditioner. The linear system arising from the discretization of the Bidomain equations is solved by the Conjugate Gradient method, preconditioned by the Multilevel Additive Schwarz preconditioner. We apply the parallel solver to the simulation of ventricular tachycardia in an idealized left ventricular geometry, affected from infarct scars. The simulations are run on 256 cores of the Marconi-KNL cluster of the italian computing center Cineca.

2 Mathematical Models

2.1 Macroscopic Mechanical Model

Let us denote the material coordinates of the undeformed or reference cardiac domain by $\mathbf{X} = (X_1, X_2, X_3)^T$, the spatial coordinates of the deformed cardiac domain by $\mathbf{x} = (x_1, x_2, x_3)^T$ and the region occupied at time t by the undeformed and deformed cardiac domains by $\widehat{\Omega}$ and $\Omega(t)$, respectively. We denote by Div and div (Grad and grad) the material and spatial divergence (gradient) of a vector (scalar), respectively. From a mechanical point of view, the cardiac tissue is modeled as a non-linear elastic material. The deformation gradient tensor $\mathbf{F}$ and its determinant J are given by

$$\mathbf{F}(\mathbf{X}, t) = \{F_{ij}\} = \left\{\frac{\partial x_i}{\partial X_j} \quad i, j = 1, 2, 3\right\}, \qquad J(\mathbf{X}, t) = det\mathbf{F}(\mathbf{X}, t).$$

The Cauchy-Green deformation tensor $\mathbf{C}$ and Lagrange-Green strain tensor $\mathbf{E}$ are

$$\mathbf{C} = \mathbf{F}^T\mathbf{F} \quad \text{and} \quad \mathbf{E} = \frac{1}{2}(\mathbf{C} - \mathbf{I}),$$

where $\mathbf{I}$ denotes the identity matrix.

Most simulation works based on cardiac electro-mechanical models, see e.g. [39–41, 59], have adopted the quasi-static mechanical regime. For an evaluation of the influence of the inertial term, we refer to [85, 103]. In the quasi-static regime, the Cauchy's equation of equilibrium in the coordinates of the deformed body, neglecting external volumetric forces, is given in term of the Cauchy stress tensor σ by

$$\text{div } \sigma = \mathbf{0}, \text{ in } \Omega(t),$$

while in the coordinates of the undeformed body it satisfies the steady-state force equilibrium equation

$$\text{Div}(\mathbf{FS}) = \mathbf{0}, \qquad \mathbf{X} \in \widehat{\Omega}, \tag{1}$$

where $\mathbf{S} = J\mathbf{F}^{-1}\sigma\mathbf{F}^{-T}$ is the second Piola-Kirchoff stress tensor. We close the quasi-static mechanical model (1) by imposing a prescribed displacement on a Dirichlet boundary $\mathbf{x}(\mathbf{X}) = \hat{\mathbf{x}}(\mathbf{X}),\ \mathbf{X} \in \partial\widehat{\Omega}_D$, a prescribed normal stress on a Neumann boundary, representing the cavitary blood pressure on the endocardial ventricular surface, i.e. $\mathbf{FS}\,\mathbf{N} = -\ p(t)\,\mathbf{N},\ \mathbf{X} \in \partial\widehat{\Omega}_N$ with $\mathbf{N}$ the outward normal to $\widehat{\Omega}$, and a free-stress condition elsewhere.

According to the so-called active-stress approach, the tensor $\mathbf{S}$ is given by the sum of a passive elastic component $\mathbf{S}^{pas}$, a volumetric component $\mathbf{S}^{vol}$ and a biochemically generated active component $\mathbf{S}^{act}$,

$$\mathbf{S} = \mathbf{S}^{pas} + \mathbf{S}^{vol} + \mathbf{S}^{act},$$

as proposed in many previous studies, see e.g. [41, 59, 67, 88, 100]. An alternative multiplicative strategy, called active-strain, for combining the passive and active properties of the cardiac tissue has been proposed in [12], see also [75, 82]. In our simulations presented in the last section of the present paper, we have adopted the former approach, because in the literature biophysical cellular models of the complex cross-bridge sarcomere dynamics are available, see e.g. [7, 47, 61, 81]. Adopting instead the active-strain approach, to our knowledge only phenomenological fiber stretch models can be used so far, see [78, 79, 82, 83].

The passive component $\mathbf{S}^{pas}$ is computed from a suitable strain energy function W^{pas} and the Green-Lagrange strain tensor $\mathbf{E}$

$$S_{ij}^{pas} = \frac{1}{2}\left(\frac{\partial W^{pas}}{\partial E_{ij}} + \frac{\partial W^{pas}}{\partial E_{ji}}\right) \quad i, j = 1, 2, 3.$$

Several strain energy functions W^{pas} have been proposed and adopted in the literature, see e.g. [21, 29, 30, 35, 58, 86].

We recall that cardiac cells are elongated cells connected end-to-end, forming the cardiac fibers, which lie on muscle sheets, running radially across the ventricular wall. Moreover, fibers are connected also side-to-side and rotate counterclockwise from epi- to endocardium, with fiber connections between adjacent sheets, see e.g. [48]. Therefore this arrangement of fibers and the laminar organization imply that, at a macroscopic level, the cardiac tissue displays orthotropic properties both of electric conduction and mechanical stress, see e.g [20, 35, 97].

Let us denote by $\mathbf{a}_l(\mathbf{x}),\ \mathbf{a}_t(\mathbf{x}),\ \mathbf{a}_n(\mathbf{x})$ a triplet of orthonormal principal axes with $\mathbf{a}_l(\mathbf{x})$ parallel to the local fiber direction, $\mathbf{a}_t(\mathbf{x})$ and $\mathbf{a}_n(\mathbf{x})$ tangent and orthogonal to the radial laminae, respectively, and both being transversal to the fiber axis. Indicating by $\hat{\mathbf{a}}_l(\mathbf{X}), \hat{\mathbf{a}}_t(\mathbf{X}), \hat{\mathbf{a}}_n(\mathbf{X})$ the unit vectors parallel and across the local fiber direction in the reference configuration, then, $\mathbf{a}_{l,t,n} = \mathbf{F}\hat{\mathbf{a}}_{l,t,n}/||\mathbf{F}\hat{\mathbf{a}}_{l,t,n}||$ gives the unit vector parallel to the local fiber in the deformed configuration and it holds $\mathbf{a}_{l,t,n} = \mathbf{F}\hat{\mathbf{a}}_{l,t,n}/\sqrt{\hat{\mathbf{a}}_{l,t,n}^T \mathbf{C}\,\hat{\mathbf{a}}_{l,t,n}^T}$.

Three examples of strain energy functions, one isotropic and the other two anisotropic, are described here below.

Isotropic Strain Energy Function In the simplest case, the myocardium is modeled as a neo-Hookean material with the strain energy function

$$W^{pas} = c(I_1 - 3), \tag{2}$$

where $I_1 = trace(\mathbf{C})$ is the first invariant of the Cauchy-Green deformation tensor $\mathbf{C}$.

Axisymmetric Strain Energy Function In this case, the myocardium is modeled as a transversely isotropic hyperelastic material, with the exponential strain energy function [29, 100]

$$\begin{aligned} W^{pas} &= \frac{1}{2} c \left(e^{Q} - 1\right), \\ Q &= b_{ll} E_{ll}^2 + b_{tn}(E_{nn}^2 + E_{tt}^2 + 2E_{nt}^2) + 2b_{lt}(E_{lt}^2 + E_{ln}^2). \end{aligned} \tag{3}$$

The Lagrange-Green strain tensor is referred here to the orthogonal local fiber coordinate system, consisting of the fiber direction (l), and two orthogonal cross fiber directions (t,n), for instance

$$E_{rs} = \hat{\mathbf{a}}_r^T \mathbf{E} \hat{\mathbf{a}}_s, \quad r, s \in \{l, t, n\}$$

$$E_{ll} = \hat{\mathbf{a}}_l^T \mathbf{E} \hat{\mathbf{a}}_l, \quad E_{tt} = \hat{\mathbf{a}}_t^T \mathbf{E} \hat{\mathbf{a}}_t, \quad E_{nn} = \hat{\mathbf{a}}_n^T \mathbf{E} \hat{\mathbf{a}}_n,$$

$$E_{ln} = \hat{\mathbf{a}}_l^T \mathbf{E} \hat{\mathbf{a}}_n, \quad E_{lt} = \hat{\mathbf{a}}_t^T \mathbf{E} \hat{\mathbf{a}}_t, \quad E_{nt} = \hat{\mathbf{a}}_n^T \mathbf{E} \hat{\mathbf{a}}_t,$$

where $\hat{\mathbf{a}}_l$, $\hat{\mathbf{a}}_t$ and $\hat{\mathbf{a}}_n$ are the unit vectors of the local fiber coordinate system in the reference configuration. The material constant c rescales the stress, b_{ll} and b_{tn} rescale the material stiffness in the fiber and the two cross fiber directions, and b_{lt} rescales the material rigidity under shear in the fiber-transverse plane. We remark that the strain energy function (3) is transversely isotropic because the parameter rescaling the terms E_{nn}^2 and E_{tt}^2 is the same. Rescaling these terms with different parameters would lead to an orthotropic strain energy function.

Orthotropic Strain Energy Function In general, the myocardium is modeled as an orthotropic hyperelastic material, with the exponential strain energy function proposed by Holzapfel-Odgen in [35]:

$$W^{pas} = \frac{a}{2b}\left(e^{b(I_1-3)} - 1\right) + \sum_{i=l,n} \frac{a_i}{2b_i}\left(e^{b_i(I_{4i}-1)^2} - 1\right) + \frac{a_{ln}}{2b_{ln}}\left(e^{b_{ln} I_{8ln}^2} - 1\right), \tag{4}$$

where $a,\ b,\ a_{(l,n,ln)},\ b_{(l,n,ln)}$ are positive material parameters and

$$I_{4l} = \hat{\mathbf{a}}_l^T \mathbf{C}\, \hat{\mathbf{a}}_l, \ I_{4n} = \hat{\mathbf{a}}_n^T \mathbf{C}\, \hat{\mathbf{a}}_n, \ I_{8ln} = \hat{\mathbf{a}}_l^T \mathbf{C}\, \hat{\mathbf{a}}_n.$$

In order to model the nearly-incompressibility of the myocardium, following [100], we add to the energy a volumetric term W^{vol}, depending on the positive bulk modulus K,

$$W^{vol} = K\left(\sqrt{\det(\mathbf{C})} - 1\right)^2, \tag{5}$$

hence the volumetric component $\mathbf{S}^{vol}$ of the second Piola-Kirchhoff stress tensor is

$$S_{ij}^{vol} = \frac{1}{2}\left(\frac{\partial W^{vol}}{\partial E_{ij}} + \frac{\partial W^{vol}}{\partial E_{ji}}\right) \quad i, j = 1, 2, 3.$$

Another possibility is to choose the volumetric energy is

$$W^{vol} = K\left(J \log J - J + 1\right).$$

see for other choices [23].

2.2 *The Excitation-Contraction Coupling Model*

The contraction of the ventricles results from the active tension generated by the model of myofilaments dynamics activated by calcium (see e.g. [37, 61, 81] for excitation-contraction coupling models). We assume that the generated active force acts only in the direction of the fiber, as in the other works [28, 67, 103]. Hence, according to [36, Ch. 10], the active Cauchy stress is expressed as

$$\boldsymbol{\sigma}^{act} = T_a \mathbf{a}_l \otimes \mathbf{a}_l,$$

where $\mathbf{a}_l$ is a unit vector parallel to the local fiber direction and $T_a 9\mathbf{x}, t)$ is the active stress related to the deformed domain.

In terms of the principal axes of the reference configuration, we obtain:

$$\mathbf{a}_l \otimes \mathbf{a}_l = \frac{\mathbf{F}\hat{\mathbf{a}}_l \otimes \mathbf{F}\hat{\mathbf{a}}_l}{||\mathbf{F}\hat{\mathbf{a}}_l||^2} = \frac{\mathbf{F}\hat{\mathbf{a}}_l\, \hat{\mathbf{a}}_l^T\, \mathbf{F}^T}{\hat{\mathbf{a}}_l^T\, \mathbf{C}\, \hat{\mathbf{a}}_l}. \tag{6}$$

Then, the corresponding second Piola-Kirchhoff active stress component is given by

$$\mathbf{S}^{act} = J\, \mathbf{F}^{-1}\, \sigma^{act}\, \mathbf{F}^{-T} = J\, \widehat{T}_a\, \frac{\hat{\mathbf{a}}_l \otimes \hat{\mathbf{a}}_l}{\hat{\mathbf{a}}_l^T \mathbf{C}\, \hat{\mathbf{a}}_l}.$$

where $\widehat{T}_a(\mathbf{X}, t) = T_a(\phi_t^{-1}(\mathbf{x}), t)$ with $\phi_t : \widehat{\Omega} \rightarrow \Omega(t)$.

Active components in the direction $\hat{\mathbf{a}}_n$ can be also considered.

The biochemically generated active tension T_a was described with models incorporating different degrees of detail of the myofilaments dynamics. Here, we briefly describe the main three models used in simulation studies.

Stretch and Stretch-Rate Independent Active Tension $\widehat{T}_a = \widehat{T}_a(\widehat{v})$ In the electro-mechanical models presented in [28, 59], the dynamics of $\widehat{T}_a(\mathbf{X}, t)$ depends only on the transmembrane potential distribution $\widehat{v}(\mathbf{X}, t)$, defined on the reference configuration, according to this simple twitch-like rule

$$\frac{\partial \widehat{T}_a}{\partial t} = \epsilon(\widehat{v})(k_{\widehat{T}_a}(\widehat{v} - v_r) - \widehat{T}_a), \tag{7}$$

where $k_{\widehat{T}_a} > 0$ controls the saturated value of $\widehat{T}_a$ for a given potential $\widehat{v}$ and a given resting potential v_r, which is about $-80\,\text{mV}$ in cardiac cells. The function $\epsilon(\widehat{v})$ controls the delay in the development and recovery of active stress with respect to $\widehat{v}$ and is given by

$$\epsilon(\widehat{v}) = \epsilon_0 + (\epsilon_\infty - \epsilon_0)\exp(-\exp(-\xi(\widehat{v} - \bar{v}))),$$

where ϵ_0, ϵ_∞, ξ are positive constants and $\bar{v} = -30\,\text{mV}$. Under this simplified assumption, the active component of the second Piola-Kirchhoff stress tensor $\mathbf{S}^{act}$ can be written as the derivative of the active energy function

$$W^{act} = \frac{1}{2}\widehat{T}_a \log(I_{4l}).$$

We remark that in general active force is not conservative, thus it can not be associated to an active strain energy function. Nevertheless, this active energy function was considered to establish the solvability of the coupled electro-mechanical model, see [2, 66].

Stretch-Rate Independent Active Tension $\widehat{T}_a = \widehat{T}_a(\widehat{Ca}_i, \lambda)$ In this model proposed in [37, 38], the active tension depends on the intracellular calcium concentration $\widehat{Ca}_i$ (one of the variables indicated with c in system (10), defined here on the reference configuration) and on the muscle fiber stretch

$$\lambda = \sqrt{I_{4l}} = \sqrt{\hat{\mathbf{a}}_l^T \mathbf{C}\, \hat{\mathbf{a}}_l}, \tag{8}$$

according to the equation

$$\widehat{T}_a = \frac{\widehat{Ca}_i^n}{\widehat{Ca}_i^n + C_{50}^n} T_a^{max}(1 + \eta(\lambda - 1)).$$

$C_{50} = 0.5\,\mu M$ is the value of the intracellular calcium concentration for 50% of T_a^{max}, $n = 3$ is a coefficient controlling the shape of the curve, $T_a^{max} = 100\,kPa$ is the maximum isometric active stress developed at $\lambda = 1$ and $\eta = 2.5$.

Stretch and Stretch-Rate Dependent Active Tension $\widehat{T}_a = \widehat{T}_a\left(\widehat{Ca}_i, \lambda, \frac{\partial\lambda}{\partial t}\right)$ In this model proposed by Land et al. [47], based on the previous models [61, 81], the dynamics of the active tension $\widehat{T}_a$ is described by the following system of ordinary differential equations

$$\begin{cases} \dfrac{\partial tr}{\partial t} = k_{tr}\left(\left(\dfrac{\widehat{Ca}_i}{Ca_{50}(1+\beta(\lambda-1))}\right)^{n_{tr}}(1-tr) - tr\right) \\ \dfrac{\partial xb}{\partial t} = k_{xb}\left(tr_{50}tr^{n_{xb}}(1-xb) - \dfrac{1}{tr_{50}tr^{n_{xb}}}xb\right) \\ \dfrac{\partial Q_i}{\partial t} = A_i\dfrac{\partial\lambda}{\partial t} - \alpha_i Q_i, \qquad i = 1, 2 \\ \widehat{T}_a = g(Q)\,h(\lambda)\,xb, \qquad Q = Q_1 + Q_2, \end{cases} \tag{9}$$

with constant parameters $0 < k_{tr} < 1, 0 < k_{xb} < 1, n_{tr} > 1, n_{xb} > 1, 0 < Ca_{50} < 1, 0 < tr_{50} < 1, \beta > 1, A_1 < 0, A_2 > 0, \alpha_1, \alpha_2 > 0$. The function $h : \mathbb{R} \to \mathbb{R}$ is continuous and defined by

$$h(x) = \begin{cases} 0 & x \leq x_0 \\ \bar{h}(x) & x_0 < x < x_1 \\ h_{max} & x \geq x_1, \end{cases}$$

with $0 < x_0 < 1$, $x_1 > 1$, $h_{max} > 0$ and $0 < \bar{h}(x) < h_{max}$ non-decreasing Lipschitz function. The function $g(Q) : \mathbb{R} \to \mathbb{R}$ is a non-decreasing, bounded, Lipschitz function.

Note that the two differential equations for Q_i, $i = 1, 2$, can be rewritten as integral equations

$$Q_i(t) = A_i \int_{-\infty}^{t} \exp(-\alpha_i(t-s))\frac{\partial\lambda}{\partial s}\,ds.$$

2.3 *The Bidomain Model*

The macroscopic Bidomain representation of the cardiac tissue is obtained by considering the superposition of two anisotropic continuous media, the intra- (i) and extra- (e) cellular media, coexisting at every point of the tissue and separated by a distributed continuous cellular membrane; see e.g. [16, 72] for a derivation of the Bidomain model from homogenization of a periodic assembling of cellular models. We remark that the Bidomain model describes the propagation of the electric impulse in the atrial and ventricular tissue. For models of the propagation of the electric impulse in the Purkinje network, see e.g. [60, 65, 99].

The anisotropy of the intra- and extracellular media, related to the macroscopic arrangement of the cardiac myocytes in the fiber structure, is described by the anisotropic conductivity tensors $\mathbf{D}_i(\mathbf{x})$ and $\mathbf{D}_e(\mathbf{x})$, respectively, defined in (11) below.

We will now introduce the parabolic-parabolic formulation of the Bidomain system on the cardiac domain $\Omega(t)$ for $t \in (0, T)$. Given an applied current per unit volume $I_{app} : \Omega(t) \to \mathbb{R}$, and initial conditions $v_0 : \Omega(0) \to \mathbb{R}$, $(\mathbf{w}_0, \mathbf{c}_0) : \Omega(0) \to \mathbb{R}^{N_w} \times \mathbb{R}^{N_c}$, find the intra- and extracellular electric potentials $u_{i,e} : \Omega(t) \to \mathbb{R}$, the transmembrane potential $v = u_i - u_e : \Omega(t) \to \mathbb{R}$, the gating and ionic concentrations variables $(\mathbf{w}, \mathbf{c}) : \Omega(t) \to \mathbb{R}^{N_w} \times \mathbb{R}^{N_c}$ such that

$$\begin{cases} c_m \dfrac{\partial v}{\partial t} - \operatorname{div}(\mathbf{D}_i \operatorname{grad} u_i) + I_{ion}(v, \mathbf{w}, \mathbf{c}) = I_{app} & \text{in } \Omega(t) \\ -c_m \dfrac{\partial v}{\partial t} - \operatorname{div}(\mathbf{D}_e \operatorname{grad} u_i) - I_{ion}(v, \mathbf{w}, \mathbf{c}) = -I_{app} & \text{in } \Omega(t) \\ \dfrac{\partial \mathbf{w}}{\partial t} - R(v, \mathbf{w}) = 0, \quad \dfrac{\partial \mathbf{c}}{\partial t} - S(v, \mathbf{w}, \mathbf{c}) = 0, & \text{in } \Omega(t) \\ \mathbf{n}^T \mathbf{D}_{i,e} \operatorname{grad} u_{i,e} = 0 & \text{in } \partial\Omega(t) \\ v(\mathbf{x}, 0) = v_0(\mathbf{x}), \quad \mathbf{w}(\mathbf{x}, 0) = \mathbf{w}_0(\mathbf{x}), \quad \mathbf{c}(\mathbf{x}, 0) = \mathbf{c}_0(\mathbf{x}) & \text{in } \Omega(0), \end{cases} \tag{10}$$

where we have assumed an insulated cardiac boundary $\partial\Omega$.

We remark that this system uniquely determines v, while the potentials u_i and u_e are defined only up to the same additive time-dependent constant related to the reference potential. We choose this potential to be the average extracellular potential in the cardiac volume by imposing $\int_{\Omega(t)} u_e \, dx = 0$. The non-linear reaction term I_{ion} and the right-hand-side functions R, S in the ODE system for the gating $\mathbf{w}$ and ionic concentrations $\mathbf{c}$ variables depend on the ionic membrane model chosen; here we will consider the ten Tusscher (TP06) membrane model [92, 93]. For the well-posedness analysis of the Bidomain model (10) coupled with complex membrane models, we refer to [98].

The conductivity tensors $\mathbf{D}_i(\mathbf{x})$ and $\mathbf{D}_e(\mathbf{x})$ at any point $\mathbf{x} \in \Omega(t)$ are defined as

$$\mathbf{D}_{i,e}(\mathbf{x}) = \sigma_t^{i,e} \, \mathbf{I} + (\sigma_l^{i,e} - \sigma_t^{i,e}) \, \mathbf{a}_t(\mathbf{x})\mathbf{a}_t^T(\mathbf{x}) + (\sigma_n^{i,e} - \sigma_t^{i,e}) \, \mathbf{a}_n(\mathbf{x})\mathbf{a}_n^T(\mathbf{x}), \tag{11}$$

where $\sigma_l^{i,e}$, $\sigma_t^{i,e}$, $\sigma_n^{i,e}$ denote the conductivity coefficients in the intra- and extracellular media measured along the directions $\mathbf{a}_l, \mathbf{a}_t, \mathbf{a}_n$.

In this work, the electric conduction of the cardiac tissue is modeled as an *axi-symmetric* anisotropic media with respect to the local finer direction, i.e. $\sigma_n^{i,e} = \sigma_t^{i,e}$. Hence, using the orthogonality of the principal axes the conductivity tensors can be written as

$$\mathbf{D}_{i,e}(\mathbf{x}) = \sigma_t^{i,e} \, \mathbf{I} + (\sigma_l^{i,e} - \sigma_t^{i,e}) \, \mathbf{a}_l(\mathbf{x}) \, \mathbf{a}_l^T(\mathbf{x}). \tag{12}$$

In order to take into account the mechanical deformation of the tissue, following [59, 67, 103] we will now introduce the parabolic-parabolic formulation of the Bidomain system on the reference cardiac domain $\widehat{\Omega}$, applying the usual *pull-back* operation.

Given an applied current per unit volume $\widehat{I}_{app} : \widehat{\Omega} \times (0, T) \to \mathbb{R}$, and initial conditions $\widehat{v}_0 : \widehat{\Omega} \to \mathbb{R}$, $\widehat{\mathbf{w}}_0 : \widehat{\Omega} \to \mathbb{R}^{N_w}$, find the intra- and extracellular potentials $\widehat{u}_{i,e} : \widehat{\Omega} \times (0, T) \to \mathbb{R}$, the transmembrane potential $\widehat{v} = \widehat{u}_i - \widehat{u}_e : \widehat{\Omega} \times (0, T) \to \mathbb{R}$, the gating and ionic concentrations variables $(\widehat{\mathbf{w}}, \widehat{c}) : \widehat{\Omega} \times (0, T) \to \mathbb{R}^{N_w} \times \mathbb{R}^{N_c}$ such that

$$\begin{cases} c_m \left(\dfrac{\partial \widehat{v}}{\partial t} - \mathbf{F}^{-T}\, \mathrm{Grad}\, \widehat{v} \cdot \mathbf{V} \right) - J^{-1}\, \mathrm{Div}(J\mathbf{F}^{-1}\widehat{\mathbf{D}}_i\mathbf{F}^{-T}\, \mathrm{Grad}\, \widehat{u}_i) + I_{ion} = \widehat{I}_{app} & \text{in } \widehat{\Omega} \\ -c_m \left(\dfrac{\partial \widehat{v}}{\partial t} - \mathbf{F}^{-T}\, \mathrm{Grad}\, \widehat{v} \cdot \mathbf{V} \right) - J^{-1}\, \mathrm{Div}(J\mathbf{F}^{-1}\widehat{\mathbf{D}}_e\mathbf{F}^{-T}\, \mathrm{Grad}\, \widehat{u}_i) - I_{ion} = -\widehat{I}_{app} & \text{in } \widehat{\Omega} \\ \hat{\mathbf{n}}^T\mathbf{F}^{-1}\widehat{\mathbf{D}}_{i,e}\mathbf{F}^{-T}\, \mathrm{Grad}\, \widehat{u}_{i,e} = 0 \text{ on } \partial\widehat{\Omega} \qquad \widehat{v}(\mathbf{X}, 0) = \widehat{v}_0(\mathbf{X}), & \text{in } \widehat{\Omega} \end{cases} \tag{13}$$

where $\mathbf{F}$ is the deformation gradient tensor, $\mathbf{V} = \frac{\partial \mathbf{x}}{\partial t}(\mathbf{X}, t)$, $I_{ion} = I_{ion}(\widehat{v}, \widehat{\mathbf{w}}, \widehat{c})$. The previous PDE system is coupled with the ODE system in the gating and concentration variables:

$$\begin{cases} \dfrac{\partial \mathbf{w}}{\partial t} - R(v, \mathbf{w}) = 0, \qquad \dfrac{\partial \mathbf{c}}{\partial t} - S(v, \mathbf{w}, \mathbf{c}) = 0, & \text{in } \Omega(t) \\ \mathbf{w}(\mathbf{x}, 0) = \mathbf{w}_0(\mathbf{x}), \qquad c(\mathbf{x}, 0) = \mathbf{c}_0(\mathbf{x}) & \text{in } \Omega(0), \end{cases}$$

We remark that this formulation of the Bidomain system, in the potential unknowns $\widehat{u}_{i,e}$, called parabolic-parabolic formulation, is equivalent to the other more popular formulation in the unknowns $\widehat{v}$, $\widehat{u}_e$, called parabolic-elliptic, because constituted by a parabolic and an elliptic PDE, see e.g. [16].

The computation of the tensors $\mathbf{F}^{-1}(\mathbf{X})\widehat{\mathbf{D}}_{i,e}(\mathbf{x})\mathbf{F}^{-T}(\mathbf{X})$ must be performed on the reference configuration $\widehat{\Omega}$. Denoting by $\hat{\mathbf{a}}_l(\mathbf{X})$ the unit vector parallel to the local fiber direction in the reference configuration, then we recall that, from (6), it holds $\mathbf{a}_l\ \mathbf{a}_l^T = \frac{\mathbf{F}\hat{\mathbf{a}}_l\hat{\mathbf{a}}_l^T\mathbf{F}^T}{\hat{\mathbf{a}}_l^T\mathbf{C}\hat{\mathbf{a}}_l}$. Considering the product $\mathbf{F}^{-1}(\mathbf{X})\widehat{\mathbf{D}}_{i,e}(\mathbf{x})\mathbf{F}^{-T}(\mathbf{X})$ and assuming that the conductivity coefficients refer to the deformed configuration, we finally obtain, in the orthotropic case (see [17]),

$$\begin{aligned} (\mathbf{F}^{-1}\widehat{\mathbf{D}}_{i,e}\mathbf{F}^{-T})(\mathbf{X}) = \sigma_t^{i,e}\, \mathbf{C}^{-1}(\mathbf{X}) + (\sigma_l^{i,e} - \sigma_t^{i,e})\, \frac{\hat{\mathbf{a}}_l(\mathbf{X})\hat{\mathbf{a}}_l^T(\mathbf{X})}{\hat{\mathbf{a}}_l^T(\mathbf{X})\mathbf{C}(\mathbf{X})\hat{\mathbf{a}}_l(\mathbf{X})} + \\ (\sigma_n^{i,e} - \sigma_t^{i,e})\, \frac{\mathbf{C}^{-1}(\mathbf{X})\hat{\mathbf{a}}_n(\mathbf{X})(\mathbf{C}^{-1}(\mathbf{X})\hat{\mathbf{a}}_n(\mathbf{X}))^T}{\hat{\mathbf{a}}_n^T(\mathbf{X})\mathbf{C}^{-1}(\mathbf{X})\hat{\mathbf{a}}_n(\mathbf{X})}, \end{aligned} \tag{14}$$

and, in the axisymmetric case,

$$(\mathbf{F}^{-1}\widehat{\mathbf{D}}_{i,e}\mathbf{F}^{-T})(\mathbf{X}) = \sigma_t^{i,e}\,\mathbf{C}^{-1}(\mathbf{X}) + (\sigma_l^{i,e} - \sigma_t^{i,e})\,\frac{\hat{\mathbf{a}}_l(\mathbf{X})\hat{\mathbf{a}}_l^T(\mathbf{X})}{\hat{\mathbf{a}}_l^T(\mathbf{X})\mathbf{C}(\mathbf{X})\hat{\mathbf{a}}_l(\mathbf{X})}. \tag{15}$$

2.4 Membrane Model and Stretch-Activated Channel Current

The ionic current in the Bidomain model (13) is given by $i_{ion} = \chi I_{ion}$, where χ is the membrane surface to volume ratio and the ionic current per unit area of the membrane surface I_{ion} is given by the sum $I_{ion}(v, \mathbf{w}, \mathbf{c}, \lambda) = I_{ion}^m(v, \mathbf{w}, \mathbf{c}) + I_{SAC}$ of the ionic term $I_{ion}^m(v, \mathbf{w}, \mathbf{c})$ given by the ten Tusscher model (TP06) [92, 93], available from the cellML depository (models.cellml.org/cellml), and a stretch-activated current (SAC) I_{SAC}, see **for its role [45].** The TP06 ionic model also specifies the functions $R_w(v, \mathbf{w})$ and $R_c(v, \mathbf{w}, \mathbf{c})$ in the ODE system, consisting of 17 ordinary differential equations modeling the main ionic currents dynamics.

The I_{SAC} current is modeled as in [62] as the sum of non-specific and specific currents $I_{SAC} = I_{SAC,n} + I_{Ko}$. The non-specific current is defined by $I_{SAC,n} = I_{SAC,Na} + I_{SAC,K}$, with

$$I_{SAC,Na} = g_{SAC}\gamma_{SL,SAC}(v - E_{Na})\left(\frac{E_R - E_K}{E_{Na} - E_R}\right), \qquad I_{SAC,K} = g_{SAC}\gamma_{SL,SAC}(v - E_K),$$

where $g_{SAC} = 4.13 \cdot 10^{-3}\,\text{mS/cm}^2$, $\gamma_{SL,SAC} = 10\max(\lambda - 1, 0)$ and $E_R = -10\,\text{mV}$. The specific stretch-dependent K^+ current is defined by

$$I_{Ko} = g_{Ko}\frac{\gamma_{SL,Ko}}{1 + exp(-(10 + v)/45)}(v - E_K),$$

where $g_{Ko} = 1.2 \cdot 10^{-2}\,\text{mS/cm}^2$ and $\gamma_{SL,Ko} = 3\max(\lambda - 1, 0) + 0.7$. We remark that the stretch activated ion channels are active only during the stretching phase of the fibers, thus we consider the constraint $\max(\lambda - 1, 0)$. For related models of stretch-activated currents, we refer e.g. to [1, 39, 40, 45, 102]. The advantage of using the SAC model described here with respect to simpler models adopted in [39, 40, 85] is that it separates SAC into non-selective and potassium selective components, which have been measured experimentally in [49].

2.5 Pressure-Volume Loop

A few studies so far have coupled the mechanics of the ventricle with Navier-Stokes equations for the blood flow inside the cavity, see [90, 91]. Most works

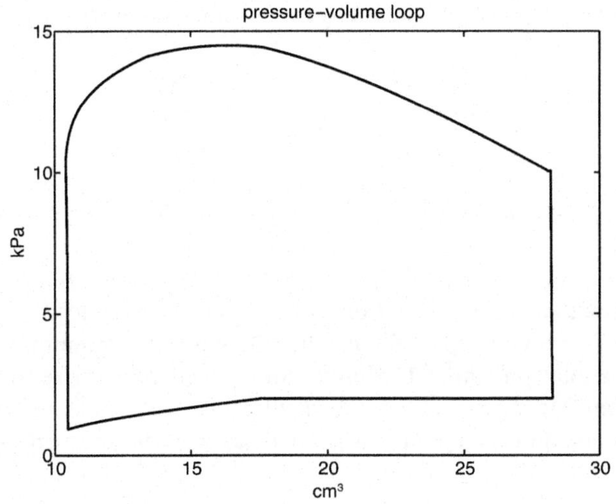

Fig. 1 Plot of pressure-volume loop

have adopted a simplified description of intracavitary blood flow, called pressure-volume loop, where pressure boundary conditions modeling the response of cavitary blood to ventricular deformation are imposed on the endocardial surface in the finite elasticity equations (1). The pressure p in the left ventricular (LV) cavity can be described by sophisticated closed-loop models as in [42]. A simpler model, adopted e.g. in [18, 24, 96], is based on the following four phases (see Fig. 1):

1. Isovolumetric LV contraction phase, where p increases from the end diastolic pressure (EDP) value of about 2 to 10 kPa;
2. Ejection phase, where the pressure-volume relationship is described by a two element Windkessel model, until the volume reduction stops;
3. Isovolumetric LV relaxation phase, where p decreases to 1 kPa;
4. Filling phase, where p increases linearly to EDP.

During the isovolumetric phases (1) and (3), the pressure is updated according to the iterative procedure introduced in [96], in order to keep the LV cavity volume constant.

3 Numerical Methods

3.1 Space and Time Discretization

Domain Geometry The cardiac domain $\widehat{\Omega} = \Omega(0)$ is the image of a cartesian periodic slab using ellipsoidal coordinates, yielding a truncated ellipsoid modeling a left ventricular geometry, described by the parametric equations

$$\begin{cases} x = a(r)\cos\theta\cos\phi & \phi_{min} \leq \phi \leq \phi_{max}, \\ y = b(r)\cos\theta\sin\phi & \theta_{min} \leq \theta \leq \theta_{max}, \\ z = c(r)\sin\theta & 0 \leq r \leq 1, \end{cases}$$

where $a(r) = a_1 + r(a_2 - a_1)$, $b(r) = b_1 + r(b_2 - b_1)$, $c(r) = c_1 + r(c_2 - c_1)$, and $a_1 = b_1 = 1.5$, $a_2 = b_2 = 2.7$, $c_1 = 4.4$, $c_2 = 5$ (all in cm) and $\phi_{min} = -\pi/2$, $\phi_{max} = 3\pi/2$, $\theta_{min} = -3\pi/8$, $\theta_{max} = \pi/8$. We will refer to the inner surface of the truncated ellipsoid ($r = 0$) as endocardium and to the outer surface ($r = 1$) as epicardium. The fibers rotate intramurally linearly with the depth for a total amount of 120° proceeding counterclockwise from epicardium to endocardium. More precisely, in a local ellipsoidal reference system $(\mathbf{e}_\phi, \mathbf{e}_\theta, \mathbf{e}_r)$, the fiber direction $\mathbf{a}_l(\mathbf{x})$ at a point $\mathbf{x}$ is given by $\mathbf{a}_l(\mathbf{x}) = \mathbf{b}_l(\mathbf{x})\cos(\beta) + \mathbf{n}(\mathbf{x})\cos(\beta)$, where

$$\mathbf{b}_l(\mathbf{x}) = \mathbf{e}_\phi \cos\alpha(r) + \mathbf{e}_\theta \sin\alpha(r), \quad \text{with} \quad \alpha(r) = \frac{2}{3}\pi(1-r) - \frac{\pi}{4}, \qquad 0 \leq r \leq 1,$$

$\mathbf{n}(\mathbf{x})$ is the unit outward normal to the ellipsoidal surface at $\mathbf{x}$ and β is the imbrication angle calibrated as in [13] given by $\beta = \arctan(\cos\alpha \ \tan\gamma)$, with $\gamma = \theta(1 - r)60/\pi$.

Space Discretization We discretize the cardiac domain with a hexahedral structured grid T_{h_m} for the mechanical model (1) and T_{h_e} for the Bidomain model (13), where T_{h_e} is a refinement of T_{h_m}, i.e. h_m is an integer multiple of h_e. We consider the variational formulations of both mechanical and bioelectrical models and then approximate all scalar and vector fields by isoparametric Q_1 finite elements in space. In all the electro-mechanical simulations, we employ an electrical mesh size $h_e = 0.01\ cm$ in order to properly resolve the sharp excitation front, while the smoother mechanical deformation allow us to use a coarse mechanical mesh of size $h_m = 0.08$ cm. We remark that, when adopting a complex membrane model such as that proposed by ten Tusscher et al. [92, 93], the depth of the steep excitation layer is about 1 mm, thus a mesh size of at least 0.01 cm is needed in order to accurately catch the velocity of the wavefront.

Time Discretization The time discretization of the electro-mechanical model is performed by the following semi-implicit splitting method, where the electrical and mechanical time steps could be different. At each time step,

a) given v^n, w^n, c^n at time t_n, solve the ODE system of the membrane model with a first order implicit-explicit (IMEX) method to compute the new w^{n+1}, c^{n+1};
b) given the calcium concentration Ca_i^{n+1}, which is included in the concentration variables c^{n+1}, solve the variational formulation of the mechanical problem (1) and the active tension system to compute the new deformed coordinates $\mathbf{x}^{n+1}$, providing the new deformation gradient tensor $\mathbf{F}_{n+1}$;

c) given w^{n+1}, c^{n+1}, $\mathbf{F}_{n+1}$ and $J_{n+1} = \det(\mathbf{F}_{n+1})$, solve the variational formulation of the Bidomain system (13) with a first order IMEX method and compute the new electric potentials v^{n+1}, u_e^{n+1} with an operator splitting method, consisting of decoupling the parabolic from the elliptic equation.

In our simulations, the electrical time step size is $\Delta_e t = 0.05\,\text{ms}$, while the mechanical times step is $\Delta_m t = 0.25\,\text{ms}$. We remark that to approximate the convective term in the variational formulation of (13), an *upwind* computation of the nodal gradient of v^n is derived, projected onto the deformation rate $\frac{\mathbf{x}^{n+1}-\mathbf{x}^n}{\Delta t_n}$ vector and then integrated against the test function. We refer to [17, 18] for more details about the numerical scheme.

3.2 Computational Kernels and Parallel Solvers

Due to the employed space and time discretization strategies, at each time step, the main computational kernels are:

i) the **mechanical solver**, which consists of solving the non-linear system deriving from the discretization of the mechanical problem (1) by a parallel Newton-GMRES-BDDC (NKBDDC) solver, see [70];
ii) the **electrical solver**, which consists of solving the two linear systems deriving from the discretization of the elliptic and parabolic equations in the Bidomain model (13) by a parallel Conjugate Gradient method preconditioned with a Multilevel Additive Schwarz preconditioner, studied in [68], and a Block Jacobi preconditioner, respectively. In the following, the two linear solvers will be denoted by elliptic and parabolic solvers, respectively, see e.g. Fig. 6.

Our parallel simulations have been performed on the Linux cluster MARCONI of CINECA laboratory using the parallel library PETSc [6] from the Argonne National Laboratory.

The next sections will describe more in detail the Multilevel Additive Schwarz and BDDC preconditioners used in the electrical and mechanical solvers, respectively.

3.3 Multilevel Additive Schwarz Preconditioners

In this section, we describe the Multilevel Additive Schwarz preconditioner used to solve the linear system arising from the Bidomain equations. Let $\Omega^k, k = 0, \ldots, \ell - 1$ be a family of ℓ nested triangulations of Ω, coarsening from $\ell - 1$ to 0, and A^k the matrix obtained by discretizing the elliptic equation of the Bidomain model (10) on Ω^k: so $\mathrm{A}^{\ell-1}$ is the matrix A_{bid} of the linear system arising from the discretization of the elliptic equation of (10) on the fine mesh. R^k are the restriction operators from

$\Omega^{\ell-1}$ to Ω^k. We decompose each grid Ω^k, for $k = 1, \dots, \ell-1$, into N_k overlapping subgrids Ω^k_m for $m = 1, \dots, N_k$, such that the overlap δ^k at level $k = 1, \dots, \ell-1$ is equal to the mesh size h^k of the grid Ω^k. Let R^k_m be the restriction operator from $\Omega^{\ell-1}$ to Ω^k_m and define $\mathrm{A}^k_m := \mathrm{R}^k_m \mathrm{A}^k \mathrm{R}^{k^T}_m$. The Multilevel Additive Schwarz (MAS) preconditioner is given by

$$B^{-1}_{MAS} := \mathrm{R}^{0^T} \mathrm{A}^{0^{-1}} \mathrm{R}^0 + \sum_{k=1}^{\ell-1} \mathrm{R}^{k^T}_m \mathrm{A}^{k^{-1}}_m \mathrm{R}^k_m.$$

The following theorem ensures the convergence of the MAS method, in particular that it is independent of the number of subdomains N_k (scalability) and the number of levels ℓ (optimality).

Theorem 1 *The condition number of the MAS operator* $T_{MAS} = B^{-1}_{MAS} A_{bid}$ *is bounded by*

$$\kappa_2(T_{MAS}) \leq C \max_{k=1,\dots,L-1} \left(1 + \frac{h_{k-1}}{\delta_k}\right),$$

where C *is a constant independent of the mesh sizes* h_k*, the number of subdomains* N_k *and the number of levels* ℓ*.*

In the general case of elliptic problems, see e.g. [94], while for the Bidomain system see [16, 68, 69].

3.4 Iterative Substructuring, Schur Complement System and BDDC Preconditioners

In this section, we describe the BDDC preconditioner used to solve the Jacobian system arising at each iteration of the Newton method applied to the non-linear mechanical system. To keep the notation simple, in the following of this section and in the next one, we will denote the reference domain by Ω instead of $\widehat{\Omega}$. Let us consider a decomposition of Ω into N non-overlapping subdomains Ω_i of diameter H_i (see e.g. Toselli and Widlund [94, Ch. 4])

$$\Omega = \bigcup_{i=1}^{N} \Omega_i,$$

and set $H = \max H_i$. As in classical iterative substructuring, we reduce the Jacobian linear system with matrix denoted by K, obtained at each Newton iteration of the mechanical solver, to the interface

$$\Gamma := \Big(\bigcup_{i=1}^{N} \partial\Omega_i\Big) \backslash \partial\Omega$$

by eliminating the interior degrees of freedom associated to basis functions with support in the interior of each subdomain, hence obtaining the Schur complement system

$$S_\Gamma x_\Gamma = g_\Gamma, \tag{16}$$

where $S_\Gamma = K_{\Gamma\Gamma} - K_{\Gamma I} K_{II}^{-1} K_{\Gamma I}$ and $g_\Gamma = f_\Gamma - K_{\Gamma I} K_{II}^{-1} f_I$ are obtained from the original Jacobian system by reordering the finite element basis functions in interior (subscript I) and interface (subscript Γ) basis functions

$$\begin{pmatrix} K_{II} & K_{I\Gamma} \\ K_{\Gamma I} & K_{\Gamma\Gamma} \end{pmatrix} \begin{pmatrix} x_I \\ x_\Gamma \end{pmatrix} = \begin{pmatrix} f_I \\ f_\Gamma \end{pmatrix}. \tag{17}$$

The Schur complement system (16) is solved iteratively by the GMRES method, where only the action of S_Γ on a vector is needed and S_Γ is never explicitly formed: instead, a block diagonal problem on interior degrees of freedom is solved while computing the matrix vector product. Once the interface solution x_Γ is computed, the internal values x_I can be recovered by solving local problems on each subdomain Ω_i. The preconditioned Schur complement system solved by GMRES is then

$$M_{\mathrm{BDDC}}^{-1} S_\Gamma x_\Gamma = M_{\mathrm{BDDC}}^{-1} f_\Gamma, \tag{18}$$

where M_{BDDC}^{-1} is the BDDC preconditioner, defined in (22) below.

Balanced Domain Decomposition by Constraints (BDDC) preconditioners where introduced by Dohrmann [22] and first analyzed by Mandel, Dohrmann and Tezaur [51, 52]. They can be regarded as an evolution of balancing Neumann-Neumann methods where all local and coarse problems are treated additively due to a choice of so-called primal continuity constraints across the interface of the subdomains. These primal constraints can be point constraints and/or averages or moments over edges or faces of the subdomains. We refer to the domain decomposition monograph by Toselli and Widlund [94, Ch. 6] for a detailed treatment of Neumann-Neumann, FETI and FETI-DP algorithms, see also [8, 25, 43, 44, 50]; the latter are closely related to BDDC methods, since they exhibit essentially the same spectrum (except for possible 0 and 1 eigenvalues) when the primal set is the same.

Subspace Decompositions Let V be the Q_1 finite element space for displacements and $V^{(i)}$ be the local discrete space defined on the subdomain Ω_i that vanish on $\partial\Omega_i \cap \partial\Omega_D$. We split the local space into a direct sum of its interior (I) and interface

(Γ) subspaces $V^{(i)} = V_I^{(i)} \bigoplus V_\Gamma^{(i)}$ and we define the associated product spaces by

$$V_I := \prod_{i=1}^{N} V_I^{(i)}, \quad V_\Gamma := \prod_{i=1}^{N} V_\Gamma^{(i)}.$$

The functions in V_Γ are generally discontinuous across Γ, while our finite element approximations are not. Therefore, we also define the subspace

$$\widehat{V}_\Gamma := \{\text{functionsof} V_\Gamma \text{thatarecontinuousacross} \Gamma\}.$$

We also need an intermediate subspace

$$\widetilde{V}_\Gamma := V_\Delta \bigoplus \widehat{V}_\Pi,$$

defined by further splitting the interface (subscript Γ) degrees of freedom into primal (subscript Π) and dual (subscript Δ) degrees of freedom, where:

a) $\widehat{V}_\Pi$ is a subspace consisting of functions which are continuous at selected *primal* variables, which can be the subdomain basis functions associated with subdomains' corners and/or edge/face basis functions with a constant value at the nodes of the associated edge/face. In order to simplify the formulas and keep the exposition compact, we assume that a change of basis has been performed and each primal variable correspond to an explicit degree of freedom; see [50].
b) $V_\Delta = \prod_{i=1}^{N} V_\Delta^{(i)}$ is the product space of the local subspaces $V_\Delta^{(i)}$ of *dual* interface functions that vanish at the primal degrees of freedom.

Restriction and Scaling Operators In order to define our preconditioners, we will need the following restriction and interpolation operators represented by matrices with elements in the set $\{0, 1\}$:

$$\begin{array}{ll} R_{\Gamma\Delta} : \widetilde{V}_\Gamma \longrightarrow V_\Delta, & R_{\Gamma\Pi} : \widetilde{V}_\Gamma \longrightarrow \widehat{V}_\Pi, \\ R_\Delta^{(i)} : V_\Delta \longrightarrow V_\Delta^{(i)}, & R_\Pi^{(i)} : \widehat{V}_\Pi \longrightarrow \widehat{V}_\Pi^{(i)}, \end{array} \tag{19}$$

where $\widehat{V}_\Pi^{(i)}$ is the local subspace of *primal* interface functions. We will also need the standard counting functions of the Neumann–Neumann methods and in particular their pseudoinverses $\delta_i^\dagger(x)$ defined at each degrees of freedom x on the interface of subdomain Ω_i by

$$\delta_i^\dagger(x) := \frac{1}{\mathcal{N}_x}, \tag{20}$$

with $\mathcal{N}_x$ the number of subdomains sharing x.

We finally define scaled local restriction operators $R^{(i)}_{D,\Delta}$ by multiplying the sole nonzero element of each row of $R^{(i)}_{\Delta}$ by $\delta_i^{\dagger}$, and we define

$$R_{D,\Gamma} := \text{ the direct sum } R_{\Gamma\Pi} \oplus R^{(i)}_{D,\Delta} R_{\Gamma\Delta}. \tag{21}$$

Choice of Primal Constraints The choice of primal degrees of freedom is fundamental for the construction of efficient BDDC preconditioners. One of the simplest choices consists of taking as primal degrees of freedom those associated with the subdomain corners. Such choice is not always sufficient to obtain scalable and fast preconditioners and this has motivated the search for richer primal sets that may yield faster preconditioners but at the expense of higher computational costs due to larger coarse problems employing also edge and/or face based primal degrees of freedom, see e.g. [44, 94].

The BDDC Preconditioner We denote by $K^{(i)}$ the local stiffness matrix restricted to subdomain Ω_i. By partitioning the local degrees of freedom into interior (I), dual (Δ), and primal (Π) degrees of freedom, $K^{(i)}$ can be written as

$$K^{(i)} = \begin{bmatrix} K^{(i)}_{II} & K^{(i)^T}_{\Gamma I} \\ K^{(i)}_{\Gamma I} & K^{(i)}_{\Gamma\Gamma} \end{bmatrix} = \begin{bmatrix} K^{(i)}_{II} & K^{(i)^T}_{\Delta I} & K^{(i)^T}_{\Pi I} \\ K^{(i)}_{\Delta I} & K^{(i)}_{\Delta\Delta} & K^{(i)^T}_{\Pi\Delta} \\ K^{(i)}_{\Pi I} & K^{(i)}_{\Pi\Delta} & K_{\Pi\Pi} \end{bmatrix}.$$

Using the scaled restriction matrices defined in (19) and (21), the BDDC preconditioner can be written as

$$M^{-1}_{\text{BDDC}} = R^T_{D,\Gamma} \widetilde{S}^{-1}_{\Gamma} R_{D,\Gamma}, \tag{22}$$

where

$$\widetilde{S}^{-1}_{\Gamma} = R^T_{\Gamma\Delta} \left(\sum_{i=1}^{N} \begin{bmatrix} 0 \; R^{(i)^T}_{\Delta} \end{bmatrix} \begin{bmatrix} K^{(i)}_{II} & K^{(i)^T}_{\Delta I} \\ K^{(i)}_{\Delta I} & K^{(i)}_{\Delta\Delta} \end{bmatrix}^{-1} \begin{bmatrix} 0 \\ R^{(i)}_{\Delta} \end{bmatrix} \right) R_{\Gamma\Delta} + \Phi S^{-1}_{\Pi\Pi} \Phi^T. \tag{23}$$

The first term in (23) is the sum of local solvers on each subdomain Ω_i, with Neumann data on the local dual nodes and with the local primal degrees of freedom constrained to vanish. The second term is a coarse solver for the primal variables, that we implemented as in [50, 71] by using the coarse matrix

$$S_{\Pi\Pi} = \sum_{i=1}^{N} R^{(i)^T}_{\Pi} \left(K^{(i)}_{\Pi\Pi} - \begin{bmatrix} K^{(i)}_{\Pi I} \; K^{(i)}_{\Pi\Delta} \end{bmatrix} \begin{bmatrix} K^{(i)}_{II} & K^{(i)^T}_{\Delta I} \\ K^{(i)}_{\Delta I} & K^{(i)}_{\Delta\Delta} \end{bmatrix}^{-1} \begin{bmatrix} K^{(i)^T}_{\Pi I} \\ K^{(i)^T}_{\Pi\Delta} \end{bmatrix} \right) R^{(i)}_{\Pi}$$

and a matrix Φ which maps primal degrees of freedom to interface variables

$$\Phi = R_{\Gamma\Pi}^T - R_{\Gamma\Delta}^T \sum_{i=1}^{N} \left[0 \; R_{\Delta}^{(i)^T} \right] \begin{bmatrix} K_{II}^{(i)} & K_{\Delta I}^{(i)^T} \\ K_{\Delta I}^{(i)} & K_{\Delta\Delta}^{(i)} \end{bmatrix}^{-1} \begin{bmatrix} K_{\Pi I}^{(i)^T} \\ K_{\Pi\Delta}^{(i)^T} \end{bmatrix} R_{\Pi}^{(i)}.$$

The columns of Φ represent the coarse basis functions defined as the minimum energy extension into the subdomains with respect to the original bilinear form and subject to the chosen set of primal constraints.

Remark 1 It is well-known that BDDC and FETI-DP preconditioners for compressible linear elasticity problems satisfy a scalable quasi-optimal condition number bound (see e.g. [94, Ch. 6.4])

$$cond(M_{\text{BDDC}}^{-1} S_{\Gamma}) \leq C\left(\frac{H}{h}\right)\left(1 + \log \frac{H}{h}\right)^2,$$

with $C(\frac{H}{h}) = \alpha$ constant for sufficiently rich coarse spaces and $C(\frac{H}{h}) = \alpha \frac{H}{h}$ for the minimal primal space spanned by the degrees of freedom defined at subdomain corners. Due to the more complex non-linear elasticity problem (1) based on an exponential strain energy function (3), we could not prove a similar GMRES bound for the convergence rate of our non-symmetric NKBDDC preconditioned operator. Nevertheless, the numerical results reported in our previous work [70] have shown that our NKBDDC method is scalable and quasi-optimal.

4 Numerical Simulations

In this final section, we apply our parallel electro-mechanical solver to the simulation of the induction of ventricular tachycardia in an idealized left ventricle affected by an infarct scar. The cardiac electro-mechanical coupling model adopted in the simulations consists of the following components:

- the quasi-static finite elasticity system, with a transversely isotropic exponential strain energy function derived from [24];
- the Bidomain system in the parabolic-elliptic formulation;
- the stretch and stretch-rate dependent active tension model proposed in [47];
- the ten Tusscher (TP06) membrane model [92, 93] for human ventricular myocytes.

4.1 Computational Domain with Scar

The computational domain consists of the truncated ellipsoid modeling a left ventricular geometry, described previously. The resulting electrical mesh consists

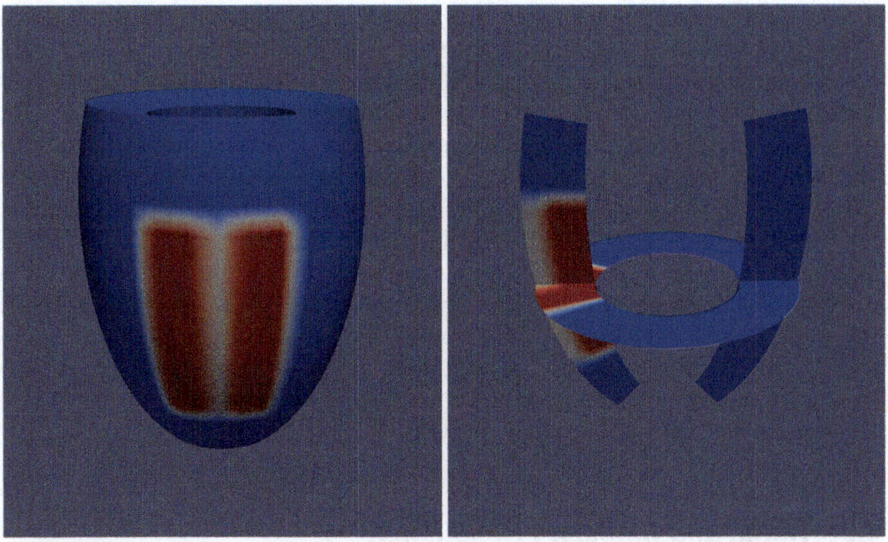

Fig. 2 Epicardial view (left) and transmural sections (right) of the idealized left ventricle computational domain with the scar. Red color denotes the scar region, white represents the border zone of the scar volume and blue indicates the normal myocardial volume

of $N_\phi \times N_\theta \times N_k = 384 \times 192 \times 48$ elements, yielding about $3.63 \cdot 10^6$ nodes, while the mechanical mesh consists of $48 \times 24 \times 6$ elements, yielding about 8400 nodes. In the left ventricular geometry, we consider the presence of a scar of dimension $2 \times 3.5 \times 1\,\text{cm}^3$ (see Fig. 2). The scar develops along the entire transmural thickness, from endocardium to epicardium. We include a border zone (BZ) region consisting of a sub-epicardial channel located at the center of the scar (see Fig. 2) of dimension $0.5 \times 3.4 \times 0.5\,\text{cm}^3$.

4.2 Electrical and Mechanical Parameters Calibration

The values of the transversely isotropic conductivity coefficients in (11), that we use in all the numerical tests for the healthy tissue, are $\sigma_l^i = 3$, $\sigma_t^i = 0.31525$, $\sigma_l^e = 2$, $\sigma_t^e = 1.3514$, all expressed in $m\Omega^{-1}\text{cm}^{-1}$. These values, coupled with the TP06 membrane model, predict conduction velocities of about 0.061 and 0.027 cm/ms for excitation wavefronts propagating along and across the fiber direction, respectively. We remark that these conduction velocities are within the physiological range. The membrane surface to volume ratio is $\chi = 10^3\,\text{cm}^{-1}$ and the membrane capacitance per unit volume is $c_m = \chi C_m$, where $C_m = 1\,\mu F/cm^2$ is the membrane capacitance per unit area. In the scar region, the conductivity coefficients and the fast sodium current conductance in the TP06 model are reduced by a factor 10. Analogously to the previous simulation studies [3, 9], the BZ is modeled by implementing a 60% reduction of fast sodium current conductance g_{Na}, 70% reduction of L-type calcium current conductance g_{CaL}, 70% and 80% reduction of the rapidly and slowly delayed rectifier potassium current conductances g_{Kr} and g_{Ks}, respectively, and 75% reduction of cross fiber intracellular conductivity coefficient σ_t^i. The parameters of the transversely isotropic strain energy function are adapted from [24]. The scar region is considered as an isotropic passive elastic material, i.e. without active stress, dropping in Eq. (4) the anisotropic terms, and increasing the parameter a by a factor 4. However, we recall that scar tissue may show a mechanical anisotropic structure due to stretch of fibroblasts and collagen fibers: for more detailed anisotropic mechanical models of scar tissue, we refer e.g. to [33, 34, 80]. The bulk modulus is $K = 200\,kPa$. In this simulation, we disregard the complete pressure-volume loop and we have assumed a fixed intracavitary blood pressure close to the diastolic value of $2\,kPa$.

4.3 Initial and Boundary Conditions

The initial conditions for the electrical model are at resting values for all the potentials and gating variables of the TP06 model, while in the Bidomain model homogeneous Neumann boundary conditions are imposed. For what concerns the mechanical boundary conditions, in order to prevent rigid body motion, we set to zero the displacement in all components only at the circumferential basal line meeting the endocardial surface, while setting to zero only the third displacement component on the rest of the basal surface. Moreover, on the endocardial surface we impose Neumann boundary conditions given by the intracavitary blood pressure, and we assume that the rest of domain is not subject to any external loading, i.e. free-stress boundary conditions are prescribed elsewhere.

4.4 Stimulation Protocol

Three stimuli of $250\,mA/cm^3$ amplitude and 1 ms duration are applied on the endocardial apical region below the scar, at a basic cycle length (BCL) of 400 ms. 290 ms after the third stimulus, an S2 stimulus of $1250\,mA/cm^3$ amplitude and 5 ms duration is applied at an epicardial site located close to the apical entrance of the BZ channel. The total simulation time is 2400 ms.

4.5 Parallel Solver Setup

The parabolic and elliptic linear systems arising from the discretization of the Bidomain equations are solved by the Conjugate Gradient (CG) method preconditioned with Block-Jacobi and 3-level Additive Schwarz preconditioners, respectively. The stopping criterion of the CG solver is the 10^{-6} reduction of the relative residual in l_2-norm. The mechanical non-linear system is solved by an inexact Newton method, using as stopping criterion the 10^{-6} reduction of the relative residual in l_2-norm. At each Newton iteration, the Jacobian linear system is solved by GMRES preconditioned with the BDDC preconditioner employing vertex and edge averages primal unknowns. The stopping criterion of the GMRES solver is a 10^{-8} reduction of the relative residual in l_2-norm. The following simulation is run on 4 nodes of the Marconi-KNL Linux cluster of CINECA, employing 64 cores for each node, for a total amount of 256 cores.

4.6 Results

After the three stimulations at BCL = 400 ms, a premature S2 stimulus delivered at an S1S2 coupling interval of 290 ms does not propagate inside the bottom entrance of the BZ channel, close to the apex of the ventricle, because the BZ tissue is still refractory, but it moves around the scar, see Fig. 3. Then it enters the BZ channel through the top entrance, close to the base of the ventricle, and, since the BZ tissue has now recovered, it propagates through the channel. Eventually the wavefront exits through the bottom gate into the healthy tissue and propagates again around the scar, leading to a stable reentry, typical of ventricular tachycardia, see Fig. 4.

We report in Fig. 5 the time evolution of transmembrane potential, extracellular potential and fiber stretch in an epicardial healthy site and in a border zone site, and of left ventricular volume. The first three action potential correspond to the normal stimulations at BCL = 400 ms, the fourth is elicited by the premature S2 stimulus, and the last two correspond to the reentrant excitation wavefront. Interestingly, the amount of contraction (see the time evolution of the fiber stretch λ) during reentry is analogous to the normal dynamics. This can be attributed to the fact that the stable reentrant excitation wavefront proceeds similar to the normal excitation wavefronts elicited by the first three beats. Only the premature excitation induced by the S2 stimulus exhibits a reduced contraction. In fact the dynamics of the left ventricular volume, which during the reentrant beats (the last two) reached the same maximum and minimum values as in the normal beats (the first three), shows a significant increase of the minimum value (the end systolic volume) only in the premature beat.

Finally, Fig. 6 shows, as function of simulation time, the CG iterations and CPU times of the parabolic solver, the CG iterations and CPU times of the elliptic solver and the Newton/GMRES iterations and CPU times of the mechanical solver. A part the normal variations during each heartbeat, corresponding to the different phases of the cardiac cycle (depolarization-repolarization, contraction-relaxation), among the six beats all the mathematical quantities remain stable, indicating that they are not influenced by the transition from the normal to the reentrant dynamics.

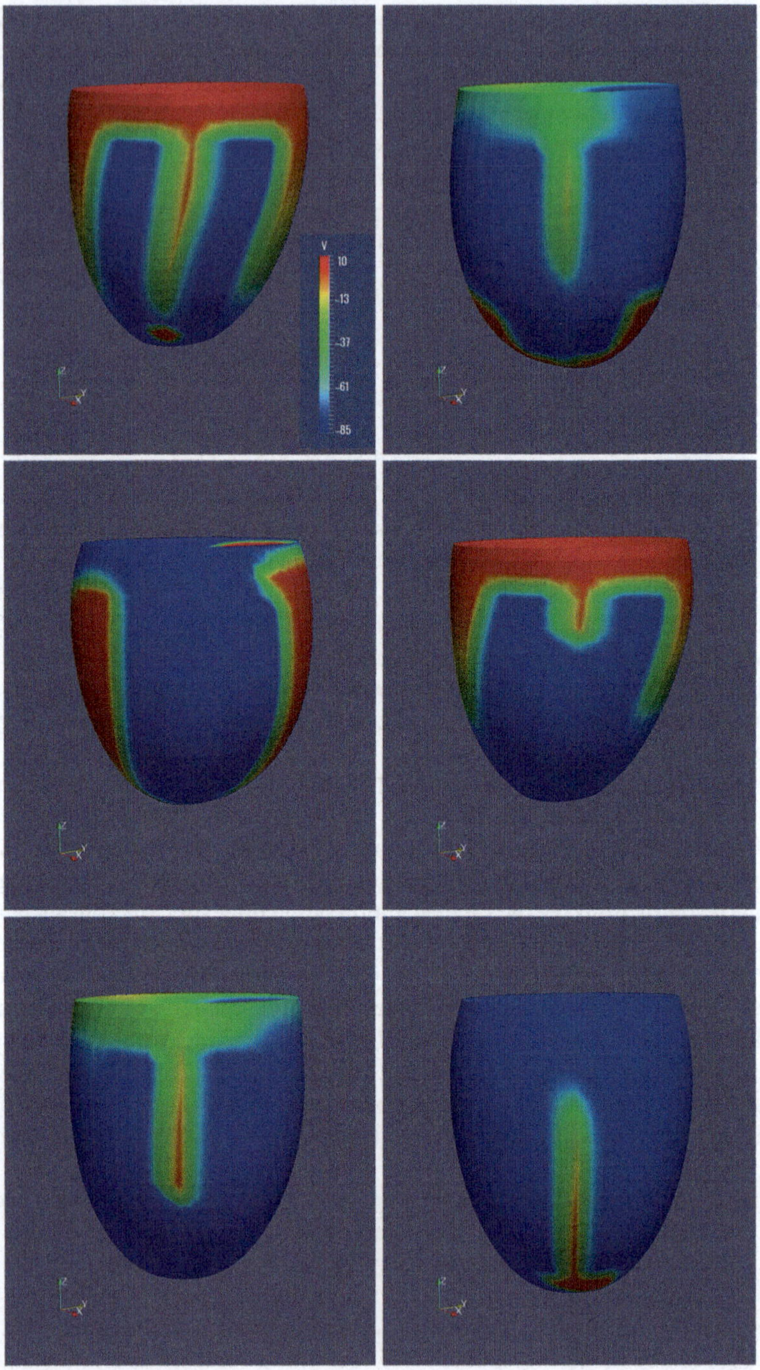

Fig. 3 Transmembrane potential snapshots on the deforming domain during reentry initiation. The six selected instants are: $t = 1085$ ms, $t = 1185$ ms (first row), $t = 1285$ ms, $t = 1385$ ms (second row) and $t = 1485$ ms, $t = 1585$ ms (third row)

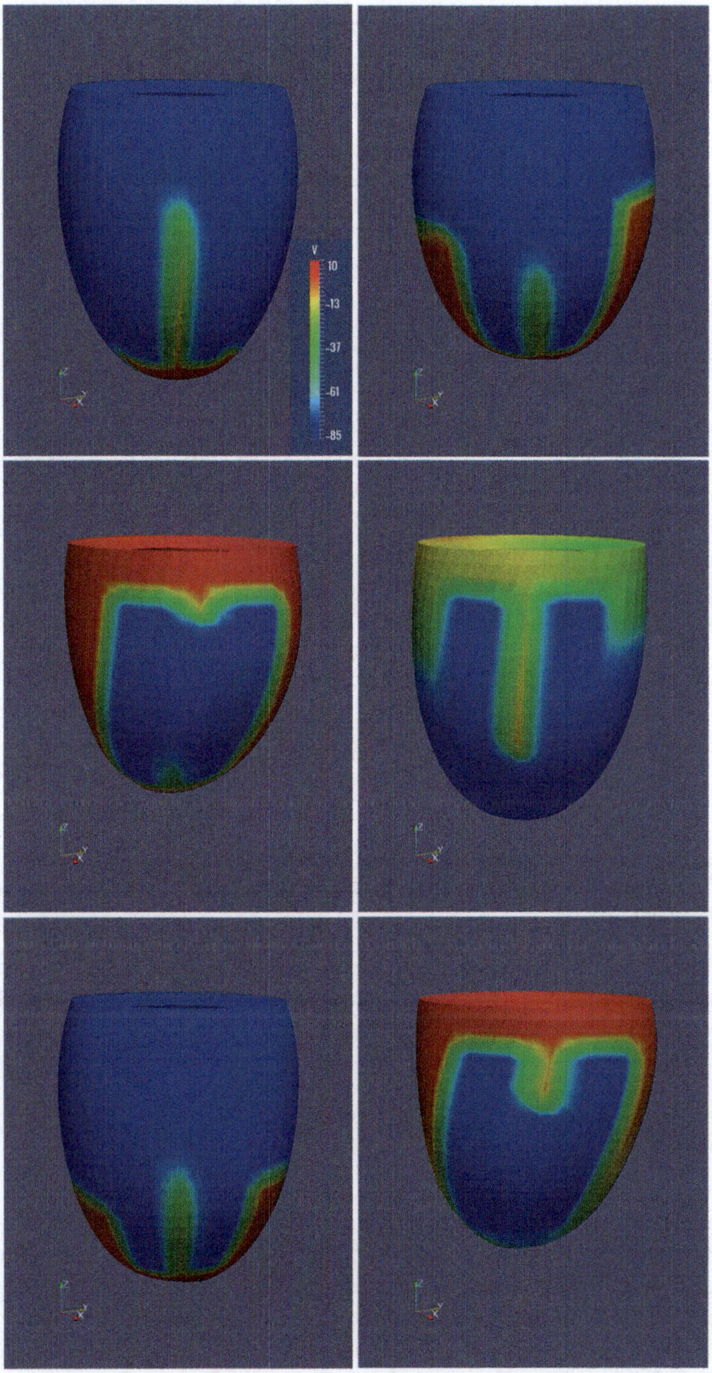

Fig. 4 Transmembrane potential snapshots on the deforming domain after reentry initiation. The six selected instants are: $t = 1600$ ms, $t = 1680$ ms (first row), $t = 1760$ ms, $t = 1920$ ms (second row) and $t = 2080$ ms, $t = 2240$ ms (third row)

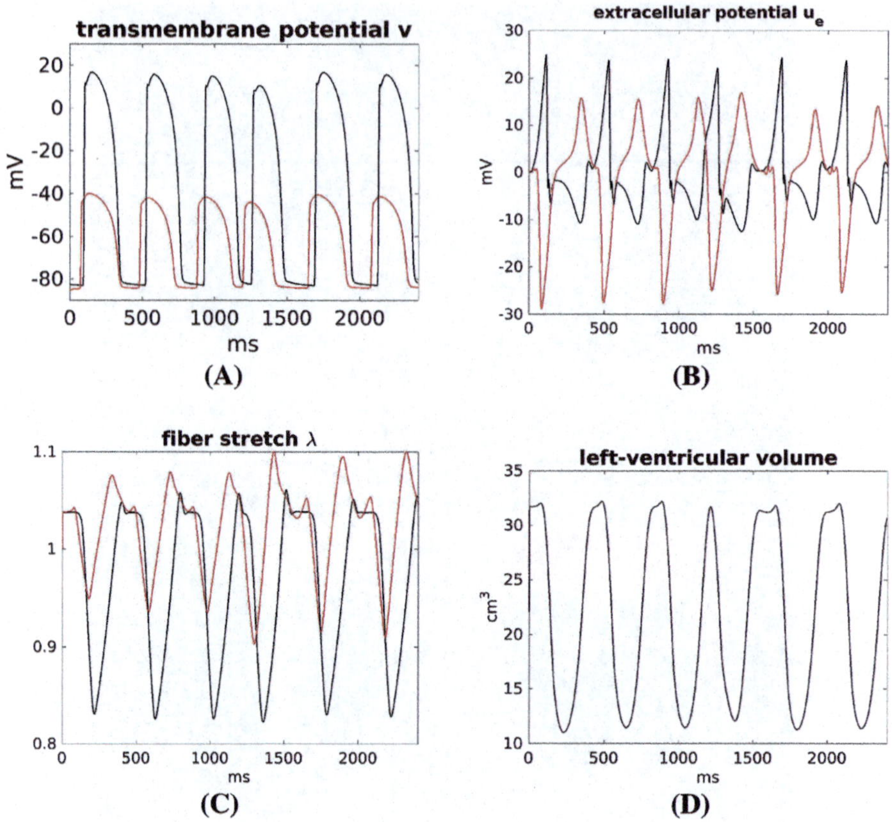

Fig. 5 Time evolution of transmembrane potential (**a**), extracellular potential (**b**) and fiber stretch (**c**) in an epicardial healthy (black line) site and in a border zone (red line) site, and of left ventricular volume (**d**)

Furthermore, the parabolic and elliptic solvers are robust with respect to the jumps in the conductivity coefficients across the scar border.

5 Conclusions

In this work, we have first reviewed the main mathematical models used to describe the cardiac electrical and mechanical interaction process, that constitute the so-called cardiac electro-mechanical coupling model. Then we have presented the numerical methods adopted to discretize the continuous model, with particular emphasis on the parallel solver used by our group to speedup the numerical simulation of the cardiac electro-mechanical activity. The results section is devoted

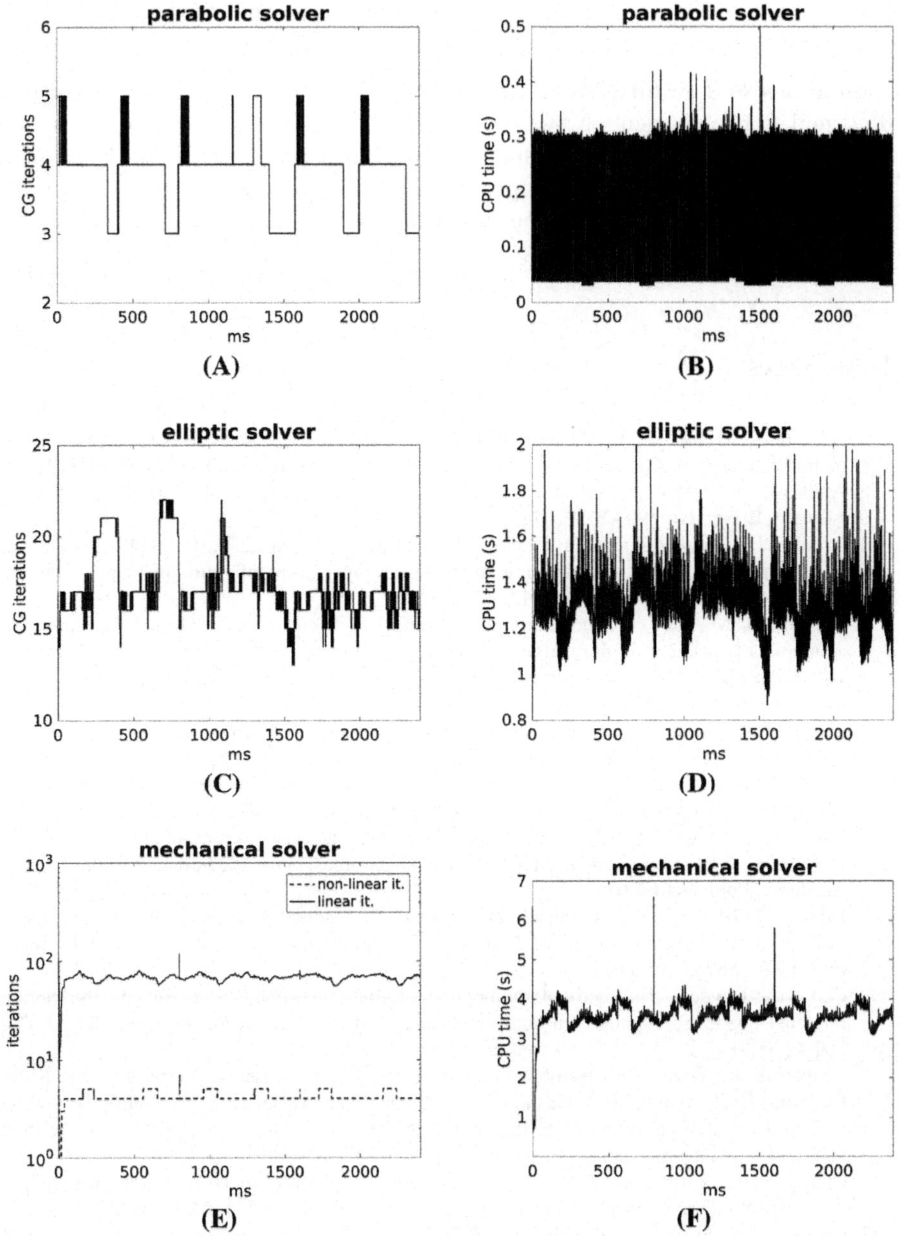

Fig. 6 Time evolution of: conjugate gradient (CG) iterations (**a**) and CPU times (**b**) of the parabolic solver (first row); CG iterations (**c**) and CPU times (**d**) of the elliptic solver (second row); Newton and GMRES iterations (**e**) and CPU times (**f**) of the mechanical solver (third row)

to show the effectiveness of the proposed solver to simulate the induction of ventricular tachycardia in a left ventricle affected from infarct scars.

Limitations and Future Works Among the limitations of the model used in the simulations, we mention the transversely isotropic tissue anisotropy and the constant left ventricular intracavitary blood pressure. Future work should be devoted to investigate the role of orthotropic tissue anisotropy and of the application of a variable endocardial pressure on the initiation and maintenance of ventricular tachycardia in presence of infact scars.

References

1. Adeniran, I., Hancox, J.C., Zhang, H.: Effect of cardiac ventricular mechanical contraction on the characteristics of the ECG: a simulation study. J. Biomed. Sci. Eng. **6**, 47–60 (2013)
2. Ambrosi, D., Pezzuto, S.: Active stress vs. active strain in mechanobiology: constitutive issues. J. Elast. **107**, 199–212 (2012)
3. Arevalo, H.J., Vadakkumpadan, F., Guallar, E., Jebb, A., Malamas, P., Wu, K.C., Trayanova, N.A.: Arrhythmia risk stratification of patients after myocardial infarction using personalized heart models. Nat. Commun. **7**, 11437 (2016)
4. Augustin, C.M., Neic, A., Liebmann, M., Prassl, A.J., Niederer, S.A., Haase, G., Plank, G.: Anatomically accurate high resolution modeling of human whole heart electromechanics: A strongly scalable algebraic multigrid solver method for nonlinear deformation. J. Comput. Phys. **305**, 622–646 (2016)
5. Austin, T.M., Trew, M.L., Pullan, A.J.: Solving the cardiac Bidomain equations for discontinuous conductivities. IEEE Trans. Biomed. Eng. **53**(7), 1265–1272 (2006)
6. Balay, S., Abhyankar, S., Adams, M.F., Brown, J., Brune, P., Buschelman, K., Dalcin, L., Eijkhout, V., Gropp, W.D., Kaushik, D., Knepley, M.G., Curfman McInnes, L., Rupp, K., Smith, B.F., Zampini, S., Zhang, H.: PETSc Web page, http://www.mcs.anl.gov/petsc, 2015
7. Bers, D.M.: Excitation-Contraction Coupling and Cardiac Contractile Force, 2nd edn. Kluwer Academic Publisher (2001)
8. Brands, D., Klawonn, A., Rheinbach, O., Schroeder, J.: Modelling and convergence in arterial wall simulations using a parallel FETI solution strategy. Comput. Methods Biomech. Biomed. Eng. **11**(5), 569–583 (2008)
9. Cabo, C., Boyden, P.: Electrical remodeling of the epicardial border zone in the canine infarcted heart: a computational analysis. Am. J. Physiol. Heart Circ. Physiol. **284**, H372–H384 (2003)
10. Chabiniok, R., Wang, V.Y., Hadjicharalambous, M., Asner, L., Lee, J., Sermesant, M., Kuhl, E., Young, A.A., Moireau, P., Nash, M.P., Chapelle, D., Nordsletten, D.A.: Multiphysics and multiscale modelling, data-model fusion and integration of organ physiology in the clinic: ventricular cardiac mechanics. Interface Focus **6**, 20150083 (2016)
11. Cherubini, C., Filippi, S., Gizzi, A., Ruiz-Baier, R.: A note on stress-driven anisotropic diffusion and its role in active deformable media. J. Theor. Biol. **430**, 221–228 (2017)
12. Cherubini, C., Filippi, S., Nardinocchi, P., Teresi, L.: An electromechanical model of cardiac tissue: constitutive issues and electrophysiological effects. Prog. Biophys. Mol. Biol. **97**, 562–573 (2008)
13. Colli Franzone, P., Guerri, L., Pennacchio, M., Taccardi, B.: Spread of excitation in 3-D models of he anisotropic cardiac tissue. II. Effects of fiber architecture and ventricular geometry. Math. Biosci. **147**, 131–171 (1998)
14. Colli Franzone, P., Pavarino, L.F.: A parallel solver for reaction-diffusion systems in computational electrocardiology. Math. Models Methods. Appl. Sci. **14**(6), 883–911 (2004)

15. Colli Franzone, P., Pavarino, L.F., Scacchi, S.: A comparison of coupled and uncoupled solvers for the cardiac Bidomain model. ESAIM: Math. Mod. Numer. Anal. **47**(4), 1017–1035 (2013)
16. Colli Franzone, P., Pavarino, L.F., Scacchi, S.: Mathematical Cardiac Electrophysiology, MSA Vol. 13. Springer, New York (2014)
17. Colli Franzone, P., Pavarino, L.F., Scacchi, S.: Bioelectrical effects of mechanical feedbacks in a strongly coupled cardiac electro-mechanical model. Math. Mod. Meth. Appl. Sci. **26**(1), 27–57 (2016)
18. Colli Franzone, P., Pavarino, L.F., Scacchi, S.: Joint influence of transmural heterogeneities and wall deformation on cardiac bioelectrical activity: A simulation study. Math. Biosci. **280**, 71–86 (2016)
19. Colli Franzone, P., Pavarino, L.F., Scacchi, S.: Effects of mechanical feedback on the stability of cardiac scroll waves: A bidomain electro-mechanical simulation study. Chaos **27**, 093905 (2017)
20. Coppola1, B.A., Omens, J.H.: Role of tissue structure on ventricular wall mechanics. Mol. Cell. Biomech. **5**(3), 183–196 (2008)
21. Costa, K.D., Holmes, J.W., McCulloch, A.D.: Modelling cardiac mechanical properties in three dimensions. Philos. Trans. R. Soc. Lond. A **359**, 1233–1250 (2001)
22. Dohrmann, C.R.: A Preconditioner for substructuring based on constrained energy minimization. SIAM J. Sci. Comput. **25**, 246–258 (2003)
23. Doll, S., Schweizerhof, K.: On the development of volumetric strain energy functions. J. Appl. Mech. **67**, 17–21 (2000)
24. Eriksson, T.S.E., Prassl, A.J., Plank, G., Holzapfel, G.A.: Influence of myocardial fiber/sheet orientations on left ventricular mechanical contraction. Math. Mech. Solids **18**, 592–606 (2013)
25. Farhat, C., Lesoinne, M., LeTallec, P., Pierson, K., Rixen, D.: FETI-DP: a dual-primal unified FETI method – part I. A faster alternative to the two-level FETI method. Int. J. Numer. Meth. Eng. **50**, 1523–1544 (2001)
26. Gerardo Giorda, L., Mirabella, L., Nobile, F., Perego, M., Veneziani, A.: A model-based block-triangular preconditioner for the Bidomain system in electrocardiology. J. Comput. Phys. **228**(10), 3625–3639 (2009)
27. Geselowitz, D.B., Miller, W.T.: A bidomain model for anisotropic cardiac muscle. Ann. Biomed. Eng. **11**, 191–206 (1983)
28. Göktepe, S., Kuhl, E.: Electromechanics of the heart - A unified approach to the strongly coupled excitation-contraction problem. Comput. Mech. **80**, 227–243 (2010)
29. Guccione, J.M., McCulloch, A.D., Waldman, L.K.: Passive material properties of intact ventricular myocardium determined from a cylindrical model. J. Biomech. Eng. **113**(1), 42–55 (1991)
30. Guccione, J.M., Costa, K.D., McCulloch, A.D.: Finite element stress analysis of left ventricular mechanics in the beating dog heart. J. Biomech. **28**, 1167–1177 (1995)
31. Gurev, V., Constantino, J., Rice, J.J., Trayanova, N.A.: Distribution of electromechanical delay in the heart: insights from a three-dimensional electromechanical model. Biophys. J. **99**, 745–754 (2010)
32. Hodgkin, A., Huxley, A.: A quantitative description of membrane current and its application to conduction and excitation in nerve. J. Physiol. **117**, 500–544 (1952)
33. Holmes, J.W., Borg, T.K., Covell, J.W.: Structure and mechanics of healing myocardial infarcts. Annu. Rev. Biomed. Eng. **7**, 223–253 (2005)
34. Holmes, J.W., Laksman, Z., Gepstein, L.: Making better scar: Emerging approaches for modifying mechanical and electrical properties following infarction and ablation. Prog. Biophys. Mol. Biol. **120**(1–3), 134–148 (2016)
35. Holzapfel, G.A., Ogden, R.W.: Constitutive modelling of passive myocardium. A structurally-based framework for material characterization. Philos. Trans. R. Soc. Lond. A **367**, 3445–3475 (2009)

36. Humphrey, J.D.: Cardiovascular Solid Mechanics, Cells, Tissues and Organs. Springer, New York (2001)
37. Hunter, P.J., McCulloch, A.D., ter Keurs, H.E.D.J.: Modelling the mechanical properties of cardiac muscle. Prog. Biophys. Mol. Biol. **69**, 289–331 (1998)
38. Hunter, P.J., Nash, M.P., Sands, G.B.: Computational electromechanics of the heart. In: Panfilov, A.V., Holden, A.V. (eds.) Computational Biology of the Heart. Wiley (1997)
39. Jie, X., Gurev, V., Trayanova, N.A.: Mechanisms of mechanically induced spontaneous arrhythmias in acute regional ischemia. Circ. Res. **106**, 185–192 (2010)
40. Keldermann, R.H., Nash, M.P., Gelderblom, H., Wang, V.Y., Panfilov, A.V.: Electromechanical wavebreak in a model of the human left ventricle. Am. J. Physiol. Heart Circ. Physiol. **299**, H134–H143 (2010)
41. Kerckhoffs, R.C.P., Bovendeerd, P.H.M., Kotte, J.C.S., Prinzen, F.W., Smits, K., Arts, T.: Homogeneity of cardiac contraction despite physiological asyncrony of depolarization: a model study. Ann. Biomed. Eng. **31**, 536–547 (2003)
42. Kerckhoffs, R.C.P., Neal, M.L., Gu, Q., Bassingthwaighte, J.B., Omens, J.H., McCulloch, A.D.: Coupling of a 3D finite element model of cardiac ventricular mechanics to lumped systems models of the systemic and pulmonic circulation. Ann. Biomed. Eng. **35**(1), 1–18 (2007)
43. Klawonn, A., Rheinbach, O.: Highly scalable parallel domain decomposition methods with an application to biomechanics. ZAMM - Z. Angew. Math. Mech. **90**(1), 5–32 (2010)
44. Klawonn, A., Widlund, O.B.: Dual-primal FETI methods for linear elasticity. Comm. Pure Appl. Math. **59**, 1523–1572 (2006)
45. Kohl, P., Sachs, F., Franz, M.R.: Cardiac Mechano-Electric Feedback and Arrhythmias: From Pipette to Patient. Elsevier Sauders (2011)
46. Krishnamurthy, A., Villongco, C.T., Chuang, J., Frank, L.R., Nigam, V., Belezzuoli, E., Stark, P., Krummen, D.E., Narayan, S., Omens, J.H., McCulloch, A.D., Kerckhoffs, R.C.P.: Patient-specific models of cardiac biomechanics. J. Comput. Phys. **244**, 4–21 (2013)
47. Land, S., Niederer, S.A., Aronsen, J.M., Espe, E.K.S., Zhang, L.L., Louch, W.E., Sjaastad, I., Sejersted, O.M., Smith, N.P.: An analysis of deformation-dependent electromechanical coupling in the mouse heart. J. Physiol. **590**, 4553–4569 (2012)
48. LeGrice, I.J., Smaill, B.H., Chai, L.Z., Edgar, S.G., Gavin, J.B., Hunter, P.J.: Laminar structure of the heart: ventricular myocyte arrangement and connective tissue architecture in the dog. Am. J. Physiol. Heart Circ. Physiol. **269**, H571–H582 (1995)
49. Li, X.T., Dyachenko, V., Zuzarte, M., Putzke, C., Preisig-Muller, R., Isenberg, G., Daut, J.: The stretch-activated potassium channel TREK-1 in rat cardiac ventricular muscle. Cardiovasc. Res. **69**, 86–97 (2006)
50. Li, J., Widlund, O.B.: FETI-DP, BDDC, and block Cholesky methods. Int. J. Numer. Meth. Eng. **66**(2), 250–271 (2006)
51. Mandel, J., Dohrmann, C.R.: Convergence of a balancing domain decomposition by constraints and energy minimization. Numer. Lin. Alg. Appl. **10**(7), 639–659 (2003)
52. Mandel, J., Dohrmann, C.R., Tezaur, R.: An algebraic theory for primal and dual substructuring methods by constraints. Appl. Numer. Math. **54**(2), 167–193 (2005)
53. Mardal, K.-A., Nielsen, B.F., Cai, X., Tveito, A.: An order optimal solver for the discretized bidomain equations. Numer. Linear Algebr. Appl. **14**(2), 83–98 (2007)
54. Munteanu, M., Pavarino, L.F.: Decoupled Schwarz algorithms for implicit discretization of nonlinear Monodomain and Bidomain systems. Math. Mod. Meth. Appl. Sci. **19**(7), 1065–1097 (2009)
55. Munteanu, M., Pavarino, L.F., Scacchi, S.: A scalable Newton-Krylov-Schwarz method for the Bidomain reaction-diffusion system. SIAM J. Sci. Comput. **31**(5), 3861–3883 (2009)
56. Murillo, M., Cai, X.: A fully implicit parallel algorithm for simulating the nonlinear electrical activity of the heart. Numer. Linear Algebr. Appl. **11**, 261–277 (2004)
57. Nardinocchi, P., Teresi, L.: Electromechanical modeling of anisotropic cardiac tissues. Math. Mech. Sol. **18**(6), 576–591 (2013)

58. Nash, M.P., Hunter, P.J.: Computational mechanics of the heart. From tissue structure to ventricular function. J. Elast. **61**, 113–141 (2000)
59. Nash, M.P., Panfilov, A.V.: Electromechanical model of excitable tissue to study reentrant cardiac arrhythmias. Prog. Biophys. Mol. Biol. **85**, 501–522 (2004)
60. Nayak, A.R., Panfilov, A.V., Pandit, R.: Spiral-wave dynamics in a mathematical model of human ventricular tissue with myocytes and Purkinje fibers. Phys. Rev. E **95**(2), 022405 (2017)
61. Niederer, S.A., Hunter, P.J., Smith, N.P.: A quantitative analysis of cardiac myocyte relaxation: a simulation study. Biophys. J. **90**, 1697–1722 (2006)
62. Niederer, S.A., Smith, N.P.: A mathematical model of the slow force response to stretch in rat ventricular myocites, Biophys. J. **92**, 4030–4044 (2007)
63. Nobile, F., Quarteroni, A., Ruiz-Baier, R.: An active strain electromechanical model of cardiac tissue. Int. J. Num. Methods Biomed. Eng. **28**, 52–71 (2012)
64. Noble, D., Rudy, Y.: Models of cardiac ventricular action potentials: iterative interaction between experiment and simulation. Philos. Trans. R. Soc. Lond. A **359**, 1127–1142 (2001)
65. Palamara, S., Vergara, C., Faggiano, E., Nobile, F.: An effective algorithm for the generation of patient-specific Purkinje networks in computational electrocardiology. J. Comput. Phys. **283**, 495–517 (2015)
66. Pathmanathan, P.J., Chapman, S.J., Gavaghan, D.J., Whiteley, J.P.: Cardiac electromechanics: the effect of contraction model on the mathematical problem and accuracy of the numerical scheme. Quart. J. Mech. Appl. Math. **63**(3), 375–399 (2010)
67. Pathmanathan, P.J., Whiteley, J.P.: A numerical method for cardiac mechanoelectric simulations. Ann. Biomed. Eng. **37**, 860–873 (2009)
68. Pavarino, L.F., Scacchi, S.: Multilevel additive Schwarz preconditioners for the Bidomain reaction-diffusion system. SIAM J. Sci. Comput. **31**(1), 420–443 (2008)
69. Pavarino, L.F., Scacchi, S.: Parallel multilevel Schwarz and block preconditioners for the Bidomain parabolic-parabolic and parabolic-elliptic formulations. SIAM J. Sci. Comp. **33**(4), 1897–1919 (2011)
70. Pavarino, L.F., Scacchi, S., Zampini, S.: Newton–Krylov–BDDC solvers for non-linear cardiac mechanics. Comput. Meth. Appl. Mech. Eng. **295**, 562–580 (2015)
71. Pavarino, L.F., Widlund, O.B., Zampini, S.: BDDC Preconditioners for spectral element discretizations of almost incompressible elasticity in three dimensions. SIAM J. Sci. Comp. **32**(6), 3604–3626 (2010)
72. Pennacchio, M., Savaré, G., Colli Franzone, P.: Multiscale modeling for the bioelectric activity of the heart. SIAM J. Math. Anal. **37**, 1333–1370 (2006)
73. Pennacchio, M., Simoncini, V.: Algebraic multigrid preconditioners for the bidomain reaction–diffusion system. Appl. Numer. Math. **59**, 3033–3050 (2009)
74. Pennacchio, M., Simoncini, V.: Fast structured AMG preconditioning for the bidomain model in electrocadiology. SIAM J. Sci. Comput. **33**, 721–745 (2011)
75. Pezzuto, S., Ambrosi, D., Quarteroni, A.: An orthotropic active-strain model for the myocardium mechanics and its numerical approximation. Eur. J. Mech. A Solids **48**(1), 83–96 (2014)
76. Plank, G., Liebmann, M., Weber dos Santos, R., Vigmond, E.J., Haase, G.: Algebraic multigrid preconditioner for the cardiac bidomain model. IEEE Trans. Biomed. Eng. **54**(4), 585–596 (2007)
77. Pullan, A.J., Buist, M.L., Cheng, L.K.: Mathematically Modelling the Electrical Activity of the Heart. World Scientific, Singapore (2005)
78. Quarteroni, A., Lassila, T., Rossi, S., Ruiz-Baier, R.: Integrated Heart–Coupling multiscale and multiphysics models for the simulation of the cardiac function. Comput. Methods Appl. Mech. Eng. **314**, 345–407 (2017)
79. Quarteroni, A., Manzoni, A., Vergara, C.: The cardiovascular system: mathematical modelling, numerical algorithms and clinical applications. Acta Numerica **26**, 365–590 (2017)
80. Richardson, W.J., Clarke, S.A., Quinn, T.A., Holmes, J.W.: Physiological implications of myocardial scar structure. Compr. Physiol. **5**(4), 1877–1909 (2015)

81. Rice, J.J., Wang, F., Bers, D.M., de Tombe, P.P.: Approximate model of cooperative activation and crossbridge cycling in cardiac muscle using ordinary differential equations. Biophys. J. **95**, 2368–2390 (2008)
82. Rossi, S., Ruiz-Baier, R., Pavarino, L.F., Quarteroni, A.: Orthotropic active strain models for the numerical simulation of cardiac biomechanics. Int. J. Num. Methods Biomed. Eng. **28**, 761–788 (2012)
83. Rossi, S., Lassila, T., Ruiz-Baier, R., Sequeira, A., Quarteroni, A.: Thermodynamically consistent orthotropic activation model capturing ventricular systolic wall thickening in cardiac electromechanics. Eur. J. Mech. A Solids **48**, 129–142 (2014)
84. Rudy, Y., Silva, J.R.: Computational biology in the study of cardiac ion channels and cell electrophysiology. Quart. Rev. Biophys. **39**(1), 57–116 (2006)
85. Sahli Costabal, F., Concha, F.A., Hurtado, D.E., Kuhl, E.: The importance of mechano-electrical feedback and inertia in cardiac electromechanics. Comput. Methods Appl. Mech. Eng. **320**, 352–368 (2017)
86. Sainte-Marie, J., Chapelle, D., Cimrman, R., Sorine, M.: Modeling and estimation of cardiac electromechanical activity. Comput. Struct. **84**, 1743–1759 (2006)
87. Scacchi, S.: A multilevel hybrid Newton-Krylov-Schwarz method for the Bidomain model of electrocardiology. Comput. Methods Appl. Mech. Eng. **200**(5–8), 717–725 (2011)
88. Smith, N.P., Nickerson, D.P., Crampin, E.J., Hunter, P.J.: Multiscale computational modelling of the heart. Acta Numerica, 371–431 (2004)
89. Sundnes, J., Lines, G.T., Mardal, K., Tveito, A.: Multigrid block preconditioning for a coupled system of partial differential equations modeling the electrical activity in the heart. Comput. Methods Biomech. Biomed. Eng. **5**, 397–409 (2002)
90. Tagliabue, A., Dedè, L., Quarteroni, A.: Fluid dynamics of an idealized left ventricle: the extended Nitsche's method for the treatment of heart valves as mixed time varying boundary conditions. Int. J. Numer. Methods Fluids **85**(3), 135–164 (2017)
91. Tagliabue, A., Dedè, L., Quarteroni, A.: Complex blood flow patterns in an idealized left ventricle: A numerical study. Chaos **27**(9), 093939 (2017)
92. K. H. W. J. ten Tusscher, D. Noble, P. J. Noble and A. Pan, V.: A model for human ventricular tissue. Am. J. Phys. Heart Circ. Physiol. **286**, H1573–H1589 (2004)
93. ten Tusscher, K.H.W.J., Panfilov, A.V.: Alternans and spiral breakup in a human ventricular tissue model. Am. J. Phys. Heart Circ. Physiol. **291**, H1088–H1100 (2006)
94. Toselli, A., Widlund, O.B.: Domain Decomposition Methods: Algorithms and Theory. Computational Mathematics, Vol. 34. Springer, Berlin (2004)
95. Tung, L.: A bidomain model for describing ischemic myocardiacl D.C. potentials. PhD dissertation, MIT, Cambridge, MA (1978)
96. Usyk, T.P., LeGrice, I.J., McCulloch, A.D.: Computational model of three-dimensional cardiac electromechanics. Comput. Visual. Sci. **4**, 249–257 (2002)
97. Usyk, T.P., Mazharia, R., McCulloch, A.D.: Effect of laminar orthotropic myofiber architecture on regional stress and strain in the canine left ventricle. J. Elast. **61**(1–3), 143–164 (2000)
98. Veneroni, M.: Reaction-diffusion systems for the macroscopic Bidomain model of the cardiac electric field. Nonlin. Anal. Real World Appl. **10**, 849–868 (2009)
99. Vergara, C., Lange, M., Palamara, S., Lassila, T., Frangi, A., Quarteroni, A.: A coupled 3D–1D numerical monodomain solver for cardiac electrical activation in the myocardium with detailed Purkinje network. J. Comput. Phys. **308**, 218–238 (2016)
100. Vetter, F.J., McCulloch, A.D.: Three-dimensional stress and strain in passive rabbit left ventricle: a model study. Ann. Biomed. Eng. **28**, 781–792 (2000)
101. Vigmond, E.J., Aguel, F., Trayanova, N.A.: Computational techniques for solving the bidomain equations in three dimensions. IEEE Trans. Biomed. Eng. **49**(11), 1260–1269 (2002)
102. Wall, S.T., Guccione, J.M., Ratcliffe, M.B., Sundnes, J.S.: Electromechanical feedback with reduced cellular connectivity alters electrical activity in an infarct injures left ventricle: a finite element model study. Am. J. Physiol. Heart Circ. Physiol. **302**, H206–H214 (2012)

103. Whiteley, J.P., Bishop, M.J., Gavaghan, D.J.: Soft tissue modelling of cardiac fibres for use in coupled mechano-electric simulations. Bull. Math. Biol. **69**, 2199–2225 (2007)
104. Zampini, S.: Balancing Neumann-Neumann methods for the cardiac Bidomain model. Numer. Math. **123**(2), 363–393 (2013)
105. Zampini, S.: Dual-primal methods for the cardiac bidomain model. Math. Models Methods Appl. Sci. **24**(4), 667–696 (2014)

Piero Colli Franzone graduated in Mathematics from the University of Pavia in 1969. He was then Researcher at the Institute of Numerical Analysis of the National Research Council (CNR) in Pavia (1971–73), Associate Professor of Mathematical Analysis at the University of Pavia (1973–81), Full Professor of Numerical Analysis at the University of Udine (1981–83) and Pavia (1983–2016), where he is now Professor Emeritus.

Luca F. Pavarino is Professor of Numerical Analysis at the Department of Mathematics of the University of Pavia. He holds a PhD (1992) and master's (1990) in mathematics from the Courant Institute, NYU, USA, and a B.S. degree (1987) in mathematics from the University of Pavia. He was a Postdoctoral Research Associate at Rice University, USA, and Professor at the University of Milan. His current research focuses on domain decomposition methods for PDEs, parallel computing, and biomathematics.

Simone Scacchi is Associate Professor of Numerical Analysis at the Department of Mathematics of the University of Milan. He received a degree in mathematics from the University of Milan (2004) and a PhD in mathematics and statistics from the University of Pavia (2008). He was a Postdoctoral Research Associate at the University of Pavia (2008) and then Assistant Professor at the University of Milan (2008–2015).

Ergotropic Effect in Cardiac Tissue After Electromagnetic and β-Adrenergic Stimulus

Lorenzo Fassina, Marisa Cornacchione, Maria Evelina Mognaschi, Giovanni Magenes, and Fabio Naro

Abstract In a murine ventricular cardiac tissue *in vitro*, via an image processing analysis, we have studied the ergotropic effect (contraction energy) after electromagnetic stimulation (frequency, 75 Hz), isoproterenol administration (10 μM), and their combination. We have found that the electromagnetic stimulation is able to counteract the β-adrenergic action of isoproterenol.

1 Introduction

A core idea of Tissue Engineering is to understand the relationships between structures and functions in mammalian cells. This information is important during the growth of tissue substitutes *in vitro*; in other words, Tissue Engineering constructs are based not only on the use of growth factors, but also on the stimuli provided by the structural context (e.g., the biomaterials with their biocompatibility and mechanical properties) and provided by the biophysical context (e.g., the forces acting onto the plasma membrane, transmitted to the cytoskeleton, and biochemically transduced).

For instance, a fluid shear stress [9, 38, 47] or ultrasounds [17] or biomaterial features [12, 19, 42] lead to the remodeling of bone matrix *in vitro*.

L. Fassina (✉) · M. E. Mognaschi · G. Magenes
Department of Electrical, Computer and Biomedical Engineering, University of Pavia, Pavia, Italy
e-mail: lorenzo.fassina@unipv.it; eve.mognaschi@unipv.it; giovanni.magenes@unipv.it

M. Cornacchione
IRCCS SDN, Naples, Italy
e-mail: marisa.cornacchione@uniroma1.it

F. Naro
Department of Anatomical, Histological, Forensic and Orthopaedic Sciences, Sapienza University of Rome, Rome, Italy
e-mail: fabio.naro@uniroma1.it

D. Boffi et al. (eds.), *Mathematical and Numerical Modeling of the Cardiovascular System and Applications*, SEMA SIMAI Springer Series 16,
https://doi.org/10.1007/978-3-319-96649-6_4

In addition, a modulation of the cellular function is well attested by the cardiomyocytes subjected to the mechanical forces induced by an electromagnetic field [3, 33]. However, the effects of the electromagnetic fields are controversial. Some works showed that basal heart rate was either decreased and coupled with arrhythmias or increased with occurrence of tachycardia [28, 44].

In the heart, the β-adrenergic receptors (βARs), associated to G proteins, play a fundamental role in regulating the cardiac function [29, 41].

In this work, we have studied the contraction movement of murine cardiomyocytes under electromagnetic and/or β-adrenergic stimulation, addressing, in particular, the ergotropic effect (contraction energy).

2 Materials and Methods

2.1 *Beating Mouse Cardiac Syncytia*

Spontaneously beating cardiac syncytia were obtained from hearts of 1- to 2-day-old CD-1® mouse pups (Charles River Laboratories Italia, Calco, Italy) (Fig. 1), as previously described [5, 45, 46] with some modifications. Briefly, beating primary cultures of murine cardiomyocytes were prepared *in vitro* as follows: the hearts were quickly excised, the atria were cut off, and the ventricles were minced and digested by incubation with 100 μg/ml type II collagenase (Invitrogen, Carlsbad, CA) and with 900 μg/ml pancreatin (Sigma-Aldrich, Milan, Italy) in ADS buffer (0.1 M HEPES, 0.1 M D-glucose, 0.5 M NaCl, 0.1 M KCl, 0.1 M $NaH_2PO_4 \bullet H_2O$, 0.1 M $MgSO_4$) for 15 min at 37 °C. The resulting cell suspension was preplated for 2 h at 37 °C to reduce the contribution of non-myocardial cells. The unattached,

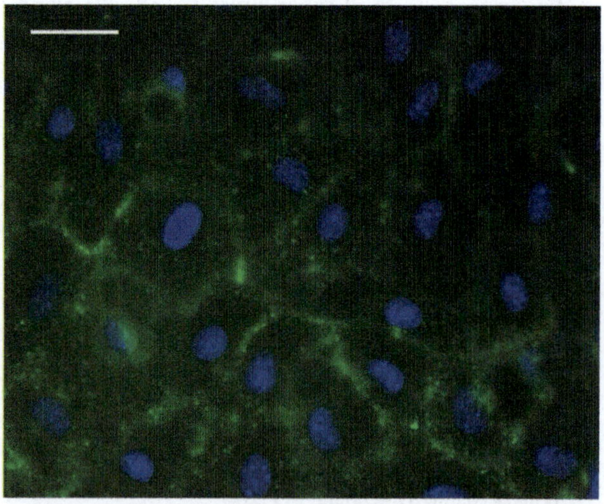

Fig. 1 Cardiac syncytium with gap junctions (green fluorescence related to connexin 43) and cellular nuclei (blue fluorescence); white marker, 20 μm

cardiomyocyte-enriched cells remaining in suspension were collected, plated onto collagen-coated 35-mm Petri dishes, and covered by DMEM containing 10% horse serum, 5% fetal bovine serum, and 1× gentamicin (Roche Molecular Biochemicals, Indianapolis, IN). About 3×10^5 cardiomyocytes were cultured in each Petri dish at 37 °C and 5% CO_2 to form a spontaneously beating cardiac syncytium (that is, a cardiac cell culture made by multilayers of contracting cardiomyocytes as in our previous works [20, 23]).

2.2 Experimental Conditions

On day 3 of culture, at a constant temperature of 37 °C and 5% CO_2, each syncytium was observed via a movie capture system (ProgRes C5, Jenoptik, Germany) in four different conditions: untreated control (CTRL), stimulated via β-adrenergic isoproterenol (ISO, 10 µM; Sigma-Aldrich, Milan, Italy), via an electromagnetic field (EMF; see below for details), via both isoproterenol and electromagnetic field (ISO+EMF). In particular, for each condition, AVI videos (duration, 20 s) of 20 beating syncytia were collected every 3 min (in general, in each video, we have systematically selected 30 spots or markers to be tracked during contraction; see below for details), permitting us to specifically study the average contraction pattern during the time interval 27–39 min.

2.3 Electromagnetic Bioreactor

The electromagnetic bioreactor used here has been previously investigated in terms of biological effects [10, 14, 16, 18, 31, 35] and in terms of numerical dosimetry and physical parameters (induced electric field, induced electric current, induced forces) [33]. The setup was based on two air-cored solenoids (Fig. 2) connected in series, placed inside a cell incubator and powered by a pulse generator (Biostim SPT from Igea, Carpi, Italy). The magnetic induction field (module, circa 3 mT; frequency, 75 Hz) was perpendicular to the seeded cells.

In particular, in our experimental setup:

- the electric current in the solenoid wire ranged from 0 to 319 mA in 1.36 ms (Fig. 2 in [33]);
- in order to optimize the spatial homogeneity of the magnetic induction field, especially in the central region where the cells were stimulated, the two solenoids were supplied by the same electric current and their dimensions and distance were comparable; the spatial homogeneity was both calculated *in silico* [33] and verified inside the cell incubator by means of Hall-effect gaussmeter;

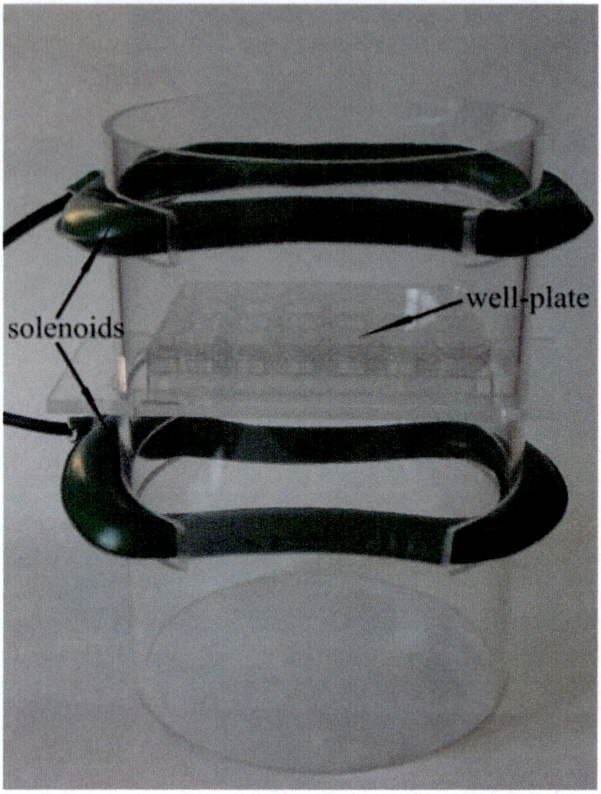

Fig. 2 Solenoids of the electromagnetic bioreactor with a culture well-plate in the central region

- the maximum electromagnetic energy density applied to the cells was about 3.18 J/m^3 and, using a thermocouple, we observed no EMF-induced heating;
- according to the Faraday-Neumann-Lenz and Lorentz laws, inside the cylindrical culture wells, the time varying and homogeneous magnetic induction generated a concentric and planar distribution of induced electric currents with corresponding induced distribution of radial mechanical forces (Fig. 10 in [33]): in the temporal range 0–1.36 ms the magnetic induction was arising, the currents clockwise, and the radial mechanical forces inwardly directed (compression), whereas, on the contrary, during the temporal range 1.36–6 ms, the magnetic induction was decreasing, the currents anticlockwise, and the radial mechanical forces outwardly directed (traction);
- during the same time interval of the electromagnetic stimulation, control cells were placed into another but identical incubator with no EMF.

2.4 *Registration of the Syncytium Movement via the Apposition of Software Markers*

By the Video Spot Tracker (VST) program, which is used to track the motion of one or more spots in an AVI video file (http://cismm.cs.unc.edu/downloads), in each video, we have systematically selected 30 spots or markers onto the first video frame, according to the same orthogonal grid [30, 34]. By starting the videos in VST, frame by frame, the program followed and registered the spatial-temporal coordinates x, y, and t for each marker, as previously described [20]. The coordinates x and y are expressed in pixel, whereas the coordinate t in s.

2.5 *Kinematics of the Beating Syncytium*

By an algorithm based on the Matlab programming language (The MathWorks, Inc., Natick, MA), frame by frame and for each marker, we have studied the kinematics of the beating cardiac syncytia, as previously described [20, 22, 23, 32, 39, 40]. In particular, in this work, we have evaluated the syncytium contraction in terms of contraction energy ($pixel^2/s^2$) (ergotropic effect).

2.6 *Statistics*

In order to compare the results between the different conditions, one-way analysis of variance (ANOVA) with *post hoc* least significant difference (LSD) test was applied, electing a significance level of 0.05. The results are expressed as mean ±95% confidence interval for the differences between means.

3 Results

In terms of ergotropy i.e. contraction energy (Fig. 3), in comparison with control (CTRL), the isoproterenol (ISO) showed a significant positive ergotropic effect ($p<0.05$) and the electromagnetic stimulation (EMF) caused a non-significant negative ergotropic action ($p>0.05$). The simultaneous use of pharmacological and physical stimulation (ISO+EMF) significantly reduced the positive ergotropic effect of ISO ($p<0.05$), giving an overall non-significant negative ergotropic action in comparison with CTRL ($p>0.05$).

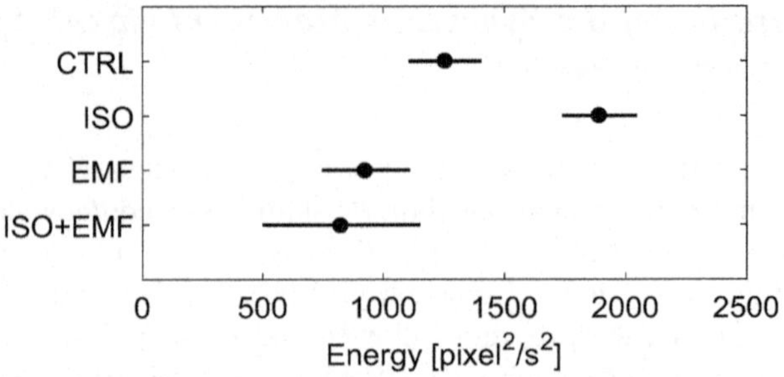

Fig. 3 Mean contraction energy (during the time interval 27–39 min). In terms of ergotropy, in comparison with control (CTRL), the isoproterenol (ISO) showed a significant positive ergotropic effect ($p<0.05$) and the electromagnetic stimulation (EMF) caused a non-significant negative ergotropic action ($p>0.05$). The simultaneous use of pharmacological and physical stimulation (ISO+EMF) significantly reduced the positive ergotropic effect of ISO ($p<0.05$), giving an overall non-significant negative ergotropic action in comparison with CTRL ($p>0.05$). The horizontal bars are the 95% confidence intervals for the differences between means according to LSD (least significant difference) statistical test: there is a statistically significant difference between the means with non-overlapping bars (n=20 syncytia for each condition)

4 Discussion

The mouse is in the center of the models of cardiovascular diseases. Using *in vitro* beating primary murine ventricular cardiomyocytes, we have studied the modification of their contraction following the mechanical forces induced by an electromagnetic field and/or a β-adrenergic stimulation (10 μM isoproterenol) [3, 33].

In particular, the electromagnetic bioreactor induced compression/traction cycles, that is, a mechanical vibration (frequency=75 Hz) onto the cells. A similar vibration (frequency=80 Hz), according to [27], is able to cause a speeding up of the cardiomyocyte relaxation and a reduced cellular tension via an increase in detachment rate of attached cross-bridges from the thin filaments of the cytoskeleton without substantial reattachment. In a preceding work [24], we then observed an accordant reduction of contraction acceleration and contractility, whereas, in the present study, we have recognized the same biological behavior in the reduction of the contraction energy (ergotropy).

In addition, in the heart, the stimulation of β_1ARs and β_2ARs enhances the contraction frequency by increasing cAMP production via activation of adenylyl cyclases [3]. Then, activated PKA phosphorylates proteins essential for contraction. β_1AR is significantly more effective than β_2AR to increase the sarcomeric shortening, because it directly activates the G_s-cAMP pathway, whereas β_2AR activates sequentially both the G_s and G_i pathways.

Moreover, by the electromagnetic bioreactor previously described [1, 2, 6, 8, 10, 11, 13–16, 18, 21, 26, 35–37, 43], our previous work indicated that an exposure to a low-frequency EMF decreases the beat frequency of neonatal murine cardiomyocytes and also the effects of the β-adrenergic stimulation [3], due to the internalization of β_2ARs.

In accordance with the cited papers, in the present work, we have showed that a low-frequency electromagnetic stimulus was able to counteract both the basal ergotropy and the β-adrenergically enhanced ergotropy, due to induced mechanical vibration [27, 33], internalization of β_2ARs [3], and inhibition of T-type calcium channels via AA/LTE4 signal [4].

5 Conclusion

While some studies arouse alarms about the low-frequency electromagnetic exposure [7, 44], this work suggests a potential use of that biophysical stimulation to modulate the β-adrenergic sensibility. In particular, a weakening of the β-adrenergic sensibility can be essential in the ischemia-reperfusion injuries [25].

Ethics Statement

All procedures involving mice were completed in accordance with the policy of the Italian National Institute of Health (Protocol nr. 118/99-A) and with the ethical guidelines for animal care of the European Community Council (Directive nr. 86/609/ECC).

CD-1® mice were obtained from the Charles River Laboratories Italia (Calco, Italy) and were housed under 12-h light/dark cycles, at constant temperature, and with food and water *ad libitum*. The mice were sacrificed by cervical dislocation.

Acknowledgements This work was supported by Research Grants from INAIL (INAIL 2010 to FN), from Sapienza University of Rome (Ateneo 2009 to FN), and from the Italian Ministry of University (FIRB 2010 RBAP109BLT_003 and PRIN 2010 KL2Y73-006 to FN).

References

1. Capelli, E., Torrisi, F., Venturini, L., Granato, M., Fassina, L., Lupo, G.F.D., Ricevuti, G.: Low-frequency pulsed electromagnetic field is able to modulate miRNAs in an experimental cell model of Alzheimer's disease. J. Healthcare Eng. **2017**, Article ID 2530270 (2017)
2. Ceccarelli, G., Bloise, N., Mantelli, M., Gastaldi, G., Fassina, L., Cusella De Angelis, M.G., Ferrari, D., Imbriani, M., Visai, L.: A comparative analysis of the *in vitro* effects of pulsed electromagnetic field treatment on osteogenic differentiation of two different mesenchymal cell lineages. Biores. Open Access **2**, 283–294 (2013)
3. Cornacchione, M., Pellegrini, M., Fassina, L., Mognaschi, M.E., Di Siena, S., Gimmelli, R., Ambrosino, P., Soldovieri, M.V., Taglialatela, M., Gianfrilli, D., Isidori, A.M., Lenzi, A., Naro, F.: β-Adrenergic response is counteracted by extremely-low-frequency pulsed electromagnetic fields in beating cardiomyocytes. J. Mol. Cell Cardiol. **98**, 146–158 (2016)

4. Cui, Y., Liu, X., Yang, T., Mei, Y.A., Hu, C.: Exposure to extremely low-frequency electromagnetic fields inhibits T-type calcium channels via AA/LTE4 signaling pathway. Cell Calcium **55**, 48–58 (2014)
5. Devic, E., Xiang, Y., Gould, D., Kobilka, B.: β-adrenergic receptor subtype-specific signaling in cardiac myocytes from β_1 and β_2 adrenoceptor knockout mice. Mol. Pharmacol. **60**, 577–583 (2001)
6. Di Barba, P., Fassina, L., Magenes, G., Mognaschi, M.E.: Shape synthesis of a well-plate for electromagnetic stimulation of cells. Int. J. Numer. Modell. Electron. Networks Devices Fields (2018, in press)
7. Elmas, O., Comlekci, S., Koylu, H.: Effects of short-term exposure to powerline-frequency electromagnetic field on the electrical activity of the heart. Arch. Environ. Occup. Health **67**, 65–71 (2012)
8. Falone, S., Marchesi, N., Osera, C., Fassina, L., Comincini, S., Amadio, M., Pascale, A.: Pulsed electromagnetic field (PEMF) prevents pro-oxidant effects of H_2O_2 in SK-N-BE(2) human neuroblastoma cells. Int. J. Radiat. Biol. **92**, 281–286 (2016)
9. Fassina, L., Visai, L., Asti, L., Benazzo, F., Speziale, P., Tanzi, M.C., Magenes, G.: Calcified matrix production by SAOS-2 cells inside a polyurethane porous scaffold, using a perfusion bioreactor. Tissue Eng. **11**, 685–700 (2005)
10. Fassina, L., Visai, L., Benazzo, F., Benedetti, L., Calligaro, A., Cusella De Angelis, M.G., Farina, A., Maliardi, V., Magenes, G.: Effects of electromagnetic stimulation on calcified matrix production by SAOS-2 cells over a polyurethane porous scaffold. Tissue Eng. **12**, 1985–1999 (2006)
11. Fassina, L., Visai, L., Cusella De Angelis, M.G., Benazzo, F., Magenes, G.: Surface modification of a porous polyurethane through a culture of human osteoblasts and an electromagnetic bioreactor. Technol. Health Care **15**, 33–45 (2007)
12. Fassina, L., Saino, E., Visai, L., Cusella De Angelis, M.G., Benazzo, F., Magenes, G.: Enhanced *in vitro* culture of human SAOS-2 osteoblasts on a sand-blasted titanium surface modified with plastic deformation. Conf. Proc. IEEE Eng Med. Biol. Soc. **2007**, 6411–6414 (2007)
13. Fassina, L., Saino, E., Visai, L., Magenes, G.: Physically enhanced coating of a titanium plasma-spray surface with human SAOS-2 osteoblasts and extracellular matrix. Conf. Proc. IEEE Eng Med. Biol. Soc. **2007**, 6415–6418 (2007)
14. Fassina, L., Saino, E., Visai, L., Silvani, G., Cusella De Angelis, M.G., Mazzini, G., Benazzo, F., Magenes, G.: Electromagnetic enhancement of a culture of human SAOS-2 osteoblasts seeded onto titanium fiber-mesh scaffolds. J. Biomed. Mater. Res. A **87**, 750–759 (2008)
15. Fassina, L., Saino, E., Visai, L., Magenes, G.: Electromagnetically enhanced coating of a sintered titanium grid with human SAOS-2 osteoblasts and extracellular matrix. Conf. Proc. IEEE Eng Med. Biol. Soc. **2008**, 3582–3585 (2008)
16. Fassina, L., Saino, E., Sbarra, M.S., Visai, L., Cusella De Angelis, M.G., Mazzini, G., Benazzo, F., Magenes, G.: Ultrasonic and electromagnetic enhancement of a culture of human SAOS-2 osteoblasts seeded onto a titanium plasma-spray surface. Tissue Eng Part C Methods **15**, 233–242 (2009)
17. Fassina, L., Saino, E., Cusella De Angelis, M.G., Magenes, G., Benazzo, F., Visai, L.: Low-power ultrasounds as a tool to culture human osteoblasts inside cancellous hydroxyapatite. Bioinorg. Chem. Appl. **2010**, Article ID 456240 (2010)
18. Fassina, L., Saino, E., Sbarra, M.S., Visai, L., Cusella De Angelis, M.G., Magenes, G., Benazzo, F.: *In vitro* electromagnetically stimulated SAOS-2 osteoblasts inside porous hydroxyapatite. J. Biomed. Mater. Res. A **93**, 1272–1279 (2010)
19. Fassina, L., Saino, E., Visai, L., Avanzini, M.A., Cusella De Angelis, M.G., Benazzo, F., Van Vlierberghe, S., Dubruel, P., Magenes, G.: Use of a gelatin cryogel as biomaterial scaffold in the differentiation process of human bone marrow stromal cells. Conf. Proc. IEEE Eng Med. Biol. Soc. **1**, 247–250 (2010)

20. Fassina, L., Di Grazia, A., Naro, F., Monaco, L., Cusella De Angelis, M.G., Magenes, G.: Video evaluation of the kinematics and dynamics of the beating cardiac syncytium: an alternative to the Langendorff method. Int. J. Artif. Organs **34**, 546–558 (2011)
21. Fassina, L., Saino, E., Visai, L., Schelfhout, J., Dierick, M., Van Hoorebeke, L., Dubruel, P., Benazzo, F., Magenes, G., Van Vlierberghe, S.: Electromagnetic stimulation to optimize the bone regeneration capacity of gelatin-based cryogels. Int. J. Immunopathol. Pharmacol. **25**, 165–174 (2012)
22. Fassina, L., Di Grazia, A., Naro, F., Aguanno, S., Cornacchione, M., Cusella De Angelis, M.G., Sardi, F., Magenes, G.: Effects of the hydrostatic pressure in *in vitro* beating cardiac syncytia in terms of kinematics (kinetic energy and beat frequency) and syncytia geometrical-functional classification. Conf. Proc. IEEE Eng Med. Biol. Soc. **2013**, 854–857 (2013)
23. Fassina, L., Magenes, G., Gimmelli, R., Naro, F.: Modulation of the cardiomyocyte contraction inside a hydrostatic pressure bioreactor: *in vitro* verification of the Frank-Starling law. Biomed. Res. Int. **2015**, Article ID 542105 (2015)
24. Fassina, L., Cornacchione, M., Pellegrini, M., Mognaschi, M.E., Gimmelli, R., Isidori, A.M., Lenzi, A., Magenes, G., Naro, F.: Model of murine ventricular cardiac tissue for *in vitro* kinematic-dynamic studies of electromagnetic and β-adrenergic stimulation. J. Healthcare Eng. **2017**, Article ID 4204085 (2017)
25. Fassina, L., Rozzi, G., Rossi, S., Scacchi, S., Galetti, M., Lo Muzio, F.P., Del Bianco, F., Colli Franzone, P., Petrilli, G., Faggian, G., Miragoli, M.: Cardiac kinematic parameters computed from video of *in situ* beating heart. Sci. Rep. **7**, Article no. 46143 (2017)
26. Icaro Cornaglia, A., Casasco, M., Riva, F., Farina, A., Fassina, L., Visai, L., Casasco, A.: Stimulation of osteoblast growth by an electromagnetic field in a model of bone-like construct. Eur. J. Histochem. **50**, 199–204 (2006)
27. Janssen, P.M.L., Schiereck, P., Honda, H., Naya, T., Koiwa, Y.: The effect of applied mechanical vibration on two different phases of rat papillary muscle relaxation. Pflügers Arch. **434**, 795–800 (1997)
28. Korpinen, L., Partanen, J., Uusitalo, A.: Influence of 50 Hz electric and magnetic fields on the human heart. Bioelectromagnetics **14**, 329–340 (1993)
29. Lohse, M.J., Engelhardt, S., Eschenhagen, T.: What is the role of β-adrenergic signaling in heart failure? Circ. Res. **93**, 896–906 (2003)
30. Mandarim-de-Lacerda, C.A.: Stereological tools in biomedical research. An. Acad. Bras. Cienc. **75**, 469–486 (2003)
31. Marchesi, N., Osera, C., Fassina, L., Amadio, M., Angeletti, F., Morini, M., Magenes, G., Venturini, L., Biggiogera, M., Ricevuti, G., Govoni, S., Caorsi, S., Pascale, A., Comincini, S.: Autophagy is modulated in human neuroblastoma cells through direct exposition to low frequency electromagnetic fields. J. Cell Physiol. **229**, 1776–1786 (2014)
32. Meraviglia, V., Wen, J., Piacentini, L., Campostrini, G., Wang, C., Florio, M.C., Azzimato, V., Fassina, L., Langes, M., Wong, J., Miragoli, M., Gaetano, C., Pompilio, G., Barbuti, A., Di Francesco, D., Mascalzoni, D., Pramstaller, P.P., Colombo, G.I., Chen, H.S., Rossini, A.: Higher cardiogenic potential of iPSCs derived from cardiac versus skin stromal cells. Front Biosci. (Landmark. Ed) **21**, 719–743 (2016)
33. Mognaschi, M.E., Di Barba, P., Magenes, G., Lenzi, A., Naro, F., Fassina, L.: Field models and numerical dosimetry inside an extremely-low-frequency electromagnetic bioreactor: the theoretical link between the electromagnetically induced mechanical forces and the biological mechanisms of the cell tensegrity. Springerplus **3**, 473 (2014)
34. Mühlfeld, C., Nyengaard, J.R., Mayhew, T.M.: A review of state-of-the-art stereology for better quantitative 3D morphology in cardiac research. Cardiovasc. Pathol. **19**, 65–82 (2010)
35. Osera, C., Fassina, L., Amadio, M., Venturini, L., Buoso, E., Magenes, G., Govoni, S., Ricevuti, G., Pascale, A.: Cytoprotective response induced by electromagnetic stimulation on SH-SY5Y human neuroblastoma cell line. Tissue Eng. Part A **17**, 2573–2582 (2011)
36. Osera, C., Amadio, M., Falone, S., Fassina, L., Magenes, G., Amicarelli, F., Ricevuti, G., Govoni, S., Pascale, A.: Pre-exposure of neuroblastoma cell line to pulsed electromagnetic field prevents H_2O_2-induced ROS production by increasing MnSOD activity. Bioelectromagnetics **36**, 219–232 (2015)

37. Pasi, F., Fassina, L., Mognaschi, M.E., Lupo, G.F.D., Corbella, F., Nano, R., Capelli, E.: Pulsed electromagnetic field with temozolomide can elicit an epigenetic pro-apoptotic effect on glioblastoma T98G cells. Anticancer Res. **36**, 5821–5826 (2016)
38. Pavalko, F.M., Norvell, S.M., Burr, D.B., Turner, C.H., Duncan, R.L., Bidwell, J.P.: A model for mechanotransduction in bone cells: the load-bearing mechanosomes. J. Cell Biochem. **88**, 104–112 (2003)
39. Rebuzzini, P., Cebral, E., Fassina, L., Redi, C.A., Zuccotti, M., Garagna, S.: Arsenic trioxide alters the differentiation of mouse embryonic stem cell into cardiomyocytes. Sci. Rep. **5** Article no. 14993 (2015)
40. Rebuzzini, P., Fassina, L., Mulas, F., Bellazzi, R., Redi, C.A., Di Liberto, R., Magenes, G., Adjaye, J., Zuccotti, M., Garagna, S.: Mouse embryonic stem cells irradiated with γ-rays differentiate into cardiomyocytes but with altered contractile properties. Mutat. Res. **756**, 37–45 (2013)
41. Rohrer, D.K., Chruscinski, A., Schauble, E.H., Bernstein, D., Kobilka, B.K.: Cardiovascular and metabolic alterations in mice lacking both β_1- and β_2-adrenergic receptors. J. Biol. Chem. **274**, 16701–16708 (1999)
42. Saino, E., Maliardi, V., Quartarone, E., Fassina, L., Benedetti, L., Cusella De Angelis, M.G., Mustarelli, P., Facchini, A., Visai, L.: *In vitro* enhancement of SAOS-2 cell calcified matrix deposition onto radio frequency magnetron sputtered bioglass-coated titanium scaffolds. Tissue Eng. Part A **16**, 995–1008 (2010)
43. Saino, E., Fassina, L., Van Vlierberghe, S., Avanzini, M.A., Dubruel, P., Magenes, G., Visai, L., Benazzo, F.: Effects of electromagnetic stimulation on osteogenic differentiation of human mesenchymal stromal cells seeded onto gelatin cryogel. Int. J. Immunopathol. Pharmacol. **24**, 1–6 (2011)
44. Savitz, D.A., Liao, D., Sastre, A., Kleckner, R.C., Kavet, R.: Magnetic field exposure and cardiovascular disease mortality among electric utility workers. Am. J. Epidemiol. **149**, 135–142 (1999)
45. Xiang, Y., Rybin, V.O., Steinberg, S.F., Kobilka, B.: Caveolar localization dictates physiologic signaling of β_2-adrenoceptors in neonatal cardiac myocytes. J. Biol. Chem. **277**, 34280–34286 (2002)
46. Xiang, Y., Naro, F., Zoudilova, M., Jin, S.L., Conti, M., Kobilka, B.: Phosphodiesterase 4D is required for β_2 adrenoceptor subtype-specific signaling in cardiac myocytes. Proc. Natl. Acad. Sci. USA **102**, 909–914 (2005)
47. Young, S.R., Gerard-O'Riley, R., Kim, J.B., Pavalko, F.M.: Focal adhesion kinase is important for fluid shear stress-induced mechanotransduction in osteoblasts. J. Bone Miner. Res. **24**, 411–424 (2009)

Lorenzo Fassina, PhD, is Assistant Professor at the Department of Electrical, Computer and Biomedical Engineering, University of Pavia (Italy). He is author of several papers in the field of tissue engineering; in particular, he is interested in bone and heart tissue engineering via biomaterials and biophysical stimuli.

Marisa Cornacchione, PhD, is postdoctoral researcher at the Department of Anatomical, Histological, Forensic and Orthopaedic Sciences, Sapienza University of Rome (Italy). She is author of several papers in the field of cell biology; in particular, she investigates the signaling pathways in cardiac cells, in relation to normal and pathological excitation/contraction coupling.

Maria Evelina Mognaschi, PhD, is Assistant Professor at the Department of Electrical, Computer and Biomedical Engineering, University of Pavia (Italy). She is author of several papers in the field of inverse problems, in particular multi-objective optimization and identification problems in electromagnetism and biological systems.

Giovanni Magenes, PhD, is Full Professor at the Department of Electrical, Computer and Biomedical Engineering, University of Pavia (Italy). He received the MS degree in electronic engineering *cum laude* from the University of Pavia and the PhD degree in biomedical engineering from the Politecnico di Milano (Italy). He was a Visiting Researcher at the Institut Nationale de la Santé et Réchèrche Medicale, Lyon (France), and a Visiting Professor at the Faculté des Sciences, Université de la Mediterraneé, Luminy (France). He is currently Full Professor of Biomedical Engineering Signal. He has been the Director of the Department of Computer Engineering and Systems Science and the Director of the Tissue Engineering Centre, and now he is the Delegate of the University of Pavia for Information and Communication Technologies. He is author of several papers in the fields of soft computing methods in biomedical applications, biomimetic sensors and systems, artificial sensorimotor systems, tissue engineering, and wearable devices.

Fabio Naro, PhD, is Associate Professor of Histology and Embryology at the Department of Anatomical, Histological, Forensic and Orthopaedic Sciences, Sapienza University of Rome (Italy). He won the Dean's Postdoctoral Fellowship at the Stanford University School of Medicine, the Award of Buzzati-Traverso Foundation, and the Fulbright Scholarship for Medical Sciences. He was Professor at the Federal Rural University of Pernambuco (Brazil). He holds institutional positions at the Sapienza University of Rome. He is Principal Investigator in numerous research projects and author of several papers in the fields of muscle cell biology and signal transduction.

Role of Electrotonic Current in Excitable Cells

Emilio Macchi, Ezio Musso, and Stefano Rossi

Abstract The aim of the present chapter is to review basic properties of electrotonic current flow in excitable cells, such as neuronal axons and cardiac tissue, during subthreshold stimulation, excitation threshold and impulse conduction. Electrotonic current is proportional to the spatial gradient of the transmembrane potential and consists of a current flow across the membrane with the effect to depolarize it. There is a close interrelationship between electrotonic current that originates from local source-sink interactions and excitation threshold. Successful impulse conduction requires that the amount of active current supplied by the membrane at the source location must be equal to or exceed the amount of electrotonic current required to excite the membrane at the sink location. Such condition is determined by the state of membrane excitability at the source, at the sink and by the degree of electrical coupling between source and sink. Conversely, conduction slowing induced by source-sink mismatch in cardiac tissue may be responsible for unidirectional conduction block and reentry, a condition leading to increased arrhythmia vulnerability, both in normal and pathological tissue. In addition to affecting impulse conduction, electrotonic current flow originating from an activation sequence locally modulates action potential repolarization, determining its duration and spatial dispersion across the tissue. Ultimately, experimental evidence is presented in support of the hypothesis of electrotonic current modulation of ventricular repolarization by two different activation sequences, sinus beat and ventricular test site drive, in normal rat heart.

E. Macchi (✉) · E. Musso
Dipartimento di Scienze Chimiche, della Vita e della Sostenibilità Ambientale, Università di Parma, Parma, Italy
e-mail: emilio.macchi@unipr.it

S. Rossi
Dipartimento di Medicina e Chirurgia, Università di Parma, Parma, Italy

D. Boffi et al. (eds.), *Mathematical and Numerical Modeling of the Cardiovascular System and Applications*, SEMA SIMAI Springer Series 16,
https://doi.org/10.1007/978-3-319-96649-6_5

1 Introduction

Considerable data has been reported concerning the spread of electrotonic current in excitable cells such as along neuronal axons and between neighboring cells in cardiac muscle. Electrotonic current is proportional to the membrane potential difference between neighboring cells and consists of a current flow across a membrane with the effect to depolarize its membrane potential. However, an accurate definition and description of electrotonic current, an important prerequisite for understanding the mechanisms of impulse initiation and conduction in excitable cells, is not readily available in the literature. In addition, experience gathered in teaching electrophysiology to Biology students suggests that, although the common significance of electrotonic current may be readily understood, still its intimate mechanism is not easily grasped. Indeed, the close relationship between electrotonic current and threshold of excitation is usually ignored in electrophysiology textbooks. Electrotonic current originates from local source-sink interactions that determine conduction of excitation in cardiac tissue. Successful propagation requires that the amount of active current supplied by the membrane at the source location must be equal to or exceed the amount of electrotonic current required to excite the membrane at the sink location. At the cellular level, this condition is determined by the state of membrane excitability at the source and the sink and by the degree of electrical coupling between source and sink. At the macroscopic scale of multicellular tissue, this condition is strongly influenced by structural properties of myocardium, such as branching of fibers or heterogeneities of electrical coupling over tissue segments. Moreover, the relevance of electrotonic current in sustaining impulse conduction extends beyond normal propagation. Over the last decades electrotonic current mismatch has been increasingly recognized as a potential substrate for abnormal rhythms and reentry, both in normal as well as pathological conditions. For example, electrotonic conduction slowing may be responsible for unidirectional conduction block and reentry, a condition leading to increased myocardial *arrhythmia vulnerability.*

In addition to affecting action potential conduction, electrotonic current modulates action potential repolarization determining its duration and spatial variation (dispersion) across the tissue. The influence of electrotonic current on the repolarization phase of a propagating action potential is commonly undervalued in spite of the relevance of electrotonic modulation by the activation sequence.

Hence, the present chapter aims to review basic properties of electrotonic current flow in excitable cells during subthreshold and threshold stimulation and during action potential conduction. We also present some experimental evidence that supports the hypothesis of electrotonic current modulation of repolarization.

2 Linear Cable Model of Excitable Cells

The linear cable, as equivalent model of excitable cells (see, e.g. [1]), represents transmembrane current i_m as the sum of two components, the capacitive (or displacement) current i_c and the ionic (or resistive) current i_{ion} (Fig. 1). Within excitable cells, current is carried by ions, primarily sodium, potassium, chloride and calcium.

Capacitive current i_c is a displacement of ionic charges on either side of cell membrane without movement across the membrane. For an ion deposition on one side of the membrane, an ion withdrawal of the same sign occurs on the opposite side. Capacitive current is much more important than one might at first suspect, because cell membrane is very thin and thus highly capacitive.

Conversely, ionic current i_{ion} consists of ions that physically cross membrane ion channels.

In brief, $i_m = i_c + i_{ion}$ with i_m, i_c, i_{ion} currents per unit length of the cell.

By definition, capacitive current $i_c = c_m \frac{\partial V_m}{\partial t}$, with c_m membrane capacitance per unit length of the cell, $V_m = V_i - V_e$ membrane potential and V_i, V_e intracellular and extracellular potential, respectively. Ionic current i_{ion} is *resistive*, i.e. according to Ohm's law, through passive ion channels, and *active* through voltage-gated, ligand-gated or mechanosensitive ion channels.

The change in potential per unit length along the intracellular (or extracellular) axial path equals the axial current times the intracellular (or extracellular) resistance/unit length (i.e., a voltage drop according to Ohm's law).

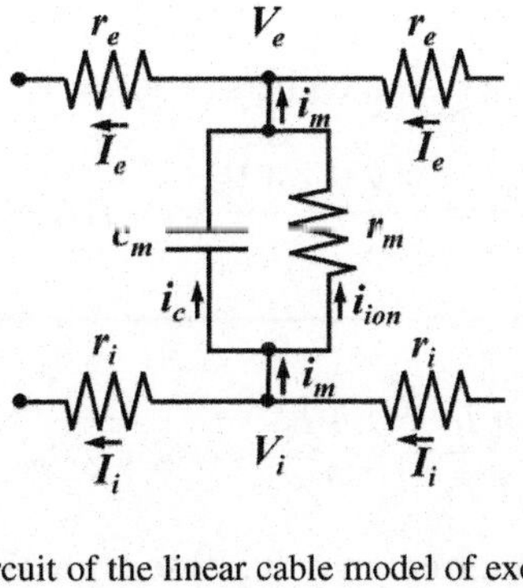

Fig. 1 Electrical equivalent circuit of the linear cable model of excitable cells. Parallel resistance r_m and capacitance c_m represent cell membrane impedance with corresponding i_{ion} and i_c current flow, respectively. Transmembrane current i_m is given by $i_m = i_{ion} + i_c$. I_i. Longitudinal extracellular and intracellular currents are I_e and I_i, extracellular and intracellular potentials are V_e and V_i and extracellular and intracellular resistances per unit length are r_e and r_i, respectively. Current flow is due to subthreshold point stimulation from an intracellular anode outside the right side of the circuit or an extracellular cathode outside the left side of the circuit

Consequently,

$$I_i = -\frac{1}{r_i}\frac{\partial V_i}{\partial x} \quad \text{and} \quad I_e = -\frac{1}{r_e}\frac{\partial V_e}{\partial x},$$

with I_i, I_e intracellular and extracellular axial current, r_i, r_e intracellular and extracellular axial resistance per unit length, respectively and the axial variable is x. The minus signs arises because we define positive longitudinal currents to be flowing in the positive x direction. The potential must *decrease* with *increasing* x for current to flow in the positive x direction because current flows from higher to lower potential values.

If a portion of intracellular axial current I_i leaves the intracellular space by outward crossing the membrane, then the axial decrease is transformed into a transmembrane current i_m and extracellular axial current I_e increases. The changes in axial current per unit length must precisely equal the transmembrane current per unit length, because the total current must be conserved, i.e.:

$$i_m = -\frac{\partial I_i}{\partial x} = \frac{\partial I_e}{\partial x}.$$

By definition, transmembrane current i_m is considered to have a positive sign when it flows across the membrane in the direction from the inside to the outside.

Hence, from the previous definitions we have:

$$i_m = \frac{1}{r_i + r_e}\frac{\partial^2 V_m}{\partial x^2}.$$

In fact, by deriving twice both members of $V_m = V_i - V_e$, we obtain at first:

$$\frac{\partial V_m}{\partial x} = \frac{\partial V_i}{\partial x} - \frac{\partial V_e}{\partial x} = -r_i I_i + r_e I_e,$$

and then:

$$\frac{\partial^2 V_m}{\partial x^2} = -r_i\frac{\partial I_i}{\partial x} + r_e\frac{\partial I_e}{\partial x} = r_i i_m + r_e i_m = (r_i + r_e)\ i_m.$$

The extracellular axial current may increase with axial distance x either due to the arrival of current that crosses the membrane (transmembrane current, i_m) or the introduction of a stimulus current from outside the preparation through inserted electrodes. In presence of impressed current, i.e. extracellular, $\frac{\partial I_e}{\partial x} = i_m + i_s$, where i_s the impressed current per unit length, positive for current entering the extracellular space via *polarizing* electrodes. The units for i_s then correspond to the same units used for i_m. In this case, it results:

$$\frac{\partial^2 V_m}{\partial x^2} = (r_i + r_e)\, i_m + r_e i_s.$$

Thus, when $i_s = 0$, i_m is directly proportional to $\frac{\partial^2 V_m}{\partial x^2}$ and depends only on r_i and r_e.

2.1 *Units of Resistance and Capacitance*

If r is the resistance (ohm) of a box of conducting material of length L (cm) and section area A (cm^2) through which ionic current flows, then $r = R \cdot L/A$, where R (ohm·cm) is the resistivity of the conducting material. The resistivity is defined as the resistance of a unit box of conducting material ($L = 1$ cm, $A = 1$ cm^2).

Hence, units of intracellular/extracellular resistance per unit length, i.e. $r_{i,e} = r/L = R_{i,e}/A$, are ohm/cm where $R_{i,e}$ is the intracellular/extracellular resistivity, respectively.

Units of membrane resistance per unit/length r_m are defined by considering a thin sheet of resistive material instead of the conducting box, with thickness l infinitesimal (δl). Hence, membrane resistivity $R_m = R \cdot \delta l$ has units ohm·cm^2. By considering the surface area A (cm^2) of a length of cell membrane L (cm), i.e. $A = s \cdot L$ with s (cm) the cell circumference, then membrane resistance per unit length is defined by $r_m = r \cdot L = R_m/s$ in units ohm·cm.

If c is the capacitance (microfarad) of a cell membrane of length L (cm), circumference s (cm) and area $A = L \cdot s$ (cm^2) through which capacitive current flows, then $c = C_m \cdot A$ where C_m (microfarad/cm^2) is the membrane capacity, defined as membrane capacitance per unit area. Accordingly, membrane capacitance per unit length is defined as $c_m = c/L = C_m \cdot s$ in units microfarad/cm.

3 Electrotonic Current and Potential

An extracellular point *current source* of strength i_s, such as a point electrode connected with the positive or negative pole of an external current generator, positioned close to an excitable cell, generates an extracellular current I_e that flows away from the positive pole (Fig. 2a, upper panel) or toward the negative pole (Fig. 2a, lower panel), respectively. A fraction of extracellular current I_e crosses cell membrane as transmembrane current i_m that transforms into intracellular current I_i flowing along the cell axis x. At steady-state, a potential gradient $\Delta V_m(x)$ exists along the cell axis where i_m crosses the membrane. Transmembrane current i_m is the *electrotonic current* and the membrane potential gradient $\Delta V_m(x)$ generated by i_m

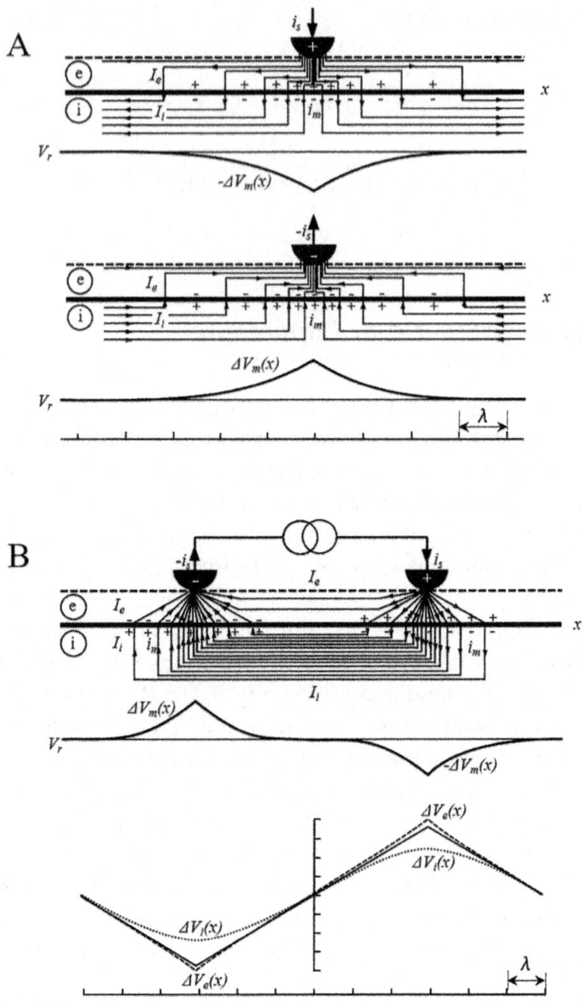

Fig. 2 Schematic diagrams showing steady-state current flow during extracellular stimulation from a current generator of i_s strength. (**a**) Upper panel: unipolar anodal stimulation. Lower panel: unipolar cathodal stimulation. (**b**) Upper panel, bipolar stimulation. Lower panel: theoretical voltage changes following bipolar stimulation; the straight line represents ΔV_e and ΔV_i, extracellular and intracellular voltage changes, respectively, at the make of the stimulus; the dashed and dotted lines represent steady-state values of ΔV_e and ΔV_i, respectively. In each diagram: *x*, longitudinal distance; λ, space constant; *e* and *i*, extracellular and intracellular space, respectively; thick horizontal line, cell membrane; dashed horizontal line, surface of extracellular conducting layer; arrowed straight lines, extracellular (I_e), intracellular (I_i) and transmembrane (i_m) current lines, respectively where density of current lines is not directly proportional to current field density; $\Delta V_m = \Delta V_i - \Delta V_e$, transmembrane voltage change; V_r, resting potential

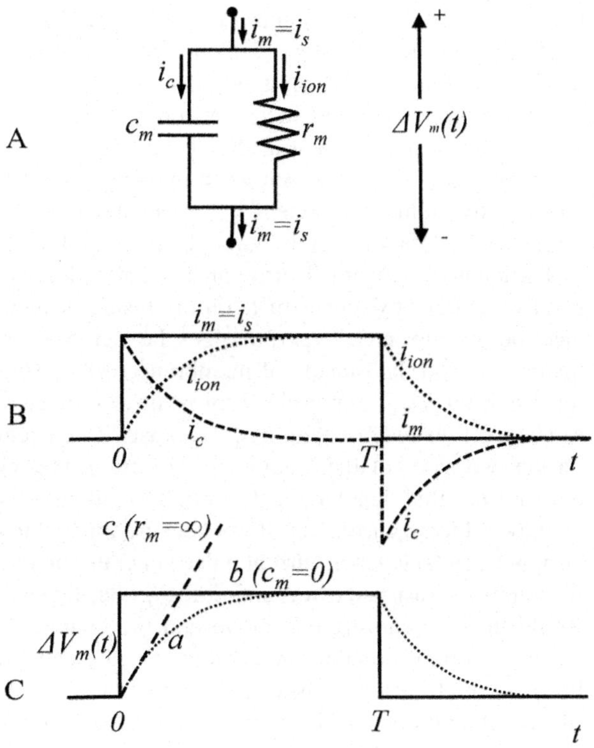

Fig. 3 (**a**) parallel resistance (r_m) and capacitance (c_m) circuit representing cell membrane impedance. (**b**) Time change of transmembrane current i_m, capacitive current i_c and ionic current i_{ion}, during and following application to the circuit shown in A of a step current pulse of strength i_s and duration T. (**c**) Time change of transmembrane voltage $\Delta V_m(t)$ for: (*a*) parallel r_m and c_m circuit; (*b*) purely r_m circuit; (*c*) purely c_m circuit, respectively, during and following application of a step current pulse of strength i_s and duration T

along the cell axis is the *electrotonic potential*. The amplitude of the electrotonic potential is highest at the source location and decreases exponentially with the distance away from the source proportionally to i_m. At distance of a few space constants λ ($\lambda = \sqrt{\frac{r_m}{r_i+r_e}}$) from the source, $i_m = 0$ so that also $\Delta V_m = 0$ and V_m is spatially constant at its resting potential V_r [1].

By representing a membrane segment δx, small enough that the transmembrane voltage is the same all across the patch, with the electrical equivalent circuit of Fig. 3a, the electrotonic current i_m that flows across the membrane is given by $i_m = i_c + i_{ion}$. Capacitive current i_c adds or subtracts positive charges on one side of the membrane and an equivalent number of negative charges on the other side and the charge change of the membrane capacitor generates the membrane potential change $\Delta V_m(t)$ that causes ionic resistive current i_{ion} to flow.

In summary, the membrane capacitor is charged or discharged by capacitive current i_c. The charge change in time on the capacitor creates membrane potential change $\Delta V_m(t)$ that generates ionic resistive current i_{ion} to flow through r_m with the same orientation of i_c. Membrane resistance per unit length r_m is passive when voltage dependent ion channels are at their resting state. In presence of a rectangular current pulse of subthreshold strength, the electrotonic potential $\Delta V_m(t)$ is an exponential function of time with time constant $\tau = r_m c_m$ (subthreshold response) (Fig. 3c, line *a*) and $i_{ion}(t) = \Delta V_m(t)/r_m$ is also an exponential function with the same time constant (Fig. 3b) (see [1]). Ionic current i_{ion} is resistive when membrane resistance per unit length r_m is passive, i.e. below activation threshold of voltage dependent channels, and does not affect the charge on the membrane surface. Thus, near the extracellular current source, transmembrane current i_m is given by $i_m = i_c + i_{ion}$: capacitive component i_c generates the electrotonic potential, and resistive ionic component i_{ion} is generated by the electrotonic potential with both i_c and i_{ion} flowing in the same direction and orientation.

Particularly, outward electrotonic current i_m depolarizes the membrane by adding positive charges by means of i_c on the inside of membrane surface, while inward electrotonic current i_m hyperpolarizes the membrane by adding positive charges by means of i_c on the outside of membrane surface (Fig. 2). In both cases, resistive ionic current i_{ion} is represented by positive ions crossing the membrane with the same direction and orientation of the capacitive current without affecting the charge on the two sides of the membrane. Hence, in the region of outward i_m, it is not outward i_{ion} that depolarizes the cell. The situation is rather the other way round: an outward ionic current i_{ion} flows because the cell is depolarized by the capacitive current i_c. In fact, a step current pulse generates an electrotonic potential ΔV_m even if membrane resistance was infinite (Fig. 3c, line *c*). The transmembrane current would then be entirely capacitive and the change in membrane potential would be simply proportional to the amount of charge applied, i.e. $\Delta V_m(t)$ would change linearly with time due to the charge accumulating on the two sides of the membrane. In case of a purely resistive circuit ($c_m = 0$), $\Delta V_m(t)$ changes simultaneously as the step current pulse (Fig. 3c, line *b*).

In excitable cells, electrotonic current can be generated by:

(a) current injection from an electrode connected with a current generator;
(b) membrane electromotive force (i.e. propagating action potential) generating *active* transmembrane ionic current i_{ion}.

4 Subthreshold Stimulation

Current injection is extracellular if both electrodes connected with the current generator are external to the cell (Fig. 2), while it is intracellular if one of the two electrodes is positioned inside the cell (Fig. 3a, electrical equivalent circuit). Furthermore, extracellular current injection is unipolar if the return current electrode is located at great distance from the current injection electrode (Fig. 2a), or bipolar if both electrodes are next to each other (Fig. 2b).

4.1 Intracellular Current Injection

In this setting, the intracellular microelectrode is connected to an anode (positive pole) or a cathode (negative pole), while the return current electrode is extracellular. In presence of an intracellular anode the current flows outward from the cell depolarizing the membrane, while in presence of an intracellular cathode the current flows into the cell hyperpolarizing the membrane. If the extracellular electrode is close to the microelectrode in the stimulated cell, current lines are mostly comprised within the inter-electrode region, while if the extracellular electrode is at a great distance, current lines display radial flow from the intracellular electrode.

Let us consider an intracellular rectangular current injection of subthreshold strength i_s, as occurs in an isolated cell or a uniformly polarized cardiac Purkinje fibre [2]. The electrical equivalent circuit is represented in Fig. 3a. At $t = 0$, current pulse onset, $i_c = i_s$ while $i_{ion} = 0$ (Fig. 3b). Subsequently, i_c decreases exponentially toward zero with time constant τ (Fig. 3b, $0 < t < T$). The charging of the capacitor generates an exponential potential change $\Delta V_m(t)$ (Fig. 3c, line *a,* $0 < t < T$) and a corresponding exponential increase in i_{ion} toward steady-state i_s (Fig. 3b, $0 < t < T$). During the current pulse, both i_c and i_{ion} flow across the membrane with the same orientation and satisfy at each instant the relationship $i_m = i_s = i_c + i_{ion}$. Under steady-state conditions, i_m is simply represented by its resistive component $i_{ion} = i_s$ since $i_c = 0$ (Fig. 3b, $0 < t < T$). At $t = T$, current pulse offset, $i_m = i_c + i_{ion} = 0$ and $i_c = -i_{ion}$, i.e. i_c and i_{ion} flow across the membrane with opposite orientation (Fig. 3b, $t > T$), discharging the capacitor c_m through the resistor r_m while the potential decreases exponentially toward zero (Fig. 3c, line *a,* $t > T$).

4.2 Extracellular Current Injection

Extracellular current injection is unipolar when one electrode, anode or cathode, is close to the cell membrane and the other electrode is at infinite distance (Fig. 2a) while it is bipolar when both electrodes, anode and cathode, are in close proximity to the cell membrane at short inter-electrode distance (Fig. 2b). Due to the higher resistance of the membrane compared to extracellular resistance, a small fraction of the injected current from an electrode crosses the membrane as electrotonic current i_m flowing along cell axis as intracellular current I_i, while the larger fraction of the injected current flows as extracellular current I_e. Electrotonic current i_m is inward at the anode and hyperpolarizes the membrane along a distance of few λ (Fig. 2a, upper panel). Electrotonic current i_m is outward at the cathode and depolarizes the membrane along a distance of few λ (Fig. 2a, lower panel). Beyond a distance of a few λ from the poles, electrotonic current $i_m = 0$ and electrotonic potential $\Delta V_m = \Delta(V_i - V_e) = \Delta V_i - \Delta V_e = 0$. In fact, beyond a distance of a few λ from the poles, ΔV_i and ΔV_e change linearly along cell axis according to Ohm's law because I_i and I_e are constant and $\Delta V_i = r_i \cdot I_i$ and $\Delta V_e = r_e \cdot I_e$ have the same amplitude and sign [3]. For bipolar current injection, Fig. 2b, lower panel, represents steady-state voltage changes ΔV_i (dotted line) and ΔV_e (dashed line) and instantaneous voltage changes at the make of the stimulus ΔV_i and ΔV_e (straight line). In case of anodal and cathodal unipolar current injection, intracellular and extracellular currents flow from the anode to infinity (Fig. 2a, upper panel) and toward the cathode from infinity (Fig. 2a, lower panel), respectively. In case of bipolar current injection both intracellular and extracellular currents flow from anode to cathode within the inter-electrode distance (Fig. 2b, upper panel).

Detailed analytical solution for the temporo-spatial response of an infinite cable to a current step is described, for example, in Jack et al. [4] and Plonsey and Barr [1].

5 Stimulation Threshold

When an excitable membrane reaches excitation threshold, it becomes itself a membrane current source or a local membrane electromotive force and, as such, can elicit a "*membrane action potential*" in absence of electrotonic current ($i_m = 0$) or a "*propagating action potential*" in presence of electrotonic current ($i_m \neq 0$). Requirements for reaching excitation threshold for the two conditions are different.

5.1 The Membrane Action Potential

Let us consider the case of a cell in which the whole membrane surface experiences the same potential changes at the same time, i.e. the membrane is clamped spatially, and there is no intracellular current flowing along cell axis. This condition occurs in an isolated cardiomyocyte, in shortened segments of Purkinje fibers or in a squid giant axon by the use of an internal silver wire which makes the squid axon membrane isopotential [2, 5]. In this setting, the membrane will become a current source when membrane potential V_m is depolarized from resting potential V_r to a threshold potential V_u (*uniform threshold potential*).

In a squid giant axon, not considering pumps and exchangers and in absence of the inward rectifier current I_{K1}, resting potential V_r is characterized by net ionic transmembrane current $i_{ion} = 0$ (electrical equilibrium) as the result of opposite ion diffusions, mainly through Na^+ and K^+ passive ion channels, respectively. In this setting,

$$i_{ion} = i_{Na} + i_K = (V_r - E_{Na}) \cdot g_{Na} + (V_r - E_K) \cdot g_K = 0,$$

with E_{Na}, E_K equilibrium potential, and g_{Na}, g_K membrane conductance for Na^+ and K^+, respectively. Extracellular cathodal current injection modifies resting potential V_r, by depolarizing V_m towards more positive values (Fig. 4). At the

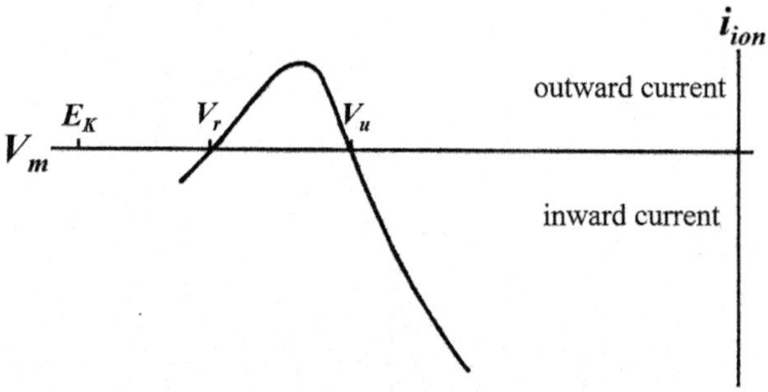

Fig. 4 Steady-state membrane current/voltage (i_{ion}/V_m) relation defining voltage threshold V_u for cable excitation by uniform polarization. E_K, K^+ equilibrium potential; V_r, resting potential

beginning of depolarization, while g_{Na} and g_K remain constant, outward K^+ current i_K increases because $V_m - E_K$ increases and inward Na^+ current i_{Na} decreases because $V_m - E_{Na}$ decreases and, as a consequence, net ionic current i_{ion} becomes outward (Fig. 4). However, when the increase in depolarization activates voltage dependent Na^+ channels, inward i_{Na} current increases due g_{Na} increase and net outward ionic current i_{ion}, after reaching a maximum, gradually decreases. With a further increase in depolarization, outward net ionic current i_{ion} becomes zero again at potential $V_m = V_u$. While V_r is a steady-state electrical equilibrium, V_u represents a highly unstable electrical equilibrium due to the counterbalance of high strength inward and outward currents through low resistance active ion channels. Electrical equilibrium instability is manifested by the i_{ion}/V_m relationship displaying negative slope (conductance) at V_u compared to positive slope at V_r (Fig. 4). Hence, if current injection is terminated at $V_m = V_u$, a small spontaneous depolarization will initiate an action potential simultaneously on every patch of membrane, while a small spontaneous repolarization will reestablish resting potential V_r. Under these conditions, the active response is a *membrane action potential* as it occurs in an isolated cell or an uniformly polarized membrane. During membrane action potential, active ionic current is entirely transformed into capacitive current $i_{ion} = -i_c$ and electrotonic current $i_m = 0$ (Fig. 5). Only during the short time interval of extracellular current injection, electrotonic current $i_m = i_c + i_{ion} \neq 0$: the stimulus current is almost entirely used to charge the local cell membrane capacitor and none flows as local circuit current. Hence, during the stimulus, both i_c and i_{ion} flow outward with the same orientation: i_c has higher strength while i_{ion} is lower due to passive high resistance ion channels (Fig. 5). Once an action potential has been initiated by passing sufficient depolarizing current i_m to exceed threshold potential V_u, the stimulus current i_m may be set to zero.

In summary, the condition for excitation of a uniformly polarized membrane is that the voltage should reach and exceed a critical value V_u, the uniform threshold potential, corresponding to net ionic current $i_{ion} = 0$. In fact, net ionic current i_{ion} is outward for V_m within the interval $V_r < V_m < V_u$, zero for $V_m = V_u$ and inward for $V_m > V_u$ during the upstroke of the action potential (Figs. 4 and 5). It is worth noticing that during a membrane action potential i_{ion}, initially inward,

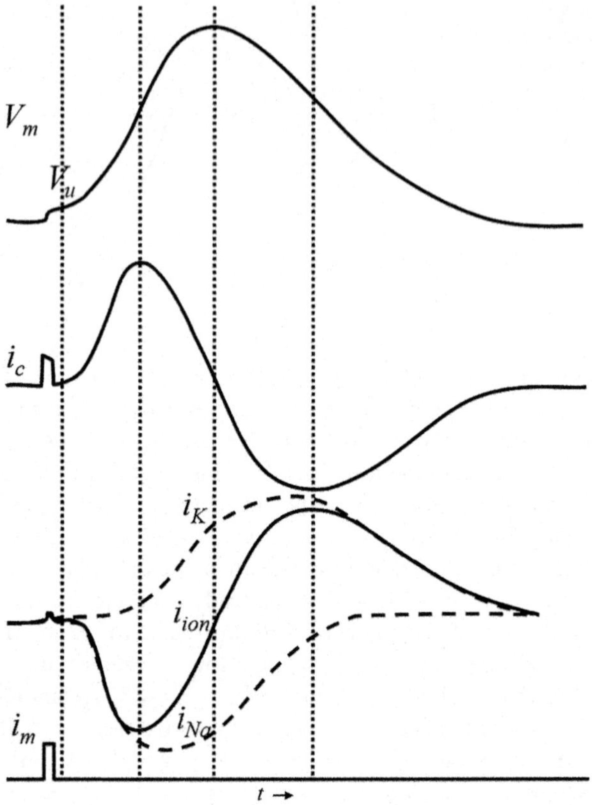

Fig. 5 Changes in voltage and currents during a membrane (uniform) action potential initiated by a brief rectangular stimulus current. The top diagram shows membrane potential change V_m as a function of time. The second diagram shows the capacity current $i_c = c_m dV_m/dt$. The initial, nearly square, wave corresponds to the applied current (see bottom diagram), most of which flows as capacity current when brief pulses are used. The third diagram shows the total ionic current (continuous curve) and its K^+ and Na^+ components (dashed curves). The bottom diagram shows the total membrane current $i_m = i_{ion} + i_c$, which is zero apart from time interval during which the stimulus is applied. The first dashed vertical line indicates when threshold V_u is reached while the following three dashed vertical lines indicate the times at which dV_m/dt is maximal, zero and minimal, respectively. This diagram may be compared with the corresponding one for a propagated action potential shown in Fig. 9. Adapted from Jack et al. [4]

reaches a minimum when dV_m/dt is maximal, is zero at the action potential peak when $dV_m/dt = 0$ and reaches a maximum when dV_m/dt is minimal.

Hence, when a uniform polarized membrane becomes an active source, i_{ion} entirely converts into i_c and the two currents flow in opposition, i.e. $i_{ion} = -i_c$.

5.2 The Propagating Action Potential

Conversely, when a cell is in contact with other cells as in cardiac tissue or long point-stimulated Purkinje fibers [2], active inward current i_{ion} generated by the cell membrane is not entirely transformed into i_c on the local membrane. Hence, $i_{ion} = -i_c + i_m$ and $i_m \neq 0$ is the electrotonic current that flows across cell membrane.

Let us consider a simple electrical model of an excitable cell as shown in Fig. 6 [4]. An external current pulse generator is connected to a model consisting of a resistor and capacitor in parallel. In addition, to represent the sodium current

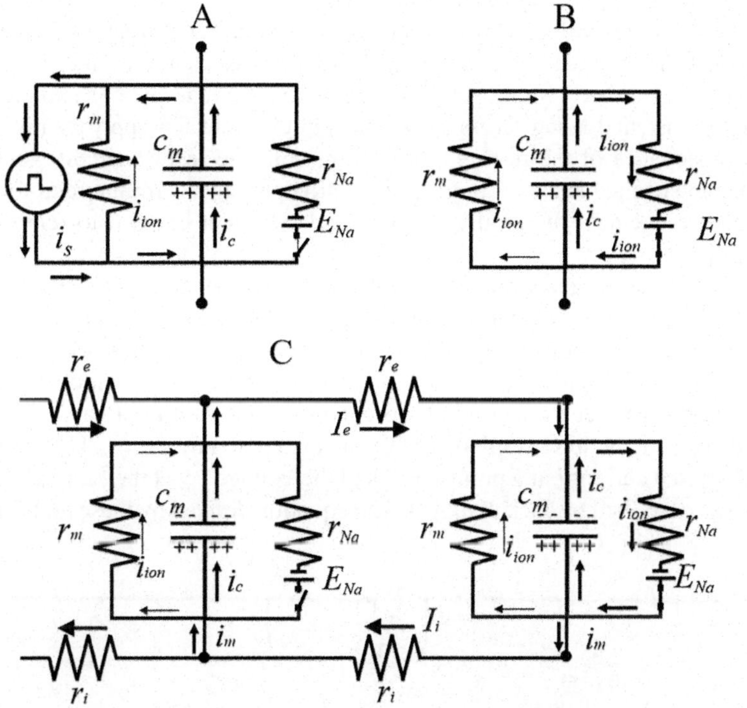

Fig. 6 Simple electrical circuits representing excitation and conduction. (**a**) an external current generator applies charge to the capacitance c_m of the model circuit. When the potential across the circuit reaches a threshold value, the switch on the Na^+ circuit closes and the Na^+ battery passes current into the cell. (**b**) the external current generator is removed. The potential change on the capacitance continues as current flows through the Na circuit and replaces the external current generator as the source of excitatory current. (**c**) the excited circuit is connected to a passive circuit and now acts like the external current generator in applying current to the passive circuit. Thus, the excitation may propagate for circuit to circuit. Since the Na^+ current in the excited circuit is the only inward current flowing, r_{Na} must be low (i.e. g_{Na} high) in order to allow sufficient Na^+ current to flow to continue charging the local capacitance and to excite the passive circuit. Hence there will be a minimal value of g_{Na} below which propagation cannot occur. Adapted from Jack et al. [4]

system, a battery E_{Na+} (the Na^+ equilibrium potential) and resistance r_{Na+} (the Na^+ active resistance) are also placed in parallel with the capacitor. A voltage-sensitive switch (i.e. the Na^+ channel 'gates'), which closes when depolarization reaches Na^+ activation threshold, controls this circuit element. When a depolarizing current i_s is applied to the inside of the model (see *Intracellular current injection*), positive charge accumulates on the inside of the membrane capacitor (Fig. 6a). The potential changes in a positive direction and, therefore, a small outward current i_{ion} also flows across the passive high membrane resistance r_m. When Na^+ activation threshold is reached, the voltage dependent switch closes and an inward current is generated across the Na^+ resistance (Fig. 6b). It is assumed that the external impressed current pulse then terminates, i.e. the stimulating current is just threshold. This condition occurs when the charge furnished by the external impressed current pulse i_s (pulse strength · pulse duration) is turned off as soon as Na^+ activation threshold is reached. The inward flow of current across the Na^+ circuit now takes over the function of the external generator in applying positive charge to the inside of the membrane capacitance. Moreover, if this circuit were connected to a similar resting circuit, the current flowing across the sodium circuit could also apply positive charge to the capacitance of the resting circuit (Fig. 6c). The resting circuit would then also become excited and so a wave of excitation propagates from circuit to circuit. Such an arrangement of serially connected circuits can be used to represent the propagation process.

The condition for generating a propagating action potential in a cable is that the voltage at the point of current injection should exceed a critical value at which the net ionic current generated by the cable as a whole becomes inward. This critical voltage should be higher than V_u, the voltage threshold for excitation by uniform polarization. In order to determine the condition for exciting a cable by current applied at one point, it is useful to consider an expression for the i_{ion}/V_m relationship for the applied current I at a point $x = 0$ and the voltage V_m at the same point $x = 0$ (Fig. 7) as described by Jack et al. [4]. The equation derived by these authors is:

$$I = 2\left(\frac{2}{r_i}\int_{V_r}^{V_m} i_{ion}\, dV\right)^{\frac{1}{2}}.$$

Since I becomes zero when the integral term is zero, the voltage threshold for cable excitation, expressed in terms of the voltage at the point of current injection, must be larger than the threshold for uniform membrane polarization V_u (Fig. 7). Moreover, provided that the $i_{ion}(V_m)$ relation is independent of time and the system is allowed to approach a steady-state before excitation occurs, the voltage threshold V_{Th} will be given by:

$$\int_{V_r}^{V_{Th}} i_{ion} dV = 0.$$

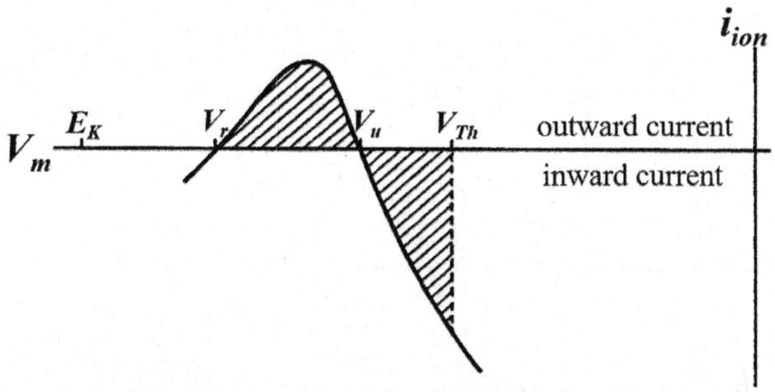

Fig. 7 Relation of voltage threshold for cable excitation V_{Th} to membrane current/voltage relation. V_r is the resting potential and V_u is the voltage threshold for excitation by uniform polarization. V_{Th} is given by the point at which the integral $\int_{V_r}^{V_{Th}} i_{ion} dV = 0$ becomes zero. Adapted from Jack et al. [4]

If there is no voltage that satisfy this equation, excitation cannot occur.

This equation may be more readily understood in terms of the corresponding curves for voltage V_m and current i_{ion} as function of distance x from the current source corresponding to threshold current applied at $x = 0$ (Fig. 8). The condition corresponding to the voltage threshold V_{Th} is that the integral of current with respect to distance should be zero so that no further current is required for the stimulation electrode, i.e.

$$\int_0^{\infty} i_{ion} dx = 0.$$

The length of fiber over which the excitation process occurs at threshold identifies the *liminal length* for excitation [5, 6]. The membrane region corresponding to the liminal length represents the *source*, delimited by V_{th} where inward i_{ion} is maximal and by V_u, where i_{ion} is zero. The membrane region where i_{ion} is outward and V_m decreases exponentially to V_r represents the *sink*, i.e. the membrane region that has to be depolarized by intracellular current flowing form the source (Fig. 8).

An important question is how safe action potential propagation is under normal and pathological conditions. Shaw and Rudy [7] provided a comprehensive approach to the computation of the so-called safety factor (*SF*) using a multi-cellular theoretical fiber, as formulated by:

$$SF = \frac{\int_A i_c \cdot dt + \int_A i_m \cdot dt}{\int_A I_i \cdot dt}.$$

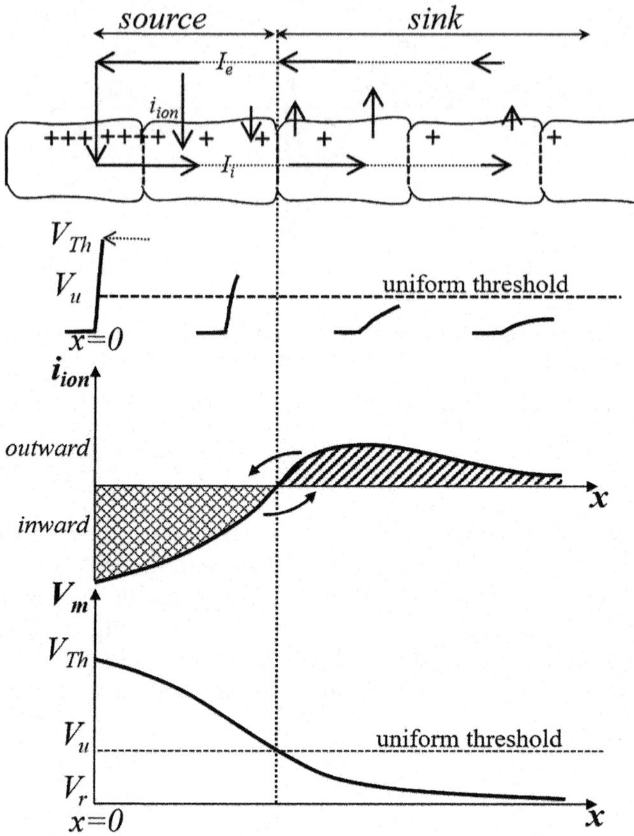

Fig. 8 Source, sink and liminal length. Upper panel: schematic diagram of currents flowing along a longitudinal sequence of cardiomyocytes induced by extracellular cathodal current injection at $x = 0$ reaching depolarization threshold. Ionic current i_{ion}, is inward in the source region where it is generated across the active membrane and outward in the sink region, i.e. electrotonic, across passive membrane that is being depolarized. Intracellular current I_i flowing from source to sink and extracellular current I_e flowing from sink to source complete the closed loop of current field lines. The source region identifies the liminal length comprised between threshold membrane potential for excitation V_{Th} at the point of current injection $x = 0$ and uniform threshold potential V_u at the point where sink region initiates. Middle panel: spatial variation in membrane current for threshold depolarization. Threshold is reached when the amount of inward current generated near the polarizing electrode is equal to the amount of (outward) repolarizing current generated by areas of membrane polarized below V_u. Lower panel: membrane potential V_m profile along source and sink, initiating at V_{Th} at the centre of the source and extending toward V_r, resting potential, where the sink ends. Adapted from Jack et al. [4]

The denominator of this equation corresponds to the electric charge flowing into a given cell during excitation and supplied by the neighboring active cell upstream (intracellular current I_i). The numerator represents the two terms into which the active ionic current i_{ion} generated by the cell divides. The first term refers to the capacitive current i_c producing the upstroke of the action potential and the second term to the current i_m that transforms into I_i flowing out of the cell to excite neighboring resting cell downstream and to propagate the action potential. The integration range A relates to the time window during which the membrane is being depolarized, i.e. foot and upstroke of the action potential. Intuitively, this definition is straight-forward: propagation is safe if SF > 1, i.e. the denominator is smaller than the numerator, that is, if the electrotonic current required to excite a given cell (denominator) is smaller than the ionic current generated by this same cell (numerator). Hence, for a uniformly propagating action potential, the denominator of SF represents the threshold charge computed as time integral at a cell point, equivalent to the voltage integral of current at the point of current injection shown in Fig. 7 or the space integral of current with respect to liminal length shown in Fig. 8.

When excitation threshold is reached, action potential initiation and propagation occurs. The currents flowing during a propagated action potential are shown in Fig. 9. This figure represents an action potential propagating from right to left, initiating far away to the right, so that the action potential is already conducting as a wave of constant shape and speed. The first sign of arrival is an exponential rise in potential V_m, corresponding to the foot of the action potential. As indicated by the current diagram at the bottom of the figure, this phase corresponds to outward current flow i_m resulting from current flowing from the active region of membrane further to the right (source). During this phase, the source is acting as an external circuit applying depolarizing current i_m to the resting region ahead, the sink. The current flow i_m is almost entirely represented by i_c since the passive membrane conductance and ionic current i_{ion} are both very small.

Along action potential profile, the curve of ionic current i_{ion} is triphasic. At the foot of the action potential, ionic current i_{ion} is initially outward and, as sodium conductance increases, i_{ion} reaches a maximum and then declines to zero at V_u (Fig. 9, vertical line *a*) becoming inward. The membrane region where i_{ion} is outward and $V_r < V_m \leq V_u$ represents the sink. The inward flow of Na^+ current further depolarizes the membrane. Threshold potential V_{Th} is reached above V_u within interval *a*–*b*. In studies by Hoffman and Cranefield [8], SF associated with a propagated action potential in normal cardiac muscle was estimated to be more than six times that of diastolic threshold. As the Na^+ conductance inactivates and the K^+ conductance increases, the inward ionic current i_{ion} decreases. However, repolarization begins (Fig. 9, vertical line *c*) before the ionic current i_{ion} becomes outward (Fig. 9, vertical line a'). The reason for this is that, in addition to adding charge to the local membrane capacity, the inward flow of ionic current i_{ion} is also supplying local circuit current flow to regions of membrane that are already repolarizing. It continues to do so beyond the peak of the action potential, and for a period of time the local capacity and ionic currents flow in the same direction (interval c–a'). Ultimately, the ionic current i_{ion} vanishes and becomes outward (Fig. 9, beyond vertical line a') and the membrane potential returns towards its resting value.

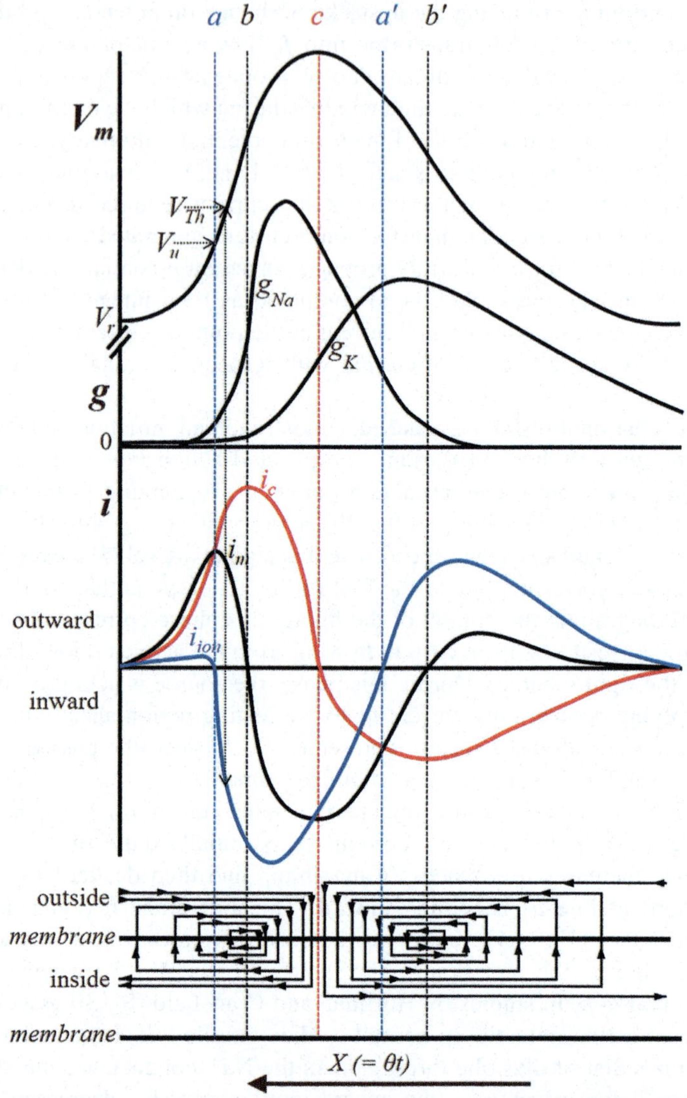

Fig. 9 The top curve shows membrane potential V_m. Immediately below are shown changes in g_{Na} and g_K conductances. The middle diagram shows the changes in i_m and its components, i_i and i_c. The bottom diagram is a schematic representation of the local circuits during propagation. The wave is propagating from right to left at uniform speed θ and the abscissa may also be regarded as time, since $x = \theta\ t$. The dashed vertical lines connect the points of the curves corresponding to zero crossing of i_{ion}, (line *a* and *a'*), zero crossing of i_m (line *b* and *b'*) and zero crossing of i_c (line *c*). Note that (i) $i_m = 0$ for i_c at maximum and V_m at maximum depolarization rate (line *a*); (ii) $i_m = 0$ for i_c at minimum and V_m at maximum repolarization rate (line *a'*); (iii) $i_c = 0$ at V_m maximum value (peak of action potential). The source and sink regions are defined by the distances comprised between $V_r - V_u$ (outward i_{ion}) and $V_u - V_{Th}$ (inward i_{ion}), respectively. Adapted from Jack et al. [4]

Along action potential profile, the curve of membrane current i_m is also triphasic, being:

- outward along the foot and initial upstroke of the action potential and becoming zero at maximum depolarization rate (Fig. 9, vertical line b, maximum of i_c curve);
- inward within interval b-b' in Fig. 9, reaching a minimum at peak of action potential (Fig. 9, vertical line c, $i_c = 0$) and becoming zero at maximum repolarization rate (Fig. 9, vertical line b', minimum of i_c curve);
- outward beyond vertical line b' in Fig. 9, where $i_m = 0$.

Interestingly, the current curves shown in this figure may all be obtained experimentally from measurements of the propagating action potential [9]. The capacity current i_c is obtained by differentiating the action potential,

$$i_c = C_m \frac{\partial V_m}{\partial t}.$$

Moreover, transmembrane current i_m may be obtained from the second derivative since, when the wave conducts with a constant shape and velocity, the time and distance scale are proportional, the constant of proportionality being θ, the conduction velocity. Hence, transmembrane current i_m becomes:

$$i_m = \frac{1}{r_i \theta^2} \frac{\partial^2 V_m}{\partial t^2}.$$

The ionic current i_{ion} may then be computed as the difference between i_m and i_c.

In summary, inward current i_m flowing across active membrane undergoing rapid depolarization and early repolarization gives rise to axial current I_i flowing in opposite directions along fiber axis. Partition point for i_m occurs at the peak of action potential profile. Axial current I_i transforms into outward electrotonic current, mainly capacitive, that sustains propagation ahead of action potential profile and delays repolarization phase behind. Hence, the amount of electrotonic current flowing behind direction of propagation modulates action potential profile, i.e. action potential duration (APD).

6 Electrotonic Modulation of Repolarization by the Activation Sequence

As compared to a cable like structure where action potential propagation is one-dimensional, propagation in cardiac muscle is three-dimensional. Electrotonic modulation of repolarization is more complex than the classic electrotonus described by cable theory for strands or axons with constant values of membrane resistance and uniformly distributed longitudinal resistance. It is well known that spatial dispersion of repolarization plays an important role in arrhythmogenesis. At tissue level, intrinsic APD heterogeneity can be modulated by electrotonic interactions between cells [10]. Regional variations of APD represent a form of electrophysiological heterogeneity in the heart. In a wide range of species, spatial gradients of APD exist from base to apex, from right to left ventricle and in the transmural plane of healthy myocardium [11, 12]. Furthermore, electrotonic interactions during repolarization can lead to dynamic modulation of APD gradients depending on the activation sequence, leading to an inverse linear relationship between activation time (AT) and APD [10, 13–20]. The dynamic nature of this modulation is due, in part, to the spatial gradient in membrane potential occurring during the repolarization phase of a propagating AP. Each cell is influenced by the electrotonic load from its neighbors such that, cells repolarizing later generate an inward electrotonic current flowing toward their earlier repolarizing neighbors with the effect of delaying repolarization phase. In homogeneous tissue, electrotonic modulation of repolarization effectively prolongs the APD of the earlier activated cells and generates gradually decreasing APDs away from the pacing site. These APD gradients are most pronounced at the pacing site, at the tissue boundaries and in directions of slow propagation [18, 21, 22].

6.1 Modulation of Activation-Recovery Interval

Experimental evidence of acute electrotonic modulation of repolarization during two different activation sequences, sinus rhythm (SR) and ventricular drive (VD), can be obtained by electrical mapping in the in situ heart. An unpublished experimental result from our laboratory is shown in Fig. 10. Unipolar potential mapping was performed from the anterior ventricular surface in rat heart by means of an 8×8 epicardial electrode array, with 1-mm resolution square mesh [23]. Unipolar stimulation at a ventricular test site was obtained from one of the array electrodes by means of 1 ms duration and twice diastolic-threshold strength cathodal current pulses. Ventricular ATs were computed from unipolar epicardial electrograms as the times of the minimum time derivative (dV/dt_{min}) during QRS complex and referenced to QRS or stimulus onset for SR and ventricular drive, respectively. Recovery times (RT) were computed as the times of the minimum of the time derivative (dV/dt_{min}) during the downslope of the T wave. Activation recovery interval (ARI), a well-validated estimate of local APD [24, 25] was obtained as difference between AT and RT. Isochrone line maps in Fig. 10 display the general time course and spatial distribution of AT, RT and ARI on the anterior ventricular epicardium during normal SR and VD at a test site located on the anterior paraseptal area.

Commonly, during normal SR in rat heart [26, 27] two breakthrough points (BTPs) characterize ventricular activation onset on the lateral right (RV) and left (LV) ventricular surfaces, respectively (not shown because outside the electrode array). Thereafter, two planar wave fronts that originate from the two BTPs prop-

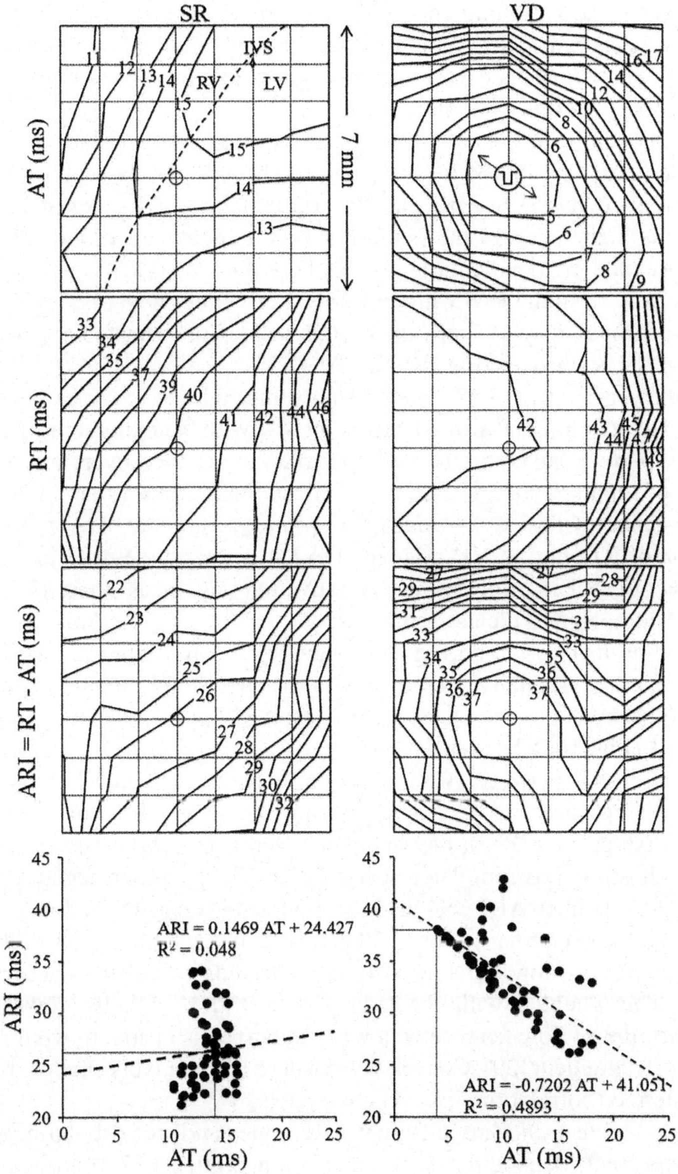

Fig. 10 Isochrone line maps during sinus rhythm (SR) and ventricular drive (VD) computed from electrograms recorded by a 8 × 8 epicardial electrode array in rat heart. AT, activation times; RT, repolarization times; ARI = RT − AT, activation recovery intervals; current pulse symbol, pacing site; circle, electrode corresponding to pacing site; double headed arrow, fiber direction at pacing site; RV, right ventricle; LV, left ventricle; IVS, interventricular septum. AT and RT isochrone numbers, milliseconds from QRS or stimulus onset during SR or VD, respectively. Bottom diagram: relationships between AT and ARI displayed by linear regression analysis at each array electrode. A good linear relationship occurs at points around pacing site. Correlation coefficient R^2 for VD is reduced by the presence of ARI at electrodes distant from pacing site, at lower right and left corners of the electrode array

agate from RV and LV along epicardial fiber direction toward the interventricular septum (IVS) (Fig. 10, SR, AT). The two wave fronts collide over the IVS and merge into a new V-shaped wave front that spreads toward the RV outflow tract, mainly across epicardial fiber direction.

RTs display uniform spatial dispersion on the anterior epicardial surface (Fig. 10, SR, RT). In particular, on the RV, recovery isochrones mainly follow the activation sequence toward the IVS with uniform parallel lines. The density of recovery isochrones is higher toward the paraseptal area indicating a pronounced recovery gradient oriented perpendicularly to the IVS. On the LV, recovery isochrones propagate away from the IVS with uniform parallel lines toward LV free wall.

ARI isochrones display uniform spatial dispersion oriented from RV to LV, similarly to RT isochrones (Fig. 10, SR, ARI).

Conversely, activation isochrones during ventricular drive were elongated, symmetric, and quasi-elliptical around the pacing site with the major axis oriented along the local sub-epicardial fiber direction (Fig. 10, VD, AT). Recovery isochrones show that onset of repolarization occurs at pacing site and that RTs are almost constant up to a distance of few λ (2–3 mm) [3] around pacing site. The different spatial distribution of RTs during VD compared to SR is due to the different electrotonic modulation of repolarization by the two activation sequences at test site: symmetric during VD compared to planar during SR.

ARI spatial distribution during VD is heterogeneous, with the longest duration at pacing site that gradually decreases with distance in all directions: longest ARI (37 ms) is observed at pacing site and shortest ARI (27 ms) lays close to the site of the latest AT (Fig. 10, VD, ARI).

The relationship between AT and ARI at each array electrode is displayed by linear regression analysis. A clear decreasing trend of ARI with AT is revealed during VD (slope $= -0.72$): electrode sites activating earliest have longer ARIs and electrode sites activating later have shorter ARIs. Similar negative correlation between AT and either ARI or APD has previously been reported in experimental models [18, 28] and humans [16]. A likely mechanism for the relationship is local electrotonic current flow between cells which tends to equalize action potential durations. This coupled with the later repolarization of cells downstream and electrotonic current flow from cells downstream to cells upstream would contribute to the AT–ARI gradient [29]. Conversely, during SR a relatively shallow relationship and an undefined correlation links ATs and ARIs.

Hence, at an electrode site, different ARI values and, correspondingly, different APDs are present, depending on the activation sequence. In particular, Fig. 10, shows that at an electrode site ARI is shorter during SR (26 ms) than during VD at the same site (38 ms). Hence, different APD values can be measured at a myocardial site depending on the different activation sequence occurring at that site, likely due to the different electrotonic modulation of repolarization. It is known that dispersion of RT is an important determinant of the vulnerability to arrhythmias following a premature activation. However, in our experiments we found that, due to electrotonic interactions, RT spatial distribution in the normal rat heart was uniform during SR and relatively homogeneous during VD. These results suggest that normal ventricular myocardium would be resistant against arrhythmias following ectopic activation, at least in the rat heart.

6.2 *Modulation of Effective Refractory Period*

In experimental and clinical electrophysiology studies, refractoriness is measured by the effective refractory period (ERP). In normal myocardium, there is a close temporal relationship between ERP and APD, and the two are often used as estimates of each other, although this relationship is altered in ischemic myocardium where the recovery of excitability lags behind full repolarization (postrepolarization refractoriness).

ERP can be measured as the interval from depolarization to the recovery of excitability when stimulation fails to induce a propagated response. The classic technique for ERP measurement is the extra-stimulus technique where a train of eight basic stimuli S1 is followed by a single premature stimulus S2 elicited at the same site (test site drive). Alternatively, ERP duration can also be measured by the extra-stimulus technique during SR, which consists of a sequence of eight sinus beats (RW, the upward deflection of the QRS complex) followed by a single premature stimulus (S). Unpublished experimental findings in rat heart show that ERP durations measured by the two protocols, S1–S2 and RW-S, are different and consistent with the difference in ARI values measured during corresponding sequences of basal ventricular activation, i.e. VD and SR. In fact, ERP is shorter when measured during RW-S than during S1–S2. Interestingly, at the same site, ERP difference is the same as ARI difference. Hence, it is hypothesized that also ERP difference is likely due to the different electrotonic modulation of repolarization by the different activation sequence. Consequently, different APDs measured at an electrode site during SR and VD would also imply correspondingly different ERPs measured at the same site. Experimental verification of electrotonic modulation of ERP by the activation sequence is shown in Fig. 11 where ERP measured at a test site during VD (S1–S2 protocol) and SR (RW-S protocol) was 90 ms and 72 ms, respectively.

Since APD changes are also function of rate, with the APD at steady-state shorter the higher the frequency, it can be argued that the different ARI (or APD) measured values are due to the different basal ventricular activation rates (S1 and SR). However, in contrast to this argument, ARI values are longer during S1 drive at a frequency higher than SR. Moreover, the difference between ARI and ERP values did not change when SR rate was matched to that of S1 drive by means of atrial drive (AD) (Fig. 11). Hence, it can be concluded that, at least for frequencies used in our experiments, electrotonic modulation of repolarization by the activation sequence is responsible for dynamic modulation of ARI and ERP gradients.

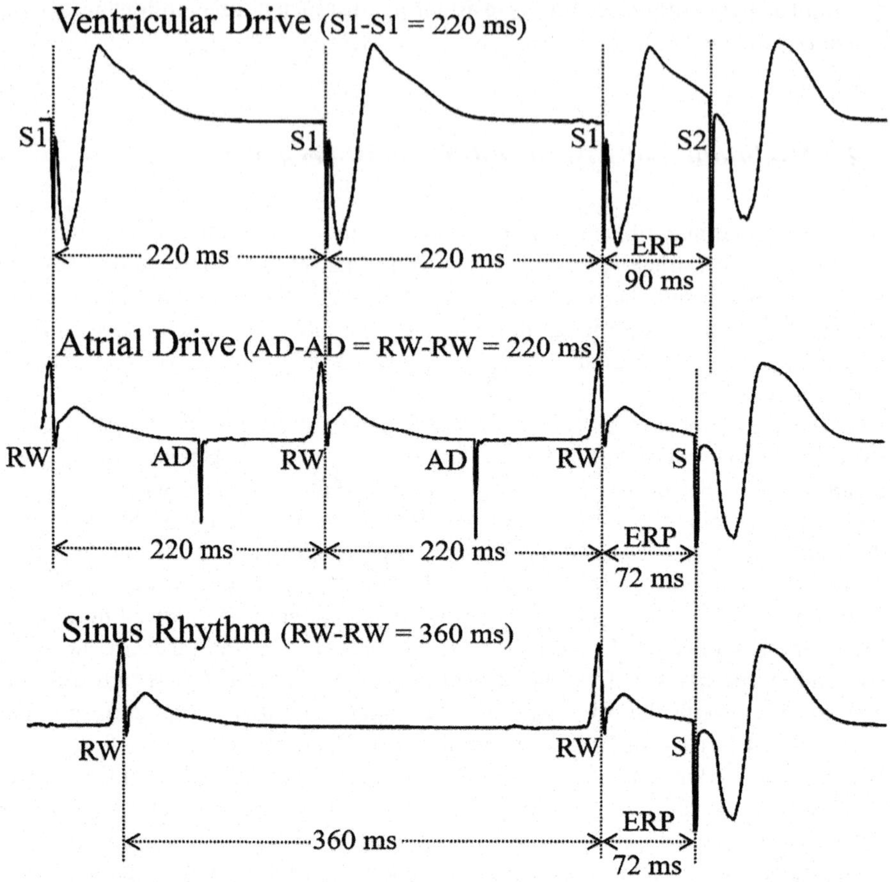

Fig. 11 Reference electrogram displaying pacing protocols for effective refractory period (ERP) measurement by premature stimulation (S2 or S) during Ventricular Drive (S1–S1), Atrial Drive (AD–AD) and Sinus Rhythm (RW–RW). Results show that ERP is longer when measured during S1–S2 than RW–S protocol, irrespective of S1–S1 or RW–RW cycle length. See text for explanation

7 Summary and General Conclusions

In summary, electrotonic current is a transmembrane current, provided by either an external current generator or an active region of cell membrane, which depolarizes membrane potential locally.

In case of an external current generator, the stimulating extracellular electrode (cathode) removes current I_e from the surrounding extracellular volume including outward transmembrane current i_m from excitable tissue. The stimulating current $i_m = i_c + i_{ion}$ depolarizes membrane potential V_m by means of capacitive component i_c that forces ionic component i_{ion} to flow across passive membrane resistance r_m.

Both i_c and i_{ion} are outward currents that flow with the same orientation across the membrane. On reaching excitation threshold at the stimulation site, action potential initiation and propagation occurs from the stimulating electrode with opposite orientation along cell axis.

In case of an active membrane region, as it occurs for a propagating action potential, the membrane itself is a current source generating inward ionic current i_{ion} whose largest fraction turns into outward capacitive current i_c depolarizing source membrane potential. It is worth noting that i_{ion} and i_c flow with opposite orientation across the active membrane region. The fraction of i_{ion} that is not used to depolarize membrane potential as i_c at the source region, i.e. $i_m = i_{ion} + i_c$, flows into the cell transforming into intracellular current I_i. Current I_i flows from the source region in opposite orientation sustained by V_m potential gradients along cell axis. The inversion point for the opposite flow occurs at the peak of the action potential profile. I_i flows downstream towards the sink region, down the slope of action potential upstroke. In the sink region, I_i transforms into transmembrane current $i_m = -\frac{\partial I_i}{\partial x} = \frac{\partial I_e}{\partial x} = \mathrm{i_c} + i_{ion}$ that depolarizes membrane potential to threshold by means of i_c greater than i_{ion}, due to high r_m value. I_i flows upstream from the source region along the repolarization slope of action potential profile transforming into outward depolarizing i_m that delays repolarization phase and prolongs APD.

A clear definition of electrotonic current and its mechanism of action is of particular importance in cardiac electrophysiology, where APD heterogeneity is in part determined by the intrinsic properties of the myocardial cells, but it has also been shown to depend on cell-to-cell coupling and activation patterns [18]. Electrotonic interactions due to the coupling of cells in tissue are capable of gradually decreasing the duration of an action potential as it propagates away from the pacing site, with the strongest effect found in the direction of small conduction velocities or for reduced coupling. Cell-to-cell coupling has also been found to modulate the rate-dependency of APD, i.e. APD restitution [30]. Since the steepness of APD restitution has been implicated in arrhythmogenesis, it is important to understand which factors can modulate this dynamical property of cardiac tissue. Epicardial potential measurements in a rat experimental model, suggest that APD is shorter at later activation times independent on basal pacing frequency. However, these findings require validation for a broad range of basal pacing frequencies.

References

1. Plonsey, R., Barr, R.C.: Bioelectricity. A Quantitative Approach, 3rd edn. Springer, Heidelberg (2007)
2. Fozzard, H.A., Schoenberg, M.: Strength-duration curves in cardiac Purkinje fibres: effects of liminal length and charge distribution. J. Physiol. **226**, 593–618 (1972)
3. Weidmann, S.: Electrical constants of trabecular muscle from mammalian heart. J. Physiol. **210**, 1041–1054 (1970)
4. Jack, J.J.B., Noble, D., Tsien, R.W.: Electric Current Flow in Excitable Cells. Clarendon Press, Oxford (1975)

5. Rushton, W.A.H.: Initiation of the propagated disturbance. Proc. R. Soc. B. **124**, 210–243 (1937)
6. Noble, D.: The relation of Rushton's 'liminal length' for excitation to the resting and active conductances of excitable cells. J. Physiol. **226**, 573–591 (1972)
7. Shaw, R.M., Rudy, Y.: Ionic mechanisms of propagation in cardiac tissue. Roles of the sodium and L-type calcium currents during reduced excitability and decreased gap junction coupling. Circ. Res. **81**, 727–741 (1997)
8. Hoffman, B.F., Cranefield, P.F.: Electrophysiology of the Heart. McGraw Hill, New York (1960)
9. Cole, K.S., Curtis, H.J.: Electric impedance of the squid giant axon during activity. J. Gen. Physiol. **22**, 649–670 (1939)
10. Laurita, K.R., Girouard, S.D., Rudy, Y., Rosenbaum, D.S.: Role of passive electrical properties during action potential restitution in intact heart. Am. J. Phys. **273**, H1205–H1214 (1997)
11. Antzelevitch, C.: Modulation of transmural repolarization. Ann. N. Y. Acad. Sci. **1047**, 314–323 (2005)
12. Wan, X., Bryant, S.M., Hart, G.: The effects of $[K^+]_o$ on regional differences in electrical characteristics of ventricular myocytes in guinea pig. Exp. Physiol. **85**, 769–774 (2000)
13. Banville, I., Gray, R.A.: Effect of action potential duration and conduction velocity restitution and their spatial dispersion on alternans and the stability of arrhythmias. J. Cardiovasc. Electrophysiol. **13**, 1141–1149 (2002)
14. Chauhan, V.S., Downar, E., Nanthakumar, K., Parker, J.D., Ross, H.J., Chan, W., Picton, P.: Increased ventricular repolarization heterogeneity in patients with ventricular arrhythmia vulnerability and cardiomyopathy: a human in vivo study. Am. J. Physiol. Heart Circ. Physiol. **290**, H79–H86 (2006)
15. Franz, M.R., Bargheer, K., Rafflenbeul, W., Haverich, A., Lichtlen, P.R.: Monophasic action potential mapping in human subjects with normal electrocardiograms: direct evidence for the genesis of the T wave. Circulation. **75**, 379–386 (1987)
16. Hanson, B., Sutton, P., Elameri, N., Gray, M., Critchley, H., Gill, J.S., Taggart, P.: Interaction of activation-repolarization coupling and restitution properties in humans. Circ. Arrhythm. Electrophysiol. **2**, 162–170 (2009)
17. Myles, R.C., Bernus, O., Burton, F.L., Cobbe, S.M., Smith, G.L.: Effect of activation sequence on transmural patterns of repolarization and action potential duration in rabbit ventricular myocardium. Am. J. Physiol. Heart Circ. Physiol. **299**, H1812–H1822 (2010)
18. Walton, R.D., Benson, A.P., Hardy, M.E., White, E., Bernus, O.: Electrophysiological and structural determinants of electrotonic modulation of repolarization by the activation sequence. Front. Physiol. **4**, 281 (2013)
19. Yuan, S., Kongstad, O., Hertervig, E., Holm, M., Grins, E., Olsson, B.: Global repolarization sequence of the ventricular endocardium: monophasic action potential mapping in swine and humans. Pacing Clin. Electrophysiol. **24**, 1479–1488 (2001)
20. Yue, A.M., Betts, T.R., Roberts, P.R., Morgan, J.M.: Global dynamic coupling of activation and repolarization in the human ventricle. Circulation. **112**, 2592–2601 (2005)
21. Joyner, R.W.: Modulation of repolarization by electrotonic interactions. Jpn. Heart J. **27**(Suppl 1), 167–183 (1986)
22. Zubair, I., Pollard, A.E., Spitzer, K.W., Burgess, M.J.: Effects of activation sequence on the spatial distribution of repolarization properties. J. Electrocardiol. **27**, 115–127 (1994)
23. Rossi, S., Buccarello, A., Ershler, P.R., Lux, R. L., Callegari, S., Corradi, D., Carnevali, L., Sgoifo, A., Miragoli, M., Musso, E., Macchi, E.: Effect of anisotropy on ventricular vulnerability to unidirectional block and reentry by single premature stimulation during normal sinus rhythm in rat heart. Am. J. Physiol. Heart Circ. Physiol. **312**, H584–H607 (2017)
24. Haws, C.W., Lux, R.L.: Correlation between in vivo transmembrane action potential durations and activation-recovery intervals from electrograms. Effects of interventions that alter repolarization time. Circulation. **81**, 281–288 (1990)

25. Millar, C.K., Kralios, F.A., Lux, R.L.: Correlation between refractory periods and activation-recovery intervals from electrograms: effects of rate and adrenergic interventions. Circulation. **72**, 1372–1379 (1985)
26. Macchi, E., Cavalieri, M., Stilli, D., Musso, E., Baruffi, S., Olivetti, G., Ershler, P.R., Lux, R.L., Taccardi, B.: High-density epicardial mapping during current injection and ventricular activation in rat hearts. Am. J. Physiol. Heart Circ. Physiol. **275**, H1886–H1897 (1998)
27. Rossi, S., Baruffi, S., Bertuzzi, A., Miragoli, M., Corradi, D., Maestri, R., Alinovi, R., Mutti, A., Musso, E., Sgoifo, A., Brisinda, D., Fenici, R., Macchi, E.: Ventricular activation is impaired in aged rat hearts. Am. J. Physiol. Heart Circ. Physiol. **295**, H2336–H2347 (2008)
28. Burgess, M.J., Steinhaus, B.M., Spitzer, K.W., Green, L.S.: Effects of activation sequence on ventricular refractory periods of ischemic canine myocardium. J. Electrocardiol. **18**, 323–329 (1985)
29. Toyoshima, H., Burgess, M.: Electrotonic interaction during canine ventricular repolarization. Circ. Res. **43**, 348–356 (1978)
30. Qu, Z.: Dynamical effects of diffusive cell coupling on cardiac excitation and propagation: a simulation study. Am. J. Physiol. Heart Circ. Physiol. **287**, H2803–H2812 (2004)

Emilio Macchi graduated with a master's in physics, Università degli Studi, Milano, in 1969. He obtained a PhD in 1973 from the Department of Physiology and Biophysics, Dalhousie University, Halifax, Nova Scotia, Canada, under the supervision of Prof. Pentti M. Rautaharju. Subsequently, he became research associate at the Istituto di Cardiologia Sperimentale, Simes-SPA, Milano, in the Electrophysiological Laboratory of Professor Bruno Taccardi (1974–1976); Full-time Researcher at the Istitituo per le Applicazioni del Calcolo "Mauro Picone," National Research Council, Rome (1976–1987); Associate Professor of General Physiology, Istituto di Fisiologia Generale, Università degli Studi, Parma (1987–1994); and Professor of Physiology, Dipartimento di Biologia Evolutiva e Funzionale and Dipartimento di Bioscienze, Università degli Studi, Parma (1994–2016). He is now Professor Emeritus at the Dipartimento di Scienze Chimiche, della Vita e della Sostenibilità Ambientale, Università di Parma. Dr. Macchi has been Visiting Professor at the Cardiovascular Research and Training Institute, University of Utah, Salt Lake City, USA, during different periods in 1986–1989. He also served as Director of the Dipartimento di Biologia Evolutiva e Funzionale, Università degli Studi, Parma (2004–2008).

Dr. Macchi's scientific activity is documented by 75 full articles/reviews/chapters in books and relates to measurement and modeling of the cardiac electric field in mathematical and animal models and in man. Independent research work in cardiac electrophysiology continuously progressed in collaboration with Prof. Bruno Taccardi and Prof. Ezio Musso. Relevant results of these studies were relationship between extracellular potential distribution and intra-cardiac electric sources, identification of the oblique dipole layer model of the activation wave front as a realistic equivalent generator presently utilized in experimental and clinical electrophysiology studies, definition of an olive-shaped intra-cavitary probe for arrhythmia detection in experimental, and clinical electrophysiology studies.

Ezio Musso completed his university studies at the University of Pavia (Italy) where he obtained the MD degree (1965) and specialization in cardiology (1971). In 1966, he started his research activity in the Electrophysiological Laboratory of Professor Bruno Taccardi in Milano. Subsequently, he became Assistant Professor of Physiology at the University of Parma, Italy (1974–1980) and Full Professor of Physiology at the same university (1980–2011) where he also served as Director of the Institute of General Physiology and/or the Department of Evolutionary and Functional Biology (1987–2011). He was Visiting Scientist/Visiting Assistant Professor of Physiology/Visiting Professor of Medicine at Laboratoires de Physiologie Comparée et de Physiologie Cellulaire associé au CNRS, Faculté de Sciences, Orsay-Paris (1969); Department of Physiology, State University of New York, Downstate Medical Center, Brooklyn-New York (1972–1975, 1976, 1977, 1981); and Cardiovascular Research Institute, New York Medical College, Valhalla-New York (2002–2004; 2005; 2006).

Dr. Musso's research in cardiac electrophysiology has focused on autonomic control of cardiac automatism, biophysical aspects and clinical applications of the cardiac electric field mechanisms of onset of arrhythmias, cardiovascular response to stress, stem cell-mediated cardiac regeneration, and mechanical and electrical competence of the regenerated heart.

His scientific activity is documented by more than 150 papers "in extenso" (the largest part of which has been published in the most reputable scientific journals) and chapters in books. His work has been funded with national, European, and NIH grants.

Stefano Rossi is assistant researcher at the Department of Medicine and Surgery, University of Parma. In 2006, he obtained a PhD in "Systemic Pathophysiology" at the Department of Evolutionary and Functional Biology, University of Parma. He was winner of a 4-year research grant for the study of cardiac electrical activity in elderly rats and a 3-year research grant for the study of atmospheric particulate effects on arrhythmogenesis in normotensive and hypertensive rats. He is also a collaborator of several research projects concerning the study of cardiac electrophysiological properties through electric epicardial mapping on different experimental models such as ischemic rats treated with cardiac stem cells, dystrophic mice, stressed rats, and hypertensive rats. Dr. Rossi is the author of 26 scientific publications in international journals with impact factor and reviewer for several international journals in the field of cardiac electrophysiology. Currently, he is involved in the study of arrhythmogenesis in normal cardiac tissue and nanotoxicity effects within the cardiovascular system.

Reduced Order Modeling for Cardiac Electrophysiology and Mechanics: New Methodologies, Challenges and Perspectives

Andrea Manzoni, Diana Bonomi, and Alfio Quarteroni

Abstract Reduced-order modeling techniques enable a remarkable speed up in the solution of the parametrized electromechanical model for heart dynamics. Being able to rapidly approximate the solution of this problem allows to investigate the impact of significant model parameters querying the parameter-to-solution map in a very inexpensive way. The construction of reduced-order approximations for cardiac electromechanics faces several challenges from both modeling and computational viewpoints, because of the multiscale nature of the problem, the need of coupling different physics, and the nonlinearities involved. Our approach relies on the reduced basis method for parametrized PDEs. This technique performs a Galerkin projection onto low-dimensional spaces built from a set of snapshots of the high-fidelity problem by the Proper Orthogonal Decomposition technique. Snapshots are obtained for different values of the parameters and computed, e.g., by the finite element method. Then, suitable hyper-reduction techniques, in particular the Discrete Empirical Interpolation Method and its matrix version, are called into play to efficiently handle nonlinear and parameter-dependent terms. In this work we show how a fast and reliable approximation of both the electrical and the mechanical model can be achieved by developing two separate reduced order models where the interaction of the cardiac electrophysiology system with the contractile muscle tissue, as well as the sub-cellular activation-contraction mechanism, are included. Open challenges and possible perspectives are finally outlined.

1 Introduction

Cardiac electromechanics refers to a model for the description of the coupling of the electrophysiology model, which describes the propagation of the signal triggering the heart contraction, and the mechanical model, which describes the contraction

A. Manzoni (✉) · D. Bonomi · A. Quarteroni
MOX, Dipartimento di Matematica, Politecnico di Milano, Milano, Italy
e-mail: andrea1.manzoni@polimi.it; diana.bonomi@polimi.it; alfio.quarteroni@polimi.it

D. Boffi et al. (eds.), *Mathematical and Numerical Modeling of the Cardiovascular System and Applications*, SEMA SIMAI Springer Series 16,
https://doi.org/10.1007/978-3-319-96649-6_6

and relaxation of the muscle tissue, including the sub-cellular activation-contraction mechanism. Solving such a coupled problem is crucial to investigate how clinically relevant processes affect different features of the heart beat [70].

Cardiac electromechanics is a challenging problem from both a mathematical and a numerical viewpoint, because of the coupling of different physical problems which take place at different spatial and temporal scales. Indeed, a model for the cardiac electrophysiology has to describe on one hand the sub cellular activity (1–100 μm) which gives rise to the cellular depolarization, on the other hand the spreading of the electrical signal through the whole myocardium (1–10 cm). The modeling of these processes yields a two-way coupled problem involving a PDE with a nonlinear reaction term and a system of nonlinear ODEs. To correctly track the propagation of the action potential, in the form of wave-front solutions, fine computational grids are needed, thus yielding large-scale algebraic problems to be solved. Moreover, the description of the mechanics of the cardiac tissue requires complex constitutive laws, characterized by an exponential strain energy function and the presence of muscular fibers and sheets, resulting in a complex highly nonlinear model. This turns into the need of assembling involved Jacobian matrices when relying, e.g., on the Newton method for the solution of nonlinear systems of equations. Furthermore, the heart muscle contracts after being electrically activated without the need of an external load, and this active behavior of the cardiac cells has to be properly taken into account when coupling the electrical and the mechanical models. To describe these processes different works have proposed more and more accurate electromechanical models [29, 31, 36, 53, 74, 80], very often yielding overwhelming computational costs.

Computational complexity is even more exacerbated if one is interested in going beyond a single, direct simulation. Indeed, when simulating cardiovascular problems, several input data affect the problem under investigation, often varying within a broad range and possibly affected by uncertainty. Addressing the impact of input variations on outputs of clinical interest is thus of paramount importance in order to (1) obtain reliable results, (2) calibrate the numerical solver and/or (3) personalize the mathematical model. In fact, model parameters have to be specifically tuned to fit subject-specific clinical data in order to take into account inter-patient variability. To correctly calibrate cardiac models and estimate the unknown input parameters, such as muscular fibers orientation or parameters affecting the signal propagation, several numerical simulations have to be carried out, thus calling for multiple queries of the parameter-to-solution map in fast and accurate ways. Beyond parameter estimation, this requirement also arises when dealing with sensitivity analysis, control and optimization, and uncertainty quantification, noteworthy classes of problems whose importance in cardiovascular modeling is growing faster and faster.

The need of solving these problems efficiently calls for the development of efficient and accurate reduced order modeling (ROM) techniques in electromechanics. These techniques are designed to provide accurate and reliable solutions to PDEs depending on several parameters at a greatly reduced computational cost. In particular, the reduced basis (RB) method replaces the original large-scale numerical problem (or high-fidelity approximation) originated by applying, e.g., a finite element (FE) method, with a reduced problem of substantially smaller dimension; this latter is generated through a projection of the high-fidelity problem upon a low-dimensional subspace, spanned by a set of high-fidelity solutions corresponding to suitably chosen parameters [41, 68].

In this work we show how to solve both the electrophysiology and the mechanical problems, when these depend on a set of parameters, in the framework of RB methods, also taking into account active mechanics triggered by the cellular depolarization. Because of the nonlinear nature of these problems, computational efficiency is obtained by combining a general-purpose technique to generate the low-dimensional subspace, such as proper orthogonal decomposition, and suitable *hyper-reduction* techniques allowing to assemble the algebraic structures required by the ROM independently of the high-fidelity arrays. This is required, e.g., when dealing with Newton iterations while solving cardiac mechanics—for which the global Jacobian matrix would have to be entirely reassembled at each Newton step—as well as time stepping in cardiac electrophysiology, for which nonlinear terms have to be evaluated at each time step, also involving the contribution from the cell model. Among recent applications of ROM techniques to problems related with the cardiovascular system, we also mention haemodynamics, for the sake of simulating blood dynamics in different flow conditions [25, 26, 54] or geometrical configurations [7, 8], also in view of the optimal design of prosthetic devices [47, 50] or parameter identification [48, 70]. Cardiac electrophysiology has also been tackled in the last decade [15, 20, 30, 35, 88], however by performing reduction only with respect to the time independent variable, thus avoiding the main difficulties related with the efficient handling of parameter-dependent problems. Recent results for parametrized problems in cardiac electrophysiology, also in view of the efficient solution of uncertainty quantification problems, can be found in [59]. Instead, regarding cardiac mechanics, this subject has only been addressed in a recent paper by the authors [14] and in [13], where also first results about one-way and two-ways coupled electromechanical problems have been obtained. The reduction of coupled problems, however, is still matter of investigation and therefore are not included in this paper.

The structure of this paper is as follows. After a brief recall (Sect. 2) on the mathematical modeling of cardiac electromechanics, we describe the high-fidelity FE approximation we start from (Sect. 3). We then introduce the key tools of the proposed ROM technique (Sect. 4): the Galerkin-POD method and suitable hyper-reduction strategies. Then, we show how to combine them to derive a ROM for both cardiac electrophysiology (Sect. 5) and mechanics (Sect. 6), independently. Numerical results dealing with patient-specific left ventricle configurations in the systolic phase are then shown (Sect. 7), and finally open critical issues and future perspectives are outlined (Sect. 8).

2 Mathematical Models

In this section we present an overview of the electromechanics mathematical models. We consider the minimal Bueno-Orovio model for describing the cellular behavior, the monodomain model for cardiac electrophysiology, and the hyperelastic Holzapfel-Ogden model for the passive ventricular mechanics, adopting an active-strain formulation to take into account active mechanics. This latter requires a dynamical system for the variable which describes fiber shortening as a function of calcium concentration (and then electrical activation).

2.1 *Cardiac Electrophysiology*

A mathematical model for cardiac electrophysiology has to include processes arising at different scale, ranging from subcellular activity, which originates the cellular depolarisation to the spreading of the signal in the whole myocardium. To model the whole heart, several works have considered continuous models, where the myocardium is approximated as a syncytium, that is a domain where the intra and extracellular spaces coexist at each point [31, 44, 61, 76]. These continuous models describe the spreading of the signal in the heart tissue and are usually coupled to a ionic model which describes the evolution of ion concentrations and ionic currents inside the cells. In this section we first introduce a general framework for the ionic models, focusing on the Bueno-Orovio model [17], and we show how to couple it to the bidomain and monodomain models, widely adopted continuous tissue models.

2.1.1 Cell Models

Cell models describe the evolution of the transmembrane potential across the cell membrane of a single cardiomyocyte. Several ionic models have been proposed, most of which are based on the well-studied Hodgkin-Huxley model [42]; see e.g. [58] for a review. They are all based on the assumption that the electrical properties of the cell membrane can be modeled as an electrical circuit, which connects in parallel a resistor and a capacitance: the latter describes the cell membrane which separates the intra and the extra cellular space, while the former models the ionic channels and pumps regulating ionic fluxes through the membrane. The conservation of currents across the cell membrane can be expressed through the relation

$$C_m \frac{\partial v}{\partial t} + I_{ion}(v, \mathbf{w}, \mathbf{c}) = I_{app},$$

where v is the transmembrane potential, C_m the membrane capacitance, I_{app} an external applied current density and I_{ion} the sum of the current densities through the membrane, that can be written using the general Hodgkin-Huxley formalism introduced in [42] as

$$I_{ion}(v, \mathbf{w}, \mathbf{c}) = \sum_{k=1}^{p} g_k(\mathbf{c}) \prod_{j=1}^{q} w_j^{p_{jk}} (v - v_k(\mathbf{c})) + I_0(v, \mathbf{w}, \mathbf{c}). \tag{1}$$

Here $\mathbf{w}$ is a vector of gating variables taking values in [0, 1] that represent the portion of open channels on the membrane, whereas $\mathbf{c}$ is a vector describing the concentration of ionic species within the cell. We denote by $g_k(\mathbf{c})$ and $v_k(\mathbf{c})$ the conductance and the Nernst equilibrium potential associated to the k-th ion, and by p_{jk} the number of sub-units composing each ionic channel, so that the ion k can flow through a ionic channel if all the sub-units forming the channel are opened. Thus, $\prod_{j=1}^{q} w_j^{p_{jk}}$ represents the probability that the ions k flow through the cellular membrane. The term $I_0(v, \mathbf{w}, \mathbf{c})$ represents possible time independent ionic fluxes. The dynamic of a single cell can thus be described in general by a ionic model under the form:

$$\begin{cases} C_m \dfrac{\partial v}{\partial t} + I_{ion}(v, \mathbf{w}, \mathbf{c}) = I_{app} & \text{in} \Omega_0 \times (0, T), \\ \dfrac{\partial \mathbf{w}}{\partial t} = \mathbf{s}(v, \mathbf{w}) & \text{in} \Omega_0 \times (0, T), \\ \dfrac{\partial \mathbf{c}}{\partial t} = \mathbf{r}(v, \mathbf{w}, \mathbf{c}) & \text{in} \Omega_0 \times (0, T), \\ \mathbf{w}(t_0) = \mathbf{w}_0, \quad \mathbf{c}(t_0) = \mathbf{c}_0 & \text{in} \Omega_0 \end{cases} \tag{2}$$

where we denote by Ω_0 the computational domain, here representing the cell. The first set of ODEs is related to the evolution of the gating variables while the second set characterizes the evolution of the ionic concentrations during the cardiac cycle. The number of equations, the functions $\mathbf{s}$ and $\mathbf{r}$ and the overall complexity depend on the considered model. In this work, we focus on the so-called minimal model introduced by Bueno-Orovio in [17], developed to reproduce physiological action potential morphologies, at moderate computational costs.

The minimal model is a four variables model, able to reproduce experimental measures characteristics of human ventricular transmembrane potential. It can be expressed as:

$$\begin{cases} C_m \dfrac{\partial v}{\partial t} + I_{fi}(v, w_1) + I_{so}(v) + I_{si}(v, w_2, w_3) = I_{app}, & \text{in} \Omega_0 \times (0, T) \\ \dfrac{\partial w_1}{\partial t} + H(v - \theta_{w_1}) \dfrac{w_1}{\tau_{w_1}^+} - [1 - H(v - \theta_{w_1})] \dfrac{w_{1,\infty} - w_1}{\tau_{w_1}^-} = 0, & \text{in} \Omega_0 \times (0, T) \\ \dfrac{\partial w_2}{\partial t} + H(v - \theta_{w_2}) \dfrac{w_2}{\tau_{w_2}^+} - [1 - H(v - \theta_{w_2})] \dfrac{w_{2,\infty} - w_2}{\tau_{w_2}^-} = 0, & \text{in} \Omega_0 \times (0, T) \\ \dfrac{\partial w_3}{\partial t} - \dfrac{[1 + tanh(k_s(v - v_s))]/2 - w_3}{\tau_{w_3}} = 0, & \text{in} \Omega_0 \times (0, T) \end{cases} \tag{3}$$

where the three currents I_{fi}, I_{si} and I_{so} represent the fast inward, the overall slow outward and the slow inward currents, respectively. In particular, we have:

$$I_{fi}(v, w_1) = w_1 \frac{(v - \theta_{w_1})(c_v - v)H(v - \theta_{w_1})}{\tau_{fi}},$$

$$I_{s0}(v) = \frac{(v - v_0)(1 - H(v - \theta_{w_2}))}{\tau_0} + \frac{H(v - \theta_{w_2})}{\tau_{so}},$$

$$I_{si}(v, w_2, w_3) = w_2 w_3 \frac{H(v - \theta_w)}{\tau_{si}},$$

where $H(\cdot)$ is the Heaviside function. The model parameters, provided in [17], allow to reproduce the action potential morphologies and the dynamics of more complex models, such as the Ten Tusscher model or the O'Hara and Rudy model [72]. The evolution of the transmembrane potential v and the gating variables w_1, w_2 and w_3 obtained with the considered parameters are reported in Fig. 1. In

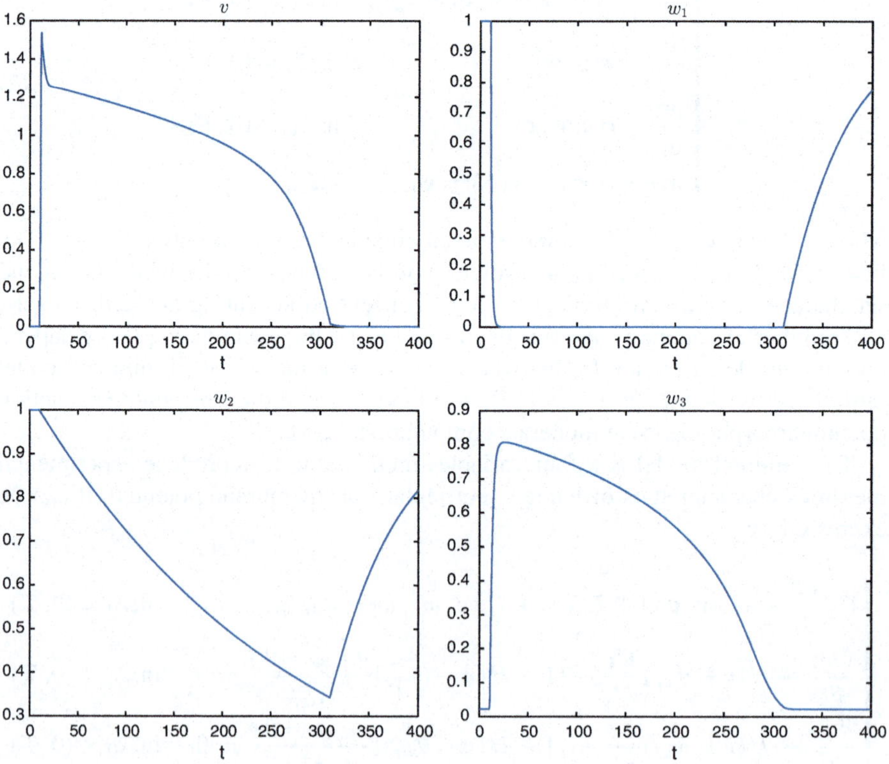

Fig. 1 From top left to bottom right: transmembrane potential v and gating variables w_1, w_2 and w_3 of the minimal model during a heart beat

particular, we observe that the variable v correctly reproduces the action potential shape characteristic of the ventricular cardiomyocytes. We notice that the variable v is dimensionless and it can be rescaled to dimensions of mV using the equations $v_{mv} = 87.5v - 84$.

Note that this model does not provide specific informations about intracellular calcium concentration. However, the variable w_3 can be assumed to be responsible of calcium dynamics, as it shows a phenomenological behavior which is really similar to the one of the calcium ions in the cardiac cells [72]. In what follows, we refer to w_3 as the calcium concentration c and denote by $\mathbf{w} = (w_1, w_2, c)$.

2.1.2 Tissue Models

To describe the propagation of the activation front in the cardiac muscle, we rely on a tissue model able to characterize the evolution of the transmembrane potential. This latter plays indeed a crucial role in the description of the heart contraction. In particular, the bidomain and the monodomain models have been widely used to study the cardiac electrophysiology (see e.g. [24, 66] and reference therein). These models arise from a homogenization process applied to the cardiac tissue and simulate the propagation of the electrical signal through the myocardium; a complete derivation of the two models can be found in [28].

The bidomain model, first proposed in [83], represents the cardiac tissue as a syncytium composed of intracellular and extracellular domains coexisting at every point of the tissue. Each domain is thus considered as a continuum, rather than a group of discrete cells connected with each other. Denoting by v_e and v_i the extracellular and the intracellular potential, respectively, and by $v = v_i - v_e$ the transmembrane potential, the bidomain model can be expressed as:

$$
\begin{cases}
C_m \dfrac{\partial v}{\partial t} + I_{ion}(v, \mathbf{w}, \mathbf{c}) - \nabla \cdot (\mathbf{D}_i \nabla v_i) = 0 & \text{in} \Omega_0 \times (0, T), \\
C_m \dfrac{\partial v}{\partial t} + I_{ion}(v, \mathbf{w}, \mathbf{c}) + \nabla \cdot (\mathbf{D}_e \nabla v_e) = I_{app} & \text{in} \Omega_0 \times (0, T), \\
\dfrac{\partial \mathbf{w}}{\partial t} = \mathbf{s}(v, \mathbf{w}), \quad \dfrac{\partial \mathbf{c}}{\partial t} = \mathbf{r}(v, \mathbf{w}, \mathbf{c}) & \text{in} \Omega_0 \times (0, T), \\
\mathbf{D}_i \nabla v_i \cdot \mathbf{n} = 0, \quad \mathbf{D}_e \nabla v_e \cdot \mathbf{n} = 0 & \text{on} \partial \Omega_0 \times (0, T), \\
v(t_0) = v_0, \quad \mathbf{w}(t_0) = \mathbf{w}_0, \quad \mathbf{c}(t_0) = \mathbf{c}_0 & \text{in} \Omega_0,
\end{cases}
\tag{4}
$$

where the Neumann boundary conditions express the condition that the cardiac tissue is electrically insulated. Here, C_m is the membrane capacitance, I_{ion} is the sum of the density currents through the membrane and I_{app} is an external applied density current. The domain Ω_0 now represents a portion of myocardium; in the case of the left ventricle, its boundary $\partial \Omega_0 = \Gamma_{endo} \cup \Gamma_{epi} \cup \Gamma_{base}$ is made by the endocardium, the epicardium and the base, respectively. The conductivity tensors $D_i \in \mathbb{R}^3$ and $D_e \in \mathbb{R}^3$ model the anisotropy of the cardiac tissue, characterized by a higher conductivity in the fiber direction, and can be expressed as:

$$\mathbf{D}_{i,e} = \sigma_f^{i,e}\mathbf{f}_0 \otimes \mathbf{f}_0 + \sigma_s^{i,e}\mathbf{s}_0 \otimes \mathbf{s}_0 + \sigma_n^{i,e}\mathbf{n}_0 \otimes \mathbf{n}_0,$$

where $\sigma_f^{i,e}$, $\sigma_s^{i,e}$ and $\sigma_n^{i,e}$ are the electrical conductivities in the intracellular and extracellular domains. Here, $\mathbf{f}_0$ denotes the fibers direction, $\mathbf{s}_0$ the sheets direction and $\mathbf{n}_0$ is orthogonal to both $\mathbf{f}_0$ and $\mathbf{s}_0$.

The bidomain model is currently the most complete mathematical model for describing the electrical signal propagation in the heart. However, it is computationally demanding, since to capture the rapid dynamics of the cellular reactions high resolutions in space and time is required. For this reason, a common approach in the literature is to reduce the bidomain equations to simpler tissue models. In particular, the monodomain model can be obtained by assuming that the intra and extracellular domains have equal anisotropy ratios, $\sigma_f^i/\sigma_f^e = \sigma_s^i/\sigma_s^e = \sigma_n^i/\sigma_n^e$. The monodomain model is not suitable for the description of pathological situations, such as, cardiac arrhythmias or fibrillation, since in these situations the extracellular domain influence the transmembrane potential and the ionic currents. However, it provides an accurate description of the cardiac tissue in physiological situations [27, 66], at reduced computational costs. From now on, we adopt the monodomain model as we are only interested, for the time being, to capture the relevant phenomena for describing the electromechanical coupling in healthy conditions.

The monodomain model reads as follows:

$$\begin{cases} C_m \dfrac{\partial v}{\partial t} + I_{ion}(v, \mathbf{w}, \mathbf{c}) - \nabla \cdot (\mathbf{D}\nabla v) = I_{app} & \text{in} \Omega_0 \times (0, T), \\ \dfrac{\partial \mathbf{w}}{\partial t} = \mathbf{s}(v, \mathbf{w}), \quad \dfrac{\partial \mathbf{c}}{\partial t} = \mathbf{r}(v, \mathbf{w}, \mathbf{c}) & \text{in} \Omega_0 \times (0, T), \\ \mathbf{D}\nabla v \cdot \mathbf{n} = 0 & \text{on} \partial\Omega_0 \times (0, T), \\ v(t_0) = v_0, \quad \mathbf{w}(t_0) = \mathbf{w}_0, \quad \mathbf{c}(t_0) = \mathbf{c}_0 & \text{in} \Omega_0, \end{cases} \tag{5}$$

where $\mathbf{D} \in \mathbb{R}^3$ is the conductivity tensor. In particular, we assume that

$$\mathbf{D} = \sigma_f \mathbf{f}_0 \otimes \mathbf{f}_0 + \sigma_s \mathbf{s}_0 \otimes \mathbf{s}_0 + \sigma_n \mathbf{n}_0 \otimes \mathbf{n}_0 \tag{6}$$

where we denote by σ_f, σ_s and σ_n the electrical conductivities in the direction $\mathbf{f}_0$, $\mathbf{s}_0$ and $\mathbf{n}_0$, respectively.

2.2 Cardiac Mechanics

The description of cardiac mechanics involves both a passive and an active contribution; besides the hyper-elastic behavior of the tissue, the active contraction of the muscular fibers has to be included in the force balance when modeling the systolic part of the cardiac cycle.

We consider a reference configuration Ω_0 and an actual configuration Ω at the current time t. We denote by $\mathbf{X}$ the position vector in Ω_0 and by $\mathbf{x}$ the position vector in Ω. We can now introduce the body deformation as the map $\boldsymbol{\varphi} : \Omega_0 \to \Omega$ from the reference to the actual configuration, such that $\mathbf{x} = \boldsymbol{\varphi}(\mathbf{X})$ for any $\mathbf{X} \in \Omega_0$, $\mathbf{x} \in \Omega$. The deformation gradient tensor $\mathbf{F}$ is defined as

$$\mathbf{F} = \frac{\partial \boldsymbol{\varphi}}{\partial \mathbf{X}}, \qquad [F_{ij}] = \frac{\partial \varphi_i}{\partial \mathbf{X}_j}, \qquad i, j = 1, 2, 3. \tag{7}$$

By denoting $\mathbf{u} : \Omega_0 \to \Omega, \mathbf{u}(\mathbf{X}) = \varphi(\mathbf{X}) - \mathbf{X}$ the displacement field, the deformation gradient tensor can be written as $\mathbf{F} = \mathbf{I} + \nabla \mathbf{u}$. We also denote by $J = det(\mathbf{F})$ the determinant of $\mathbf{F}$ and by $\mathbf{C} = \mathbf{F}^T\mathbf{F}$ the left Cauchy-Green strain tensor.

2.2.1 Passive Mechanics

We first provide a description of the passive ventricular mechanics, by recalling the hyperelastic model proposed by Holzapfel and Ogden in [43]. Cardiac deformations can be modeled by considering the myocardium as orthotropic, hyperelastic, and incompressible with passive properties characterized by means of an exponential strain energy function.

The equations of motion for the cardiac tissue express the balance of linear momentum in material coordinates, which reads as

$$\rho_0 \frac{d^2\mathbf{u}}{dt^2} - \nabla_0 \cdot \mathbf{P}_p = \mathbf{b}_0,$$

where ρ_0 is the tissue density and $\mathbf{b}_0$ are the body forces. Here, $\mathbf{P}_p$ is the first (passive) Piola(-Kirchhoff) tensor, which is related to the surface tractions $\mathbf{t}_0$ through the relation $\mathbf{t}_0 = \mathbf{P}_p\mathbf{n}$, where $\mathbf{n}$ is the normal to the boundary of the reference domain. As usual in cardiac mechanics literature (see e.g. [29, 37, 72]), inertial forces can be neglected, since they are about two orders of magnitude smaller than other terms [86], thus obtaining the quasi-static problem

$$-\nabla_0 \cdot \mathbf{P}_p = \mathbf{b}_0. \tag{8}$$

We impose Neumann boundary conditions on the endocardium ($\Gamma_N = \Gamma_{endo}$) to model the effect of blood pressure,[1] and Robin boundary conditions on the epicardium and on the base ($\Gamma_R = \Gamma_{epi} \cup \Gamma_{base}$); the boundaries are reported for a patient-specific left ventricle geometry in Fig. 4. We also neglect the body forces $\mathbf{b}_0$

[1]More specifically, we would have $\mathbf{g} = p_{endo}(t)\mathbf{n}$ where $\mathbf{n}$ is the unit normal vector to the boundary, and $p_{endo} = p_{endo}(t)$ is the external load applied by the fluid at the endocardium wall, which in this context is assumed to be prescribed.

because their contribution is negligible [72]. In conclusion, the cardiac deformation $\mathbf{u}$ solves:

$$\begin{cases} -\nabla_0 \cdot \mathbf{P}_p(\mathbf{u}) = \mathbf{0} & \text{in} \Omega \\ \mathbf{P}_p(\mathbf{u})\mathbf{n} = \mathbf{g} & \text{on} \Gamma_{endo} \\ \mathbf{P}_p(\mathbf{u})\mathbf{n} + \alpha \mathbf{u} = \mathbf{0} & \text{on} \Gamma_{epi} \cup \Gamma_{base}. \end{cases} \tag{9}$$

The myocardium is considered as an hyperelastic material: there exists a strain energy function $\mathscr{W} : \Omega_0 \to \mathbb{R}$ related to the Piola tensor through the relation

$$\mathbf{P}_p(\mathbf{u}) = \frac{\partial \mathscr{W}(\mathbf{u})}{\partial \mathbf{F}}. \tag{10}$$

The description of the cardiac muscle mechanics faces a number of difficulties. Indeed, the myocardium is non-homogeneous and it is composed by several layers; moreover, fibers have different orientation in each layer and rotate across the heart wall, featuring a complex mechanical characterization. To model this complex behavior we consider the orthotropic model proposed by Holzapfel and Ogden in [43], characterized by a simple invariant-based formulation. This model hinges upon the idea that for an orthotropic, incompressible material the strain energy density function can be written as

$$\mathscr{W} = \mathscr{W}_1(\mathscr{I}_1) + \mathscr{W}_{4,\mathbf{f}_0}(\mathscr{I}_{4,\mathbf{f}_0}) + \mathscr{W}_{4,\mathbf{s}_0}(\mathscr{I}_{4,\mathbf{s}_0}) + \mathscr{W}_{8,\mathbf{f}_0\mathbf{s}_0}(\mathscr{I}_{8,\mathbf{f}_0\mathbf{s}_0}),$$

where $\mathbf{f}_0$, $\mathbf{s}_0$ are the two (fibers and sheets respectively) preferred directions and $\mathscr{I}_1$, $\mathscr{I}_{4,\mathbf{f}_0}$, $\mathscr{I}_{4,\mathbf{s}_0}$, $\mathscr{I}_{8,\mathbf{f}_0\mathbf{s}_0}$ are invariants of the right Cauchy-Green strain tensor,

$$\mathscr{I}_1 = tr(\mathbf{C}), \quad \mathscr{I}_{4,\mathbf{f}_0} = \mathbf{f}_0 \cdot \mathbf{C}\mathbf{f}_0, \quad \mathscr{I}_{4,\mathbf{s}_0} = \mathbf{s}_0 \cdot \mathbf{C}\mathbf{s}_0, \quad \mathscr{I}_{8,\mathbf{f}_0\mathbf{s}_0} = \mathbf{f}_0 \cdot \mathbf{C}\mathbf{s}_0,$$

respectively. In particular, we have

$$\mathscr{W}_1(\mathscr{I}_1) = \frac{a}{2b}\left[e^{b(\mathscr{I}_1-3)} - 1\right], \qquad \mathscr{W}_{4,\mathbf{f}_0}(\mathscr{I}_{4,\mathbf{f}_0}) = \frac{a_f}{2b_f}\left[e^{b_f(\mathscr{I}_{4,\mathbf{f}_0}-1)^2} - 1\right],$$

$$\mathscr{W}_{4,\mathbf{s}_0}(\mathscr{I}_{4,\mathbf{s}_0}) = \frac{a_s}{2b_s}\left[e^{b_s(\mathscr{I}_{4,\mathbf{s}_0}-1)^2} - 1\right], \qquad \mathscr{W}_{8,\mathbf{f}_0\mathbf{s}_0}(\mathscr{I}_{8,\mathbf{f}_0\mathbf{s}_0}) = \frac{a_{fs}}{2b_{fs}}\left[e^{b_{fs}\mathscr{I}^2_{8,\mathbf{f}_0\mathbf{s}_0}} - 1\right]. \tag{11}$$

The coefficients of the Holzapfel-Ogden constitutive law are taken from [33] and are reported in Table 1.

Table 1 Parameters of the Holzapfel-Ogden model

$a = 3.33$ kPa	$a_f = 18.47$ kPa	$a_s = 2.481$ kPa	$a_{fs} = 0.417$ kPa
$b = 8.023$	$b_f = 16.026$	$b_s = 11.120$	$b_{fs} = 11.436$

In order to describe myocardium deformations, we consider a quasi-incompressible formulation [39, 77], which offers several advantages with respect to a full incompressible one, from both a modeling and a numerical viewpoint. Indeed, taking into account limited volumetric changes is possible according to experimental evidence since the volume of cardiac tissue can vary until 7% during systolic contraction [6]. Moreover, a quasi-incompressible formulation leads to a simpler numerical problem with respect to a full incompressible one [65]. The adopted formulation can be obtained by introducing a multiplicative decomposition $\mathbf{F} = \mathbf{F}_{iso}\mathbf{F}_{vol}$ of the deformation gradient tensor, where we impose $det(\mathbf{F}_{iso}) = 1$ and $det(\mathbf{F}_{vol}) = J$. This formulation leads to an additive decomposition of the isotropic part $\mathscr{W}_1$ of the strain energy function, which reads as

$$\mathscr{W}_1 = \mathscr{W}_{1,iso} + \mathscr{W}_{vol} = \frac{a}{2b}\left[e^{b(J^{-\frac{2}{3}}\mathscr{I}_1-3)} - 1\right] + \frac{\kappa}{2}J\mathbf{F}^{-T}(J - J^{-1}) \tag{12}$$

where $\kappa > 0$ denotes the Bulk modulus, which measures the material resistance to a uniform compression.

2.2.2 Active Mechanics

The cardiomyocytes of the heart muscle contract after being electrically activated, without the need of an external load. This behavior can be modeled by including the active contraction of the muscular fibers in the force balance (9); however, this is a challenging task because muscular contraction occurring at the macroscale is caused by release of energy at the microscale, inside each cardiomyocyte. Different approaches have been investigated in order to obtain accurate mathematical description of the active mechanics; the most popular ones are the active stress [37, 52, 53, 62, 78] and the active strain approaches; see [4, 73] for numerical comparisons. Both strategies allow to couple electrophysiology and mechanics, defining a modified first Piola-Kirchhoff tensor $\mathbf{P}$ which involves a passive component describing the stress required to obtain a given deformation of the passive myocardium, and an active component denoting the tension generated by the depolarization of the propagating electrical signal that provides the internal active forces responsible for the contraction. Thus, Eq. (9) becomes

$$\begin{cases} -\nabla_0 \cdot \mathbf{P}(\mathbf{u}(t); t, c) = \mathbf{0} & \text{in}\,\Omega \\ \mathbf{P}(\mathbf{u}(t); t, c)\,\mathbf{n} = \mathbf{g} & \text{on}\,\Gamma_{endo} \\ \mathbf{P}(\mathbf{u}(t); t, c)\,\mathbf{n} + \alpha\mathbf{u}(t) = \mathbf{0} & \text{on}\,\Gamma_{epi} \cup \Gamma_{base}. \end{cases} \tag{13}$$

This leads to a coupled electromechanical problem, where the electrical solution affects cardiac deformations. Here we focus on the active strain approach [3, 23, 51, 81], which is based on a multiplicative decomposition of the deformation gradient tensor, under the form

$$\mathbf{F}(\mathbf{u}, t) = \mathbf{F}_e(\mathbf{u})\mathbf{F}_a(t).$$

$\mathbf{F}_e$ describes the elastic deformation of the myocardium and $\mathbf{F}_a(t)$ is the anelastic deformation due to the fibers contraction. The active strain decomposition is based on the idea that fibers inside the muscle contract and shorten; the deformation $\mathbf{F}_a$ can thus be seen as a prescribed distortion of the microstructure, whereas the deformation at the macroscale $\mathbf{F}_e$ is needed to ensure compatibility of $\mathbf{F}$.

In particular, the anelastic deformation takes the form

$$\mathbf{F}_a(t) = I + \gamma_f(t)\mathbf{f}_0 \otimes \mathbf{f}_0 + \gamma_s(t)\mathbf{s}_0 \otimes \mathbf{s}_0 + \gamma_n(t)\mathbf{n}_0 \otimes \mathbf{n}_0,$$

where $\mathbf{n}_0$ is a vector normal to $\mathbf{f}_0$ and $\mathbf{s}_0$; γ_f, γ_s and γ_n are time-dependent coefficients describing the cell shortening respectively in the $\mathbf{f}_0$, $\mathbf{s}_0$ and $\mathbf{n}_0$ directions. The fibers shortening γ_f can be computed from the following evolution law

$$\begin{cases} \mu_A \dot{\gamma_f} = f_A(c) + \dfrac{2\mathscr{I}_{4,f}}{(1+\gamma_f)^3} - 2\mathscr{I}_{4,f}|_{c=c_0} & \text{in}\Omega_0 \times (0,T), \\ \gamma_f(0) = \gamma_{f,0} & \text{in}\Omega_0, \end{cases} \tag{14}$$

where $c(t)$ is the calcium concentration. We remark that the anelastic deformation $\mathbf{F}_a$ depends on the calcium concentration through the coefficients γ_f, γ_s and γ_n. Here, $f_A(c(t)) = \alpha(c(t) - c_0)^2 R_{FL}(\mathscr{I}_{4,f})$, where we assume $\alpha = -2.5$ [72], and R_{FL} is the sarcomere force-length relationship of the cardiac cells given by

$$R_{FL}(\mathscr{I}_{4,f}) = \chi_{[SL_{min}, SL_{max}]}(\mathscr{I}_{4,f}) \left\{ \frac{c_0}{2} + \sum_{n=1}^{3} [c_n \sin(n\mathscr{I}_{4,f} l_0) + d_n \cos(n\mathscr{I}_{4,f} l_0)] \right\};$$

here, l_0 represents the initial length of a single contractile unit (sarcomere) and we assume $l_0 = 1.95\,\mu\text{m}$, whereas the coefficients c_n and d_n are parameters of a truncated Fourier series fitted to match the experimental length-force relations reported in [79]; see [75] for further details. Moreover, $\chi_{[SL_{min}, SL_{max}]}(\cdot)$ is the characteristic function of the interval $[SL_{min}, SL_{max}]$, which represent the minimum and maximum sarcomere length, respectively; here we assume $SL_{min} = 0.87\,\mu\text{m}$ $SL_{max} = 1.33\,\mu\text{m}$ [72]. Here c_0 represents the calcium concentration at the end of the diastolic phase.

The other two coefficients γ_s and γ_n can be directly derived from the expression of γ_f, relying on an orthotropic activation model, as

$$\gamma_s = \kappa_f \gamma_f, \quad \gamma_n = \frac{1}{(1+\gamma_f)(1+\gamma_s)} - 1.$$

The parameter κ_f allows to correctly describe the thickening occurring during myocardial contraction in the sheets direction. In the mechanical equations, the Piola tensor takes the following form

$$\mathbf{P} = det(\mathbf{F}_a)\frac{\partial \mathscr{W}(\mathbf{F}_e)}{\partial \mathbf{F}_e}\mathbf{F}_a^{-T}$$

leading to the following full mechanical problem, where we highlight the dependence on $\gamma_f(t)$:

$$\begin{cases} -\nabla_0 \cdot \mathbf{P}(\mathbf{u}(t); \gamma_f(t)) = \mathbf{0} & \text{in} \Omega \\ \mathbf{P}(\mathbf{u}(t); \gamma_f(t))\,\mathbf{n} = \mathbf{g} & \text{on} \Gamma_{endo} \\ \mathbf{P}(\mathbf{u}(t); \gamma_f(t))\,\mathbf{n} + \alpha\mathbf{u}(t) = \mathbf{0} & \text{on} \Gamma_{epi} \cup \Gamma_{base}. \end{cases} \tag{15}$$

We point out that in the active strain approach the solution of the mechanical problem depends on the calcium concentration c, rather than on the transmembrane potential. However, we need to compute the solution of the full electrical problem in order to characterize the mechanical displacement, since c is coupled to the transmembrane potential v, e.g., in the monodomain model (5).

A schematic representation of the model describing the electromechanical coupling is reported in Fig. 2.

2.3 *Parameters of Interest*

ROM techniques allow to efficiently solve the problems introduced so far in different scenarios in order assess the effect of clinically relevant parameters on their solutions by taking into account a possible inter-patient variability. We denote by $\boldsymbol{\mu} \in \mathscr{P} \subset \mathbb{R}^p$ the set of p selected parameters.

In particular, we are interested in analyzing how the electrical conductivities—σ_f, σ_s and σ_n introduced in (5)—and the fibers orientation $\mathbf{f}_0$ affect heart contraction. Electrical conductivities significantly influence the propagation of the electrical signal and, consequently, the displacement of the cardiac muscle; fibers orientation highly varies among subjects and can have a crucial impact on the correct torsion and shortening of the ventricle. We will restrict our attention to these parameters in this work; note that fibers' orientation affects both the electrophysiology and the mechanics since it directly enters in the monodomain Eq. (5) as well as in the constitutive law (11). Instead, electrical conductivities affects primarily the monodomain equation through the conductivity tensor $\mathbf{D}$, and only indirectly the myocardium displacement. Additional parameters of interest are, for instance, the (isotropic) coefficient $a(\boldsymbol{\mu})$ introduced in (11), related to the stiffness of the cardiac muscle, as well as the Bulk modulus κ defined in (12), related to the material incompressibility; more can be found in [13, 14].

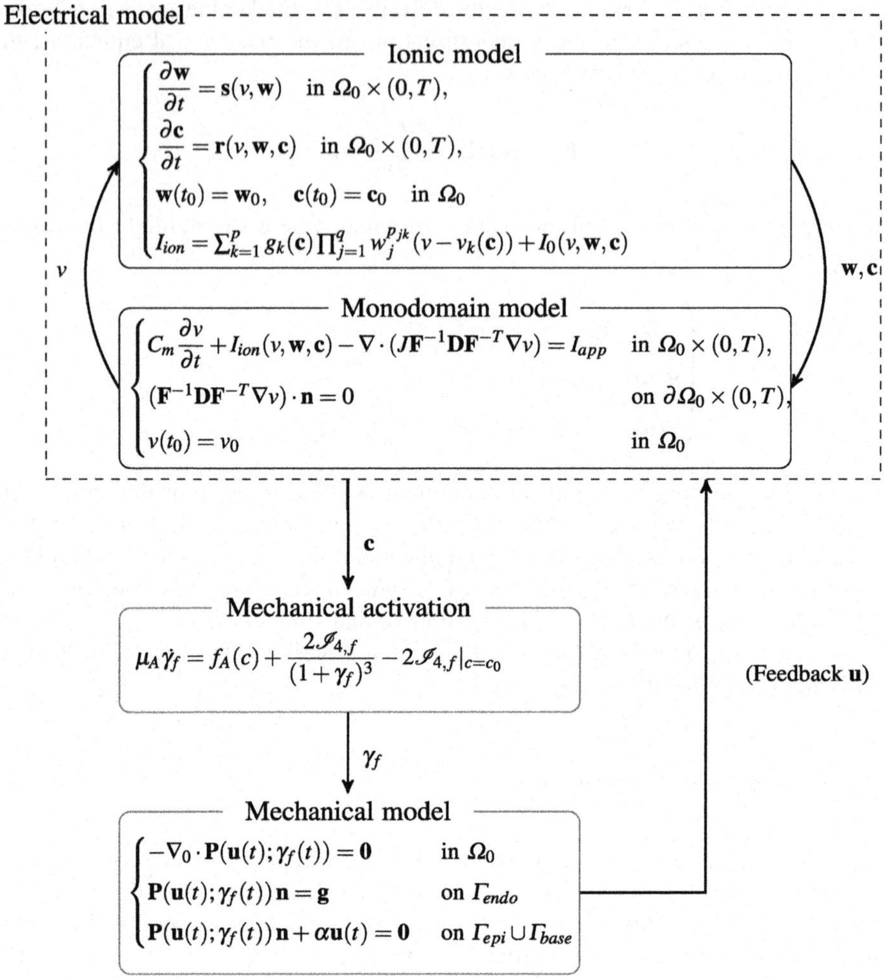

Fig. 2 Schematical representation of the cardiac electromechanical model

Regarding electrophysiology, we thus have $\mathbf{D} = \mathbf{D}(\boldsymbol{\mu})$ and the parametrized electrical model reads as:

$$
\begin{cases}
C_m \dfrac{\partial v(\boldsymbol{\mu})}{\partial t} + I_{ion}(v(\boldsymbol{\mu}), \mathbf{w}(\boldsymbol{\mu})) - \nabla \cdot (\mathbf{D}(\boldsymbol{\mu}) \nabla v(\boldsymbol{\mu})) = I_{app} & \text{in} \Omega_0 \times (0, T), \\
\dfrac{\partial \mathbf{w}(\boldsymbol{\mu})}{\partial t} = \mathbf{s}(v(\boldsymbol{\mu}), \mathbf{w}(\boldsymbol{\mu})) & \text{in} \Omega_0 \times (0, T), \\
\mathbf{D}(\boldsymbol{\mu}) \nabla v(\boldsymbol{\mu}) \cdot \mathbf{n} = 0 & \text{on} \partial\Omega_0 \times (0, T), \\
v(t_0) = v_0 \qquad w(t_0) = w_0 & \text{in} \Omega_0.
\end{cases}
\tag{16}
$$

If $\mathbf{D}$ is $\boldsymbol{\mu}$-dependent, also the variables v and $\mathbf{w}$ will depend on parameters. Since the activation equation depends on the solution of the electrical problem, we also have $\gamma_f = \gamma_f(\boldsymbol{\mu})$, so that (14) reads as

$$\mu_A \dot{\gamma_f}(\boldsymbol{\mu}) = f_A(c(\boldsymbol{\mu})) + \frac{2\mathscr{I}_{4,f}}{(1+\gamma_f(\boldsymbol{\mu}))^3} - 2\mathscr{I}_{4,f}|_{c=c_0} \quad \text{in}\Omega_0 \times (0,T).$$

As for the mechanical problem, we have $\mathbf{u} = \mathbf{u}(\boldsymbol{\mu})$, either if we consider parameters directly affecting mechanics or parameters which directly enter only in the electrophysiology, so that (13) modifies as:

$$\begin{cases} -\nabla_0 \cdot \mathbf{P}(\mathbf{u}(t,\boldsymbol{\mu}); \gamma_f(t,\boldsymbol{\mu})) = \mathbf{0} & \text{in}\Omega \\ \mathbf{P}(\mathbf{u}(t,\boldsymbol{\mu}); \gamma_f(t,\boldsymbol{\mu}))\,\mathbf{n} = \mathbf{g} & \text{on}\Gamma_{endo} \\ \mathbf{P}(\mathbf{u}(t,\boldsymbol{\mu}); \gamma_f(t,\boldsymbol{\mu}))\,\mathbf{n} + \alpha\mathbf{u}(t,\boldsymbol{\mu}) = 0 & \text{on}\Gamma_{epi} \cup \Gamma_{base}. \end{cases} \tag{17}$$

In this work, we neglect the influence of the blood in the ventricular chamber, thus taking $\mathbf{g} = \mathbf{0}$—that is, $p_{endo}(t) = 0$, by assuming that no information on the blood inside the ventricle are available. Note that the problem is not trivial: indeed, even if external loads are zero, stress (and then deformation) originates because of the presence of a term depending on the fiber shortening γ_f.

3 Full-Order Model: Finite Element Method

Before addressing the reduction of the electrophysiology and the mechanical problems, we sketch their finite element (FE) approximation [31, 38, 82, 84], which the reduced order model is built on, and plays the role of full-order model (FOM). For the sake of notation the dependence on the parameter vector $\boldsymbol{\mu}$ is understood in this section.

3.1 Electrical Model

After deriving the weak formulation of problem (16), we introduce the discretization $\mathscr{T}_{ep}$ of the domain Ω_0 and the finite dimensional spaces Z_h and Q_h with $dim(Z_h) = N_z < \infty$ and $dim(Q_h) = N_q < \infty$, respectively, for the approximation of the potential v and the ionic variables $\mathbf{w} \in \mathbb{R}^d$, $d = 3$. We denote by $\{\varphi_i\}_{i=1}^{N_z}$, $\varphi_i \in Z_h$ and $\{\psi_i\}_{i=1}^{N_q}$, $\psi_i \in Q_h$, their FE bases, so that the approximated potential v_h and ionic variables $\mathbf{w}_h$ can be expressed under the form

$$v_h(\mathbf{x},t) = \sum_{i=1}^{N_z} \mathrm{v}_i(t)\varphi_i(\mathbf{x}) \quad \text{and} \quad \mathbf{w}_h(\mathbf{x},t) = \begin{pmatrix} w_{1,h}(\mathbf{x},t) \\ w_{2,h}(\mathbf{x},t) \\ c_h(\mathbf{x},t) \end{pmatrix} = \sum_{i=1}^{N_q} \mathbf{w}_i(t) \circ \boldsymbol{\psi}_i{}^d(\mathbf{x}); \tag{18}$$

here $\boldsymbol{\psi}_i{}^d$ are the basis functions of the space Q_h^d and the operator $\circ$ is the element-wise vector product. We denote the components of $\mathbf{w}_i \in \mathbb{R}^3$, $i = 1, \ldots, N_q$, as $\mathrm{w}_{1,i}$, $\mathrm{w}_{2,i}$, c_i, respectively. The FE approximation of the monodomain model (5) turns into the following nonlinear system of algebraic equation

$$\begin{cases} \displaystyle\sum_{i=1}^{N_z} \Big(C_m \frac{\partial v_i}{\partial t}(\varphi_i, \varphi_j) + v_i a(\varphi_i, \varphi_j) \Big) + I\Big(\sum_{i=1}^{N_z} v_i\varphi_i, \sum_{i=1}^{N_q} \mathbf{w}_i \circ \boldsymbol{\psi}_i{}^d; \varphi_j\Big) = 0 & \forall j = 1, \ldots, N_z, \\ \displaystyle\sum_{i=1}^{N_q} \frac{\partial \mathbf{w}_i}{\partial t}(\boldsymbol{\psi}_i, \boldsymbol{\psi}_j) = \Big(\mathbf{s}\big(\sum_{i=1}^{N_z} v_i\varphi_i, \sum_{i=1}^{N_q} \mathbf{w}_i \circ \boldsymbol{\psi}_i{}^d\big), \boldsymbol{\psi}_j\Big) & \forall j = 1, \ldots, N_q \end{cases} \tag{19}$$

by defining the bilinear form a and the functional I as

$$a(v,z) = \int_{\Omega_0} \mathbf{D}\nabla v \cdot \nabla z d\Omega_0, \qquad I(v, \mathbf{w}; z) = \big(I_{ion}(v, \mathbf{w}) - I_{app}, z\big),$$

respectively, with

$$(v, w) = \int_{\Omega_0} vw d\Omega_0, \qquad (\mathbf{v}, \mathbf{w}) = \int_{\Omega_0} \mathbf{v} \cdot \mathbf{w} d\Omega_0.$$

Here we restrict ourselves to the case of linear (P1) finite elements, so that $N_z = N_q = N_h^e$. We denote by $\mathbf{V} = (\mathrm{v}_1, \ldots, \mathrm{v}_{N_h^e})^T$, $\mathbf{W} = (\mathbf{W}_1, \mathbf{W}_2, \mathbf{C})$, where $\mathbf{W}_1 = (\mathrm{w}_{1,1}, \ldots, \mathrm{w}_{1,N_h^e})^T$, $\mathbf{W}_2 = (\mathrm{w}_{2,1}, \ldots, \mathrm{w}_{2,N_h^e})^T$ and $\mathbf{C} = (\mathrm{c}_1, \ldots, \mathrm{c}_{N_h^e})^T$. Moreover, $\mathbf{M}_{ij} = (\varphi_j, \varphi_i)$, $\mathbf{M}^d_{q,ij} = (\boldsymbol{\psi}_j, \boldsymbol{\psi}_i)$ and $\mathbf{K}_{ij} = a(\varphi_j, \varphi_i)$ denote the mass matrices and the stiffness matrix associated to problem (19), respectively. To treat the nonlinear term I, we rely on the so-called *ionic current interpolation* (ICI) method, which introduces a linear interpolation of the ionic currents,

$$I\Big(\sum_{i=1}^{N_h^e} v_i\varphi_i, \sum_{i=1}^{N_h^e} \mathbf{w}_i \circ \boldsymbol{\psi}_i{}^d; \varphi_j\Big) \approx \int_{\Omega_0} I(v_j, \mathbf{w}_i)\varphi_j(\mathbf{x})\varphi_j(\mathbf{x}) = \mathbf{I}(\mathbf{V}, \mathbf{W}) \tag{20}$$

and makes the assembling of the ionic currents term straightforward, only requiring a matrix-vector multiplication, $\mathbf{I}(\mathbf{V}, \mathbf{W}) = \mathbf{M}\tilde{\mathbf{I}}(\mathbf{V}, \mathbf{W})$, where $\tilde{\mathbf{I}}(\mathbf{V}, \mathbf{W}) = (I(v_1, \mathbf{w}_1), \ldots, I(v_{N_h^e}, \mathbf{w}_{N_h^e}))^T$. A further enhancement of this procedure can be obtained by considering a lumped version of the ICI strategy (L-ICI) where the ionic currents are interpolated nodally or, equivalently, the mass matrix arising in the ICI method is lumped, thus yielding $\mathbf{I}(\mathbf{V}, \mathbf{W}) = \mathbf{M}_L \tilde{\mathbf{I}}(\mathbf{V}, \mathbf{W})$. We remark that a possible alternative to the ICI method is the so-called *state variable interpolation* (SVI) method, in which the transmembrane potential and the variables of the ionic model are computed on the quadrature points and then used to evaluate the ionic current in the monodomain equation. The SVI method turns out to be more accurate than the ICI, since it does not approximate the nonlinear term with piecewise linear functions. However, the ICI method represents a reasonable trade-off between accuracy and computational cost; a detailed comparison between the two approaches can be found in [63]. Moreover, the ICI method better fits with the matrix formulation of the hyper-reduction techniques we have exploited, thus making the coupling between the monodomain equation and the ionic model more efficient when dealing with the reduced order model for electrophysiology. We also point out that when using coarse meshes, the L-ICI method underestimates the propagation velocity of the electrical signal. However, this problem can be solved by artificially increasing the electrical conductivities σ_f, σ_s and σ_n in the L-ICI method, as done in [72].

The nonlinear function $\mathbf{s}$ appearing in the ionic model is approximated as

$$\Big(\mathbf{s}\big(\sum_{i=1}^{N_h^e} v_i \varphi_i, \sum_{i=1}^{N_h^e} \mathbf{w}_i \circ \boldsymbol{\psi}_i{}^d\big), \boldsymbol{\psi}_j\Big) \approx \sum_{i=1}^{N_h^e} \mathbf{s}(v_i, \mathbf{w}_i)(\boldsymbol{\psi}_i{}^d, \boldsymbol{\psi}_j{}^d) = \mathbf{S}(\mathbf{V}, \mathbf{W}), \tag{21}$$

so that $\mathbf{S}(\mathbf{V}, \mathbf{W}) = \mathbf{M}_q^d \tilde{\mathbf{S}}(\mathbf{V}, \mathbf{W})$, where $\tilde{\mathbf{S}}(\mathbf{V}, \mathbf{W}) = (\mathbf{s}(v_1, \mathbf{w}_1), \ldots, \mathbf{s}(v_{N_h^e}, \mathbf{w}_{N_h^e}))^T$. Since the ionic variables can be computed at each node independently, the second equation of (19) yields to a system of N_h^e uncoupled ODEs, $\dot{\mathbf{W}} - \tilde{\mathbf{S}}(\mathbf{V}, \mathbf{W}) = 0$.

The spatial FE discretization of (19) thus leads to the following system:

$$\begin{cases} C_m \mathbf{M}\dot{\mathbf{V}} + \mathbf{K}\mathbf{V} + \mathbf{M}_L \tilde{\mathbf{I}}(\mathbf{V}, \mathbf{W}) = 0 \\ \dot{\mathbf{W}} - \tilde{\mathbf{S}}(\mathbf{V}, \mathbf{W}) = 0 \end{cases} \tag{22}$$

Let us now introduce a time discretization of (22), denoting by $\Delta t_e = T/n_t$ the time step, with $t_n = n\Delta t_e$, $n = 1, \ldots, n_t$; the superscript n denotes a quantity evaluated at time t_n. We adopt a forward Euler scheme to solve the ODEs representing the ionic model, with a sufficiently small Δt to preserve the stability of the method. To discretize in time the monodomain equation, we rely on a semi-implicit scheme; in particular, we treat the diffusion term implicitly and the reaction term explicitly. This strategy allows us to solve, at each time step, a linear system instead of a nonlinear one, as it would have been instead required by an implicit method.

We finally assume a weak coupling between all the electrophysiology fields, that is we consider an explicit algorithm to compute the solution of (22), solving sequentially two separate problems in order to compute the ionic variables $\mathbf{w}$ and the potential v. In conclusion, we obtain the following system: given $\mathbf{V}^0$, $\mathbf{W}^0$, for

$n = 0, \dots, n_t - 1$ compute

$$\begin{cases} \mathbf{W}^{n+1} = \mathbf{W}^n + \Delta t \tilde{\mathbf{S}}(\mathbf{V}^n, \mathbf{W}^n) \\ \left(\frac{C_m}{\Delta t}\mathbf{M}_L + \mathbf{K}\right)\mathbf{V}^{n+1} = \frac{C_m}{\Delta t}\mathbf{M}_L \mathbf{V}^n + \mathbf{M}_L \tilde{\mathbf{I}}(\mathbf{V}^n, \mathbf{W}^{n+1}), \end{cases} \tag{23}$$

where mass lumping is commonly used (note the presence of the matrix $\mathbf{M}_L$) to reduce possible oscillations near the wave-front [46, 67].

3.2 Activation Equation

Regarding the activation Eq. (14), we first perform a Taylor series expansion around $\gamma_f = 0$ of the quantity

$$\mathscr{F}(\gamma_f) = \frac{2\mathscr{I}_{4,f}}{(1+\gamma_f)^3} = \sum_{j=0}^{\infty} -1^j (j+1)(j+2)\mathscr{I}_{4,f}\gamma_f^j \tag{24}$$

appearing at the right-hand side of the ODE in (14). Since $\mathscr{F}(\mathbf{u}, 0) = 2\mathscr{I}_{4,f}|_{c=c_0}$, we can approximate the ODE under the form

$$\mu_A \dot{\gamma_f} = f_A(c) + \sum_{j=1}^{M} -1^j (j+1)(j+2)\mathscr{I}_{4,f}\gamma_f^j =: F(c, \gamma_f); \tag{25}$$

the choice $M = 5$ ensures that $||\mathscr{F}(\mathbf{u}, \gamma_f) - \sum_{j=1}^{M} -1^j (j+1)(j+2)\mathscr{I}_{4,f}\gamma_f^j|| < 0.005$, see [72]. Similarly to the discretization of the ionic model (21), the spatial semidiscretization reads

$$\mu_A \mathbf{M}_q \dot{\mathbf{G}}_f = \mathbf{F}(\mathbf{C}, \mathbf{G}_f) := \sum_{i=1}^{N_h^e} F(c_i, \gamma_{f_i})(\psi_i, \psi_j),$$

where linear finite elements have been chosen to approximate γ_f. Here $\mathbf{G}_f = (\gamma_{f_1}, \dots, \gamma_{f_N})^T \in \mathbb{R}^{N_h^e}$ and $\mathbf{C} \in \mathbb{R}^{N_h^e}$ denote the vectors of degrees of freedom (dofs) related to γ_f and to the calcium variable c, respectively. The fully discretized problem, obtained by adopting the forward Euler scheme for the time discretization, then reads as follows: given $\mathbf{G}_f^0$, for $n = 0, \dots, n_t - 1$ compute

$$\mathbf{M}_q \mathbf{G}_f^{n+1} = \mathbf{M}_q \mathbf{G}_f^n + \frac{\Delta t}{\mu_A}\mathbf{F}(\mathbf{C}^{n+1}, \mathbf{G}_f^n). \tag{26}$$

3.3 Mechanical Model

We now turn to the FE approximation of the mechanical subproblem (15), which is a fully nonlinear, quasi-static problem since it depends on time through the coupling with the electrical subproblem. We rely on the Newton method to solve, at a given time instant, such a nonlinear problem. Performing a Newton step at the continuous level around a generic displacement $\hat{\mathbf{u}}(t)$ yields a weak problem under the following form to be solved: find $\delta\mathbf{u}(t)$ such that

$$J_{\hat{\mathbf{u}}(t)}(\delta\mathbf{u}(t), \mathbf{v}) = -(R_{\hat{\mathbf{u}}(t)}, \mathbf{v}) \quad \forall \mathbf{v} \in V, \ t \in [0, T), \tag{27}$$

where

$$J_{\hat{\mathbf{u}}(t)}(\delta\mathbf{u}(t), \mathbf{v}) = \int_{\Omega_0} \left[\frac{\partial \mathbf{P}}{\partial \mathbf{F}}(\hat{\mathbf{u}}(t)) : \nabla\delta\mathbf{u}(t)\right] : \nabla\mathbf{v}d\Omega + \int_{\Gamma_R} \alpha\delta\mathbf{u}(t) \cdot \mathbf{v}d\sigma$$

and

$$(R_{\hat{\mathbf{u}}(t)}, \mathbf{v}) = \int_{\Omega_0} \mathbf{P}(\hat{\mathbf{u}}(t)) : \nabla\mathbf{v}d\Omega + \int_{\Gamma_R} \alpha\hat{\mathbf{u}}(t) \cdot \mathbf{v}d\sigma - \int_{\Gamma_N} \mathbf{g} \cdot \mathbf{v}d\sigma.$$

This problem arises after integrating (15) by parts over Ω_0 and linearizing the resulting problem; recall that $\mathbf{F}$ is the deformation gradient tensor defined in (7).

The FE approximation of problem (27) over a suitable triangulation $\mathcal{T}_m$ of the domain Ω_0 can be obtained in a straightforward way; in this work, we use linear (P1) finite elements to approximate the mechanical displacement, denoting by N_h^m the dimension of the FE space for the approximated displacement. We then obtain the following algebraic form of the Newton problem: for each t_n, $n = 0, \ldots, N_t - 1$, given $\mathbf{U}^{(0)}(t_n)$, for every $k \geq 1$ we search $\delta\mathbf{U}(t_n)$ satisfying

$$\begin{cases} \mathbf{J}(\mathbf{U}^{(k-1)}(t_n))\delta\mathbf{U}^{(k)}(t_n) = -\mathbf{R}(\mathbf{U}^{(k-1)}(t_n)), \\ \mathbf{U}^{(k)}(t_n) = \mathbf{U}^{(k-1)}(t_n) + \delta\mathbf{U}^{(k)}(t_n) \end{cases} \tag{28}$$

until $||\mathbf{R}(\mathbf{u}_h^k(t_n))||_2 < \varepsilon$, being $\varepsilon > 0$ a small fixed tolerance. Here, for any $\mathbf{U} \in \mathbb{R}^{N_h^m}$,

$$\begin{aligned} [\mathbf{J}(\mathbf{U}(t))]_{ij} &= J_{\mathbf{u}(t)}(\boldsymbol{\varphi}_j, \boldsymbol{\varphi}_i), \qquad i, j = 1, \ldots, N_h^m \\ [\mathbf{R}(\mathbf{U}(t))]_i &= (R_{\mathbf{u}(t)}, \boldsymbol{\varphi}_i), \qquad i = 1, \ldots, N_h^m \end{aligned} \tag{29}$$

are the components of the Jacobian $\mathbf{J}(\mathbf{U}) \in \mathbb{R}^{N_h^m \times N_h^m}$ and the residual $\mathbf{R}(\mathbf{U}) \in \mathbb{R}^{N_h^m}$ evaluated at $\mathbf{U} \in \mathbb{R}^{N_h^m}$; $\{\boldsymbol{\varphi}_i, i = 1, \ldots, N_h^m\}$ denote the (vector) basis functions of the FE space for the displacement and $\mathbf{U}(t) = (u_1(t), \ldots, u_{N_h^m}(t))^T$ the vector of

Algorithm 1 FOM for the (one-way) coupled electromechanical model

INPUT: v_0, $\mathbf{w}_0$, H_0 and $\mathbf{u}_0$
OUTPUT: $\mathbf{u}$

1: **for** $m = 0, \ldots, T/\Delta t_m$ **do**
2: **for** $n = mD, \ldots, (m+1)D - 1$ **do** solve
3:

$$\mathbf{W}^{n+1} = \mathbf{W}^n + \Delta t \tilde{\mathbf{S}}(\mathbf{V}^n, \mathbf{W}^n) \qquad \text{(ionic model)}$$

$$\left(\frac{C_m}{\Delta t}\mathbf{M}_L + \mathbf{K}\right)\mathbf{V}^{n+1} = \frac{C_m}{\Delta t}\mathbf{M}_L \mathbf{V}^n + \mathbf{M}_L \tilde{\mathbf{I}}(\mathbf{V}^n, \mathbf{W}^{n+1}) \qquad \text{(monodomain model)}$$

$$\mathbf{M}_q \mathbf{G}_f^{n+1} = \mathbf{M}_q \mathbf{G}_f^n + \frac{\Delta t}{\mu_A}\mathbf{F}(\mathbf{C}^{n+1}, \mathbf{G}_f^n) \qquad \text{(activation model (active strain))}$$

4: **end for**
5: Interpolate $\mathbf{G}_f^{n+1}$ on the mesh $\mathcal{T}_m$ using the RBF strategy
6: **while** $||\mathbf{R}(\mathbf{U}^{(k-1)}(t^{m+1}); \mathbf{G}_f^{n+1})||_{L^2} < \varepsilon$ **do** solve (Newton step)
7:

$$\begin{cases} \mathbf{J}(\mathbf{U}^{(k-1)}(t^{m+1}); \mathbf{G}_f^{n+1})\delta\mathbf{U}^{(k)}(t^{m+1}) = -\mathbf{R}(\mathbf{U}^{(k-1)}(t^{m+1}); \mathbf{G}_f^{n+1}), \\ \mathbf{U}^{(k)}(t^{m+1}) = \mathbf{U}^{(k-1)}(t^{m+1}) + \delta\mathbf{U}^{(k)}(t^{m+1}) \end{cases} \tag{30}$$

8: **end while**
9: **end for**

dofs of its high-fidelity approximation $\mathbf{u}(t)$—here denoted, with a slight abuse of notation, in the same way as the displacement at the continuous level.

A segregated algorithm is finally chosen for the solution of the electromechanical problem, in which the governing equations are solved sequentially—that is, segregated from one another; see Algorithm 1. This approach is shown to be appropriate when considering a model which is independent on the fibers stretch or which depends on the stretch but not on the stretch-rate [62]. In particular:

- we consider two different time steps for the electrical and the mechanical problem. The electrophysiology requires a significantly small time-step $\Delta t_e = T/n_t$ in order to correctly capture the propagation front of the electrical potential. Since the mechanical displacement is slower (of a factor ranging between 10 and 100) than the electrical signal propagation, for the mechanical model it is sufficient to consider a time step $\Delta t_m = T/N_t$ small enough to guarantee the convergence of the Newton method. In particular, we set them so that $\Delta t_m = D\Delta t_e$, that is, $n_t = DN_t$. Since we are discretizing the activation equation with the forward Euler method, we need to solve (26) using the time step Δt_e of the electrical model to guarantee the stability of the overall numerical scheme. Therefore, D time steps for the electrical subproblem (23) and the activation Eq. (26) are evaluated between two time instants in which the mechanical subproblem is solved;
- different meshes $\mathcal{T}_{ep}$, $\mathcal{T}_m$ are employed for the electrical and mechanical problems, respectively; the latter problem requires less mesh-size restrictions than the former and can thus be solved on a coarser mesh [69]. This implies the need of

transferring information between the two meshes: to impose the activation in the mechanical problem we need to evaluate the solution of the electrophysiology problem on $\mathscr{T}_m$, whereas the electromechanical feedback (which here is not considered) would require to evaluate the solution of the mechanical problem on $\mathscr{T}_{ep}$. This inter-grid transfer is performed by means of a rescaled localized radial basis functions (RBF) interpolation technique introduced in [32]; see also [9, 72].

4 Reduced Order Modeling Techniques

The reduced basis (RB) method allows to speed up the approximation of a parameter-dependent PDE in the case multiple evaluations of its solution are required for several values of the parameter $\boldsymbol{\mu} \in \mathscr{P}$. The basic idea of the RB method is to seek the solution of a problem in a subspace of much smaller dimension than the one, N_h, of the FOM space. During the offline stage, the parameter domain is explored, and a set of high-fidelity solutions (snapshots) is computed to generate a low dimensional RB space of dimension $N \ll N_h$. This space can be built by means of either a greedy algorithm (if suitable a posteriori error estimators are available and can be efficiently evaluated) or, in more general situations, proper orthogonal decomposition (POD) technique; see Sect. 4.1. Then, during the online phase, for each new value of $\boldsymbol{\mu}$, the RB approximation is rapidly computed by combining (possibly few) arrays stored offline, whose complexity must no longer depend on N_h.

The technique is well-established for linear PDEs (both of elliptic and parabolic type) showing an affine dependence on $\boldsymbol{\mu}$; see, e.g., [68] for an in-depth presentation of the methodology. Under the assumption of affine parametric dependence, the differential operators and data can be expressed as a linear combination of $\boldsymbol{\mu}$-independent forms (which can thus be precomputed) weighted by $\boldsymbol{\mu}$-dependent coefficients, which can be inexpensively evaluated. The RB method in its classical formulation, however, is no longer efficient when dealing with nonlinear (and/or nonaffinely parametrized) problems, unless we employ suitable hyper-reduction techniques to perform system approximation in addition to solution-space reduction; in the nonlinear case, these two operations shall be performed at the same time.

We first address the issue of solution-space reduction, which is achieved by means of the POD technique. We consider the case of a nonlinear stationary problem, a class which the mechanical subproblem fits in, for the sake of general exposition. We postpone the case of time-dependent nonlinear problems—albeit treated in a semi-implicit way, thus requiring the solution of a linear system at each time step—to Sect. 5, where we address the reduction of the monodomain equation. For the sake of exposition, we formulate everything in a purely algebraic form.

4.1 POD-Galerkin Method

Let us consider the following, abstract, nonlinear $\boldsymbol{\mu}$-dependent algebraic system

$$\mathbf{R}(\mathbf{U}(\boldsymbol{\mu});\boldsymbol{\mu}) = \mathbf{0}. \tag{31}$$

and the associated Newton method: given $\mathbf{U}^{(0)}(\boldsymbol{\mu}) \in \mathbb{R}^{N_h}$, for $k \geq 1$, find $\delta\mathbf{U}(\boldsymbol{\mu}) \in \mathbb{R}^{N_h}$ s.t.

$$\begin{cases} \mathbf{J}(\mathbf{U}^{(k-1)}(\boldsymbol{\mu});\boldsymbol{\mu})\delta\mathbf{U}^{(k)}(\boldsymbol{\mu}) = -\mathbf{R}(\mathbf{U}^{(k-1)}(\boldsymbol{\mu});\boldsymbol{\mu}), \\ \mathbf{U}^{(k)}(\boldsymbol{\mu}) = \mathbf{U}^{(k-1)}(\boldsymbol{\mu}) + \delta\mathbf{U}^{(k)}(\boldsymbol{\mu}) \end{cases} \tag{32}$$

and iterate until $||\mathbf{R}(\mathbf{U}^{(k)}(\boldsymbol{\mu});\boldsymbol{\mu})||_2 < \varepsilon$, being $\varepsilon > 0$ a small, given tolerance. As before, $\mathbf{J}(\mathbf{U}(\boldsymbol{\mu});\boldsymbol{\mu}) \in \mathbb{R}^{N_h \times N_h}$ denotes the $\boldsymbol{\mu}$-dependent Jacobian matrix (with linearization around $\mathbf{U}(\boldsymbol{\mu})$) and $\mathbf{R}(\mathbf{U}(\boldsymbol{\mu});\boldsymbol{\mu}) \in \mathbb{R}^{N_h}$ the $\boldsymbol{\mu}$-dependent residual vector.

In the case of mechanical problems characterized by complex nonlinear constitutive laws, the computational burden in solving $(32)_1$ is represented by the assembling of the Jacobian matrix, which can consume almost the entire CPU time required by each Newton step. To reduce the computational complexity of problem (32), the RB method seeks, for any $\boldsymbol{\mu} \in \mathscr{P}$, an approximation of $\mathbf{U}^{(k)}(\boldsymbol{\mu})$ given by a linear combination of (possibly few) basis functions,

$$\mathbf{U}^{(k)}(\boldsymbol{\mu}) \approx \mathbf{Z}\mathbf{u}_N^{(k)}(\boldsymbol{\mu}), \quad \forall k \geq 1,\ N \ll N_h, \tag{33}$$

where $\mathbf{u}_N^{(k)}(\boldsymbol{\mu}) \in \mathbb{R}^N$ and $\mathbf{Z} \in \mathbb{R}^{N_h \times N}$ is a matrix whose $N \ll N_h$ columns contained the nodal values of the RB functions. Problem (32) is then replaced by the following: given $\mathbf{u}_N^{(0)}(\boldsymbol{\mu}) \in \mathbb{R}^N$, for $k \geq 1$, find $\delta\mathbf{u}_N \in \mathbb{R}^N$ s.t.

$$\begin{cases} \mathbf{Z}^T\mathbf{J}(\mathbf{Z}\mathbf{u}_N^{(k-1)}(\boldsymbol{\mu});\boldsymbol{\mu})\mathbf{Z}\delta\mathbf{u}_N(\boldsymbol{\mu}) = -\mathbf{Z}^T\mathbf{R}(\mathbf{Z}\mathbf{u}_N^{(k-1)}(\boldsymbol{\mu});\boldsymbol{\mu}), \\ \mathbf{u}_N^{(k)}(\boldsymbol{\mu}) = \mathbf{u}_N^{(k-1)}(\boldsymbol{\mu}) + \delta\mathbf{u}_N(\boldsymbol{\mu}), \end{cases} \tag{34}$$

and iterate until $||\mathbf{Z}^T\mathbf{R}(\mathbf{Z}\mathbf{u}_N^{(k)}(\boldsymbol{\mu});\boldsymbol{\mu})||_2 < \varepsilon_{RB}$, being $\varepsilon_{RB} > 0$ a small, given tolerance. If (34) converges, $\mathbf{Z}\mathbf{u}_N(\boldsymbol{\mu})$ can be regarded as an approximation of $\mathbf{U}(\boldsymbol{\mu})$ in the RB space, with $\mathbf{u}_N = \lim_{k\to\infty}\mathbf{u}_N^{(k)}$. Problem $(34)_1$ is obtained by requiring that the Galerkin projection over V_N of the FOM residual computed on the ansatz (33) vanishes, where V_N has to be intended as the space spanned by the columns of $\mathbf{Z}$.

The POD technique—through the so-called method of snapshots—can be used to compute the reduced basis $\mathbf{Z}$ (and, as we shall see in Sect. 4.2, for the construction of both DEIM and MDEIM bases). In the case of a stationary, $\boldsymbol{\mu}$-dependent nonlinear problem, POD performs the singular value decomposition of a matrix

$$\mathbf{S} = [\mathbf{U}^{(1)}(\boldsymbol{\mu}_1) \quad \mathbf{U}^{(2)}(\boldsymbol{\mu}_1) \quad \dots \quad \mathbf{U}^{(1)}(\boldsymbol{\mu}_{n_s}) \quad \mathbf{U}^{(2)}(\boldsymbol{\mu}_{n_s}) \quad \dots]$$

of snapshots of the high-fidelity problem and returns an orthonormal basis of the RB space made by the first N right singular vectors of S. Here snapshots are represented by Newton steps obtained for n_s parameter vectors $\boldsymbol{\mu}_i \in \mathscr{D}$, $i = 1, \dots, n_s$, randomly sampled over $\mathscr{P}$; more ad-hoc strategies, such as e.g. *latin hypercube sampling* or *sparse grid*, could be exploited especially for high-dimensional parameter spaces.

Hence, from the factorization $\mathbf{S} = \mathbf{Z}\boldsymbol{\Sigma}\boldsymbol{\Lambda}^T$, where $\boldsymbol{\Lambda} = diag(\sigma_1, \sigma_2, \dots)$ is the matrix of singular values of $\mathbf{S}$, the POD basis $\mathbf{Z}_N$ of dimension $N \leq n_s$ is obtained by collecting the first N columns of $\mathbf{Z}$ (i.e. the first N left singular vectors), corresponding to the first N (largest) singular values; we can set the basis dimension N as the minimum integer such that

$$\frac{\sum_{i=1}^{N} \sigma_i^2}{\sum_{i=1}^{n_s} \sigma_i^2} \geq 1 - \varepsilon_{POD},$$

given a suitable, small tolerance $\varepsilon_{POD} > 0$. The reduced basis provided by POD is optimal, in the sense that it minimizes the sum of the squared distances between each snapshot and the corresponding projection onto the subspace. When denoting by $\mathbf{Z}$ the POD basis, its dimension N will be understood.

4.2 Hyper-Reduction Techniques

When using Newton iterations to solve nonlinear problems, assembling the ROM for any new parameter would require to assemble (also in the online phase) the FOM arrays first and then to project them onto the reduced space, thus calling into play high-fidelity arrays at the online stage, too. This issue is even more relevant when dealing with fully nonlinear problems, for which the global Jacobian matrix has to be entirely reassembled at each Newton step.

Hyper-reduction (or system approximation) techniques aim to recover an approximate affine structure of nonlinear terms to guarantee an efficient offline-online decomposition. The archetypical hyper-reduction technique introduced in the RB framework is the empirical interpolation method (EIM), developed in [10, 49] to approximate nonaffinely parametrized functions. Its discrete variant, the so-called Discrete EIM (DEIM), was originally proposed in [21] to efficiently deal with nonlinear problems, but has also been applied to nonaffinely parametrized linear operators. More recently, a *matrix version* of DEIM (MDEIM) has been developed [19, 87], and then further explored in [55] to approximate the full-order parametrized operators in a purely algebraic way.

4.2.1 DEIM for Residual Approximation

For the problem at hand, at each Newton step, DEIM [21] allows to efficiently express the residual vector as a linear combination of (possibly few) $\boldsymbol{\mu}$-independent terms so that the $\boldsymbol{\mu}$-dependent weights of this combination can be efficiently computed by solving an interpolation problem. In particular, we project the residual vector $\mathbf{R}(\mathbf{Z}\mathbf{u}_N^{(k)}(\boldsymbol{\mu}); \boldsymbol{\mu})$ onto a low-dimensional subspace spanned by a basis $\boldsymbol{\Phi}_R \in \mathbb{R}^{N_h \times m_R}$ such that, $\forall k \geq 1$

$$\mathbf{R}(\mathbf{Z}\mathbf{u}_N^{(k)}(\boldsymbol{\mu}); \boldsymbol{\mu}) \approx \mathbf{R}_m(\mathbf{Z}\mathbf{u}_N^{(k)}(\boldsymbol{\mu}); \boldsymbol{\mu}) = \boldsymbol{\Phi}_R \boldsymbol{\theta}_R(\mathbf{Z}\mathbf{u}_N^{(k)}(\boldsymbol{\mu}), \boldsymbol{\mu}) \tag{35}$$

where $\boldsymbol{\theta}_R(\mathbf{Z}\mathbf{u}_N^{(k)}(\boldsymbol{\mu}), \boldsymbol{\mu}) \in \mathbb{R}^{m_R}$ is a coefficient vector to be determined. In particular:

- the basis $\boldsymbol{\Phi}_R$ can be computed (once for all) by performing POD on a set of snapshots

$$\mathbf{S}_R = \{\mathbf{R}(\mathbf{Z}\mathbf{u}_N^{(k)}(\boldsymbol{\mu}_i); \boldsymbol{\mu}_i) \quad i = 1, \ldots, n_s, k = 1, \ldots\}.$$

  To obtain the residual snapshots $\mathbf{R}(\mathbf{Z}\mathbf{u}_N^{(k)}(\boldsymbol{\mu}_i); \boldsymbol{\mu}_i)$, we need to solve the reduced problem (34) for different values of $\boldsymbol{\mu}$ and, at each Newton iteration, to store the computed residual vectors;
- the coefficient vector $\boldsymbol{\theta}_R(\mathbf{Z}\mathbf{u}_N^{(k)}(\boldsymbol{\mu}), \boldsymbol{\mu})$ can be evaluated for each new value of $\boldsymbol{\mu}$ by imposing m_R interpolation constraints on a subset $\wp = [\wp_1, \ldots, \wp_{m_R}]$ of entries of $\mathbf{R}(\mathbf{Z}\mathbf{u}_N^{(k)}(\boldsymbol{\mu}); \boldsymbol{\mu})$ (the so-called magic points, introduced in [49]), selected by the DEIM algorithm, see Algorithm 2. For ease of notation, we introduce the matrix

$$\mathbf{P} = [\mathbf{e}_{\wp_1}, \cdots, \mathbf{e}_{\wp_{m_R}}] \in \mathbb{R}^{N_h \times m_R}, \tag{36}$$

  where $\mathbf{e}_{\wp_i} = [0, \cdots, 0, 1, 0, \cdots, 0]^T \in \mathbb{R}^{N_h}$ is the $\wp_i$-th column of the identity matrix $I \in \mathbb{R}^{N_h \times N_h}$, for $i = 1, \cdots, m_R$. The coefficient vector $\boldsymbol{\theta}_R(\mathbf{Z}\mathbf{u}_N^{(k)}(\boldsymbol{\mu}), \boldsymbol{\mu})$

Algorithm 2 DEIM algorithm (as originally proposed in [21])

INPUT: $\boldsymbol{\Phi} = [\boldsymbol{\phi}_1, \ldots, \boldsymbol{\phi}_m] \in \mathbb{R}^{N_h \times m}$ made by linearly independent columns
OUTPUT: $\wp = [\wp_1, \ldots, \wp_m] \in \mathbb{R}^m$

```
1: ℘_1 = maxpos{φ_1}
2: Φ = [φ_1], P = [e_℘1]
3: for k = 2, ..., m do
4:     Solve (P^T Φ)c = (P^T φ_k)
5:     r = φ_k − Φc
6:     ℘_k = maxpos{r}
7:     Φ ← [Φ  φ_k], P ← [P  e_℘k]
8: end for
```

is then obtained as the solution of

$$\mathbf{P}^T \boldsymbol{\Phi}_R \boldsymbol{\theta}_R(\mathbf{Z}\mathbf{u}_N^{(k)}(\boldsymbol{\mu}), \boldsymbol{\mu}) = \mathbf{P}^T \mathbf{R}(\mathbf{Z}\mathbf{u}_N^{(k)}(\boldsymbol{\mu}); \boldsymbol{\mu});$$

$\mathbf{P}^T \boldsymbol{\Phi}_R$ and $\mathbf{P}^T \mathbf{R}(\mathbf{Z}\mathbf{u}_N^{(k)}(\boldsymbol{\mu}); \boldsymbol{\mu})$ are the restrictions of $\boldsymbol{\Phi}_R$ and $\mathbf{R}(\mathbf{Z}\mathbf{u}_N^{(k)}(\boldsymbol{\mu}); \boldsymbol{\mu})$ to the subset of indices $\wp$, respectively.

The approximation of the reduced residual vector in (34) can be obtained by projecting (35) onto the reduced space yielding

$$\mathbf{Z}^T \mathbf{R}(\mathbf{Z}\mathbf{u}_N^{(k)}(\boldsymbol{\mu}); \boldsymbol{\mu}) \approx \mathbf{Z}^T \boldsymbol{\Phi}_R (\mathbf{P}^T \boldsymbol{\Phi}_R)^{-1} \mathbf{P}^T \mathbf{R}(\mathbf{Z}\mathbf{u}_N^{(k)}(\boldsymbol{\mu}); \boldsymbol{\mu}) := \mathbf{R}_{N,m}(\mathbf{Z}\mathbf{u}_N^{(k)}(\boldsymbol{\mu}); \boldsymbol{\mu}). \tag{37}$$

All the quantities appearing in (37) which do not depend on $\boldsymbol{\mu}$ can be precomputed offline; in the online stage we only need to assemble $\mathbf{P}^T \mathbf{R}(\mathbf{Z}\mathbf{u}_N^{(k)}(\boldsymbol{\mu}); \boldsymbol{\mu})$, which is the restriction of the residual to the subset of DEIM nodes. In the FE context, this restriction can be computed by simply integrating the residual only on the quadrature points belonging to those mesh elements which provide a non-zero contribution to the entries $\wp$; this set of elements is usually referred to as *reduced mesh* [18].

4.3 Jacobian Approximation

An affine approximation of the reduced Jacobian matrix $\mathbf{J}_N(\mathbf{Z}\mathbf{u}_N^{(k)}(\boldsymbol{\mu}); \boldsymbol{\mu})$ can be obtained by relying on either the DEIM algorithm or a MDEIM alternative technique.

The classical DEIM approach to tackle nonlinear problems (see e.g. [11, 22, 45, 64]) computes the reduced Jacobian $\mathbf{J}_N(\mathbf{Z}\mathbf{w}_N(\boldsymbol{\mu}); \boldsymbol{\mu})$, for any $\mathbf{w}_N \in \mathbb{R}^N$, as the derivative of the reduced approximated residual vector (i.e. the right-hand side of (37)),

$$\mathbf{J}_{N,m}(\mathbf{Z}\mathbf{w}_N(\boldsymbol{\mu}); \boldsymbol{\mu}) = \frac{\partial \mathbf{R}_{N,m}(\mathbf{Z}\mathbf{w}_N(\boldsymbol{\mu}); \boldsymbol{\mu})}{\partial \mathbf{w}_N} = \mathbf{Z}^T \boldsymbol{\Phi}_R (\mathbf{P}^T \boldsymbol{\Phi}_R)^{-1} \mathbf{P}^T \mathbf{J}(\mathbf{Z}\mathbf{w}_N(\boldsymbol{\mu}); \boldsymbol{\mu}) \mathbf{Z}. \tag{38}$$

As for the residual vector, we can precompute the $\boldsymbol{\mu}$-independent quantities offline, while online we have to assemble $\mathbf{P}^T \mathbf{J}(\mathbf{Z}\mathbf{w}_N(\boldsymbol{\mu}); \boldsymbol{\mu}) \in \mathbb{R}^{m_R \times N_h}$, that is the restriction of the Jacobian matrix to the rows which correspond to the indices in $\wp$. Consequently, we need to assemble online, at each Newton step, a matrix of dimension $m_R \times N_h$, which still depends on the dimension N_h of the FE problem, which is unfeasible when m_R becomes large. Note that since the reduced Jacobian matrix is obtained as the derivative of the reduced residual, DEIM yields the application of the *exact* Newton method (i.e. with the exact reduced Jacobian matrix) although on an approximated version of problem (34).

When a large number m_R of DEIM terms is obtained, a matrix version of DEIM (MDEIM) can be employed to perform hyper-reduction of the Jacobian matrices arising in (34). The idea is to directly approximate the reduced Jacobian $\mathbf{Z}^T\mathbf{J}(\mathbf{Z}\mathbf{u}_N^{(k)}(\boldsymbol{\mu});\boldsymbol{\mu})\mathbf{Z}$ by relying on a different basis than the one used for the residual. This yields a quasi-Newton method since the reduced Jacobian matrix is not the exact derivative of the reduced residual; nevertheless, since the $\boldsymbol{\mu}$-dependent Jacobian matrix usually varies in a significantly smaller range compared to the residual vector, few (much less than m_R) terms are required.

MDEIM provides an approximation of the Jacobian matrix $\mathbf{J}(\mathbf{Z}\mathbf{u}_N^{(k)}(\boldsymbol{\mu});\boldsymbol{\mu}) \in \mathbb{R}^{N_h\times N_h}$ under the form

$$\mathbf{J}(\mathbf{Z}\mathbf{u}_N^{(k)}(\boldsymbol{\mu});\boldsymbol{\mu}) \approx \mathbf{J}_m(\mathbf{Z}\mathbf{u}_N^{(k)}(\boldsymbol{\mu});\boldsymbol{\mu}) = \sum_{i=1}^{m_J} \theta_J^i(\boldsymbol{\mu})\mathbf{J}^i, \tag{39}$$

being $\{\mathbf{J}^i \in \mathbb{R}^{N_h\times N_h},\ i = 1,\dots,m_J\}$ a set of $\boldsymbol{\mu}$-independent matrices that can be computed once for all and $\boldsymbol{\theta}_J(\boldsymbol{\mu}) = (\theta_J^1(\boldsymbol{\mu}),\cdots,\theta_J^{m_J}(\boldsymbol{\mu}))^T$ a coefficient vector. This approximation is obtained by defining $j(\mathbf{Z}\mathbf{u}_N^{(k)}(\boldsymbol{\mu});\boldsymbol{\mu}) = \mathrm{vec}(\mathbf{J}(\mathbf{Z}\mathbf{u}_N^{(k)}(\boldsymbol{\mu});\boldsymbol{\mu}) \in \mathbb{R}^{N_h^2}$ as the vector obtained by stacking all the columns of $\mathbf{J}(\mathbf{Z}\mathbf{u}_N^{(k)}(\boldsymbol{\mu});\boldsymbol{\mu})$, and approximating $j(\mathbf{Z}\mathbf{u}_N^{(k)}(\boldsymbol{\mu});\boldsymbol{\mu})$ by its DEIM counterpart

$$j(\mathbf{Z}\mathbf{u}_N^{(k)}(\boldsymbol{\mu});\boldsymbol{\mu}) \approx j_m(\mathbf{Z}\mathbf{u}_N^{(k)}(\boldsymbol{\mu});\boldsymbol{\mu}) = \boldsymbol{\Phi}_J\boldsymbol{\theta}_J(\boldsymbol{\mu}), \quad \boldsymbol{\Phi}_J = (\boldsymbol{\phi}_1,\dots,\boldsymbol{\phi}_n)^T \in \mathbb{R}^{N_h^2\times m_J}.$$

Then, the matrices $\mathbf{J}^i$ can be computed transforming each column $\boldsymbol{\phi}_i \in \mathbb{R}^{N_h^2}$ of $\boldsymbol{\Phi}_J$ into a matrix $\mathbf{J}^i \in \mathbb{R}^{N_h\times N_h}$ by reverting the vec operation, as $\mathbf{J}^i = \mathrm{vec}^{-1}(\boldsymbol{\phi}_i)$, so that $\mathbf{J}_m(\mathbf{Z}\mathbf{u}_N^{(k)}(\boldsymbol{\mu});\boldsymbol{\mu}) = \mathrm{vec}^{-1}(j_m(\mathbf{Z}\mathbf{u}_N^{(k)}(\boldsymbol{\mu});\boldsymbol{\mu}))$. The basis $\boldsymbol{\Phi}_J$ and the coefficient vector $\boldsymbol{\theta}_J(\boldsymbol{\mu})$ are determined following the same procedure used for the residual vectors, relying on a set of snapshots $\mathbf{S}_J = \{\mathbf{J}(\mathbf{Z}\mathbf{u}_N^{(k)}(\boldsymbol{\mu}_i);\boldsymbol{\mu}_i),\ i = 1,\dots,n_s,\ k = 1,\dots\}$, evaluated on the reduced solution. Finally, the reduced Jacobian matrix in (34) can be approximated as

$$\mathbf{Z}^T\mathbf{J}(\mathbf{Z}\mathbf{u}_N^{(k)}(\boldsymbol{\mu});\boldsymbol{\mu})\mathbf{Z} \approx \sum_{i=1}^{m_J} \theta_J^i(\boldsymbol{\mu})\mathbf{Z}^T\mathbf{J}^i\mathbf{Z}. \tag{40}$$

4.3.1 Efficient Assembling on a Reduced Mesh

By using DEIM and MDEIM as above, the POD solution-space reduction is first performed, while the bases associated to the DEIM and MDEIM approximations of **R** and **J** are computed at a later time. The major drawback of this strategy is that problem (34) must be solved n_s times. To avoid this, we can rely on an *intermediate*

Algorithm 3 ROM construction (stationary, nonlinear problem)

INPUT: n_s combinations of parameters $\{\boldsymbol{\mu}_1, \ldots, \boldsymbol{\mu}_{n_s}\}$
OUTPUT: $\mathbf{Z}, \boldsymbol{\Phi}_J, \boldsymbol{\Phi}_R$

```
for i = 1, ..., n_s do
    Solve problem (32) for μ_i
    At each Newton iteration k:
        S_U ← [S_U   U^(k)(μ_i)],        S_J ← [S_J   vec(J(U^(k−1)(μ_i), μ_i))]
end for
Z = POD(S_U; ε), Φ_J = POD(S_J; ε_J)
for i = 1, ..., n_s do
    Solve problem (41) for μ_i
    At each Newton iteration k:
        S_R ← [S_R   R(Zu_N^(k−1)(μ_i), μ_i)]
end for
Φ_R = POD(U_R; ε_R).
```

problem where the Jacobian matrix is replaced by its MDEIM approximation and the residual is exact: given $\mathbf{u}_{N,m}^{(0)} \in \mathbb{R}^N$, at each Newton step we search $\delta\mathbf{u}_{N,m}^{(k)} \in \mathbb{R}^N$, $k \geq 1$ satisfying

$$\begin{cases} \displaystyle\sum_{i=1}^{m_J} \theta_i^J(\boldsymbol{\mu})\mathbf{Z}^T\mathbf{J}^i\mathbf{Z}\,\delta\mathbf{u}_{N,m}(\boldsymbol{\mu}) = -\mathbf{Z}^T\mathbf{R}(\mathbf{Z}\mathbf{u}_{N,m}^{(k-1)}(\boldsymbol{\mu})) \\ \mathbf{u}_{N,m}^{(k)}(\boldsymbol{\mu}) = \mathbf{u}_{N,m}^{(k-1)}(\boldsymbol{\mu}) + \delta\mathbf{u}_{N,m}(\boldsymbol{\mu}) \end{cases} \tag{41}$$

and iterate until $||\mathbf{Z}^T\mathbf{R}(\mathbf{Z}\mathbf{u}_N^{(k)}(\boldsymbol{\mu}); \boldsymbol{\mu})||_2 < \varepsilon_{RB}$. Solving this problem is significantly faster than solving problem (34), since $\mathbf{J}$ is assembled only onto the reduced mesh. Moreover, problem (41) is very fast to solve since it requires almost the same effort of the full hyper-reduced problem. The complete procedure for the ROM construction is reported in Algorithm 3.

5 RB Methods for Cardiac Electrophysiology

We now apply the techniques introduced in the previous section to the electrical problem. The parametrized version of problem (23) reads: given $\boldsymbol{\mu} \in \mathscr{D}$, $\mathbf{V}^0(\boldsymbol{\mu})$ and $\mathbf{W}^0(\boldsymbol{\mu})$, for each $n = 0, \ldots, n_t - 1$ solve

$$\begin{cases} \mathbf{W}^{n+1}(\boldsymbol{\mu}) = \mathbf{W}^n(\boldsymbol{\mu}) + \Delta t\tilde{\mathbf{S}}(\mathbf{V}^n(\boldsymbol{\mu}), \mathbf{W}^n(\boldsymbol{\mu})) \\ \left(\dfrac{C_m}{\Delta t}\mathbf{M}_L + \mathbf{K}(\boldsymbol{\mu})\right)\mathbf{V}^{n+1}(\boldsymbol{\mu}) = \dfrac{C_m}{\Delta t}\mathbf{M}_L\mathbf{V}^n(\boldsymbol{\mu}) + \mathbf{M}_L\tilde{\mathbf{I}}(\mathbf{V}^n(\boldsymbol{\mu}), \mathbf{W}^{n+1}(\boldsymbol{\mu})). \end{cases} \tag{42}$$

A ROM for the monodomain Eq. $(42)_1$, exploits the POD technique for the construction of the reduced space, and on the DEIM technique to approximate the nonlinear terms. Moreover, we exploit the MDEIM technique to efficiently recover an affine decomposition for the $\boldsymbol{\mu}$-dependent stiffness matrix, and use the DEIM technique to reduce the computational cost associated to the ionic model $(42)_2$.

We consider a vector of physical parameters $\boldsymbol{\mu} \in \mathscr{P}$ as in Eq. (16) affecting the conductivity tensor $\mathbf{D} = \mathbf{D}(\boldsymbol{\mu})$ (and possibly the initial data); this yields a $\boldsymbol{\mu}$-dependent stiffness matrix $\mathbf{K}(\boldsymbol{\mu}) \in \mathbb{R}^{N_h^e \times N_h^e}$; in the case we considered a geometrical parametrization of the domain where the problem is set, also the mass matrix $\mathbf{M}_L$ would be $\boldsymbol{\mu}$-dependent. Moreover, for the sake of simplicity we do not consider the case of parameters affecting the ionic terms; see [59] for more about this subject.

5.1 *Monodomain Equation*

We rely on the POD-Galerkin method recalled in Sect. 4.1: for each value of $\boldsymbol{\mu} \in \mathscr{P}$, we approximate the FE discretization of the potential at time t_n as

$$\mathbf{V}^n(\boldsymbol{\mu}) \approx \mathbf{Z}_e \mathbf{V}^n_{N^e}(\boldsymbol{\mu}), \quad n = 0, \dots, n_t, \tag{43}$$

where $\mathbf{V}_{N^e}(\boldsymbol{\mu}) \in \mathbb{R}^{N^e}$ denotes the reduced transmembrane potential and $\mathbf{Z}_e$ denotes the matrix whose columns span the RB space for the monodomain equation. By substituting (43) in $(23)_1$ and projecting the resulting residual onto the reduced space spanned by the columns of $\mathbf{Z}_e$, we obtain the following problem: given $\boldsymbol{\mu} \in \mathscr{P}$ and $\mathbf{V}^0_{N^e}(\boldsymbol{\mu})$, for each $n = 0, \dots, n_t - 1$ solve

$$\begin{aligned}\mathbf{Z}_e^T\Big(\frac{C_m}{\Delta t}\mathbf{M}_L + \mathbf{K}(\boldsymbol{\mu})\Big)\mathbf{Z}_e\mathbf{V}^{n+1}_{N^e}(\boldsymbol{\mu}) &= \frac{C_m}{\Delta t}\mathbf{Z}_e^T\mathbf{M}_L\mathbf{Z}_e\mathbf{V}^n_{N^e}(\boldsymbol{\mu}) \\ &\quad + \mathbf{Z}_e^T\mathbf{M}_L\tilde{\mathbf{I}}(\mathbf{Z}_e\mathbf{V}^n_{N^e}(\boldsymbol{\mu}), \mathbf{W}^{n+1}(\boldsymbol{\mu})).\end{aligned} \tag{44}$$

Since the matrix $\mathbf{Z}_e^T\mathbf{K}(\boldsymbol{\mu})\mathbf{Z}_e$ depends on $\boldsymbol{\mu}$, it has to be reassembled for each new value of $\boldsymbol{\mu}$; to avoid this, we replace $\mathbf{K}(\boldsymbol{\mu})$ by an approximate affine expansion

$$\mathbf{K}(\boldsymbol{\mu}) \approx \sum_{i=1}^{m_K} \theta_K^i(\boldsymbol{\mu})\mathbf{K}^i \tag{45}$$

obtained through the MDEIM technique. Here $\{\mathbf{K}^i \in \mathbb{R}^{N_h^e \times N_h^e}, i = 1, \dots, m_K\}$ are a set of $\boldsymbol{\mu}$-independent matrices that can be computed once for all and $\boldsymbol{\theta}_K(\boldsymbol{\mu}) = [\theta_K^1(\boldsymbol{\mu}), \cdots, \theta_K^{m_K}(\boldsymbol{\mu})]$ a vector of coefficients to be evaluated for any $\boldsymbol{\mu} \in \mathscr{P}$. In this way, no matter which kind of $\boldsymbol{\mu}$-dependence is considered in the expression of the diffusion tensor (6), the ROM stiffness matrix can be assembled using a set of precomputed quantities; in particular, if $\boldsymbol{\mu}$ also includes the maximum fibers rotation angle θ_{max}, the vector defining the fiber direction $\mathbf{f}_0(\boldsymbol{\mu})$ cannot be written under affine form; see [13]. We point out that $\mathbf{K}(\boldsymbol{\mu})$ is time-independent, hence it has to be assembled only once for each value of $\boldsymbol{\mu}$; in particular, during the online stage, we can compute $\boldsymbol{\theta}_K(\boldsymbol{\mu})$ and the corresponding approximation (45) of $\mathbf{K}(\boldsymbol{\mu})$ when selecting a new $\boldsymbol{\mu}$, and then use the computed quantities for all time instants.

Using approximation (45), problem (44) becomes: given $\boldsymbol{\mu} \in \mathscr{P}$ and $\mathbf{V}^0_{N^e}(\boldsymbol{\mu})$, for each $n = 0, \ldots, n_t - 1$ solve

$$\left[\frac{C_m}{\Delta t}\mathbf{Z}_e^T\mathbf{M}_L\mathbf{Z}_e + \sum_{i=1}^{m_K}\theta_K^i(\boldsymbol{\mu})\mathbf{Z}_e^T\mathbf{K}_i\mathbf{Z}_e\right]\mathbf{V}^{n+1}_{N^e}(\boldsymbol{\mu}) = \frac{C_m}{\Delta t}\mathbf{Z}_e^T\mathbf{M}_L\mathbf{Z}_e\mathbf{V}^n_{N^e}(\boldsymbol{\mu}) + \mathbf{Z}_e^T\mathbf{M}_L\tilde{\mathbf{I}}(\mathbf{Z}_e\mathbf{V}^n_{N^e}(\boldsymbol{\mu}), \mathbf{W}^{n+1}(\boldsymbol{\mu})). \tag{46}$$

Since the term describing the ionic current is nonlinear, we replace it by its DEIM approximation. We introduce a basis $\boldsymbol{\Phi}_I \in \mathbb{R}^{N_h^e \times m_I}$, and express

$$\tilde{\mathbf{I}}(\mathbf{Z}_e\mathbf{V}^n_{N^e}(\boldsymbol{\mu}), \mathbf{W}^{n+1}(\boldsymbol{\mu})) \approx \boldsymbol{\Phi}_I\boldsymbol{\theta}_I(\mathbf{V}^n_{N^e}(\boldsymbol{\mu}), \mathbf{W}^{n+1}(\boldsymbol{\mu})). \tag{47}$$

The basis $\boldsymbol{\Phi}_I \in \mathbb{R}^{N_h^e \times m_I}$ has to be precomputed during the offline phase by performing a POD on a set of snapshots $\{\tilde{\mathbf{I}}(\mathbf{Z}_e\mathbf{V}^n_{N^e}(\boldsymbol{\mu}_i), \mathbf{W}^{n+1}(\boldsymbol{\mu}_i)), i = 1, \ldots, n_s\}$; instead, the coefficient vector $\boldsymbol{\theta}_I(\mathbf{V}^n_{N^e}(\boldsymbol{\mu}), \mathbf{W}^{n+1}(\boldsymbol{\mu})) \in \mathbb{R}^{m_I}$ has to be computed online, by solving the interpolation problem

$$\mathbf{P}_I^T\boldsymbol{\Phi}_I\boldsymbol{\theta}_I(\mathbf{V}^n_{N^e}(\boldsymbol{\mu}), \mathbf{W}^{n+1}(\boldsymbol{\mu})) = \mathbf{P}_I^T\tilde{\mathbf{I}}(\mathbf{Z}_e\mathbf{V}^n_{N^e}(\boldsymbol{\mu}), \mathbf{W}^{n+1}(\boldsymbol{\mu})), \tag{48}$$

where $\mathbf{P}_I^T = [\mathbf{e}_{\wp_{I,1}}, \cdots, \mathbf{e}_{\wp_{I,m_R}}]^T \in \mathbb{R}^{N_h^e \times m_I}$ is the restriction matrix to the set of DEIM indices $\boldsymbol{\wp}_I$ defined in (36). The DEIM approximation of (47) is thus

$$\tilde{\mathbf{I}}(\mathbf{Z}_e\mathbf{V}^n_{N^e}(\boldsymbol{\mu}), \mathbf{W}^{n+1}(\boldsymbol{\mu})) \approx \boldsymbol{\Phi}_I(\mathbf{P}_I^T\boldsymbol{\Phi}_I)^{-1}\mathbf{P}_I^T\tilde{\mathbf{I}}(\mathbf{Z}_e\mathbf{V}^n_{N^e}(\boldsymbol{\mu}), \mathbf{W}^{n+1}(\boldsymbol{\mu})). \tag{49}$$

To obtain the second term of the right hand side of (44) we can simply project (49) onto the reduced space spanned by $\mathbf{Z}_e$, thus getting

$$\mathbf{Z}_e^T\mathbf{M}_L\tilde{\mathbf{I}}(\mathbf{V}^n_{N^e}(\boldsymbol{\mu}), \mathbf{W}^{n+1}(\boldsymbol{\mu})) \approx \mathbf{Z}_e^T\mathbf{M}_L\boldsymbol{\Phi}_I(\mathbf{P}_I^T\boldsymbol{\Phi}_I)^{-1}\mathbf{P}_I^T\tilde{\mathbf{I}}(\mathbf{Z}_e\mathbf{V}^n_{N^e}(\boldsymbol{\mu}), \mathbf{W}^{n+1}(\boldsymbol{\mu})). \tag{50}$$

We point out that the matrices $\mathbf{Z}_e^T\mathbf{M}_L\boldsymbol{\Phi}_I \in \mathbb{R}^{N^e \times m_I}$ and $\mathbf{P}_I^T\boldsymbol{\Phi}_I \in \mathbb{R}^{m_I \times m_I}$ can be precomputed during the offline stage, since they are $\boldsymbol{\mu}$-independent. For the sake of system approximation, during the online stage we only need to evaluate (only once) the restriction of $\tilde{\mathbf{I}}$ to the indices $\boldsymbol{\wp}_I$, that is, $\mathbf{P}_I^T\tilde{\mathbf{I}}(\mathbf{Z}_e\mathbf{V}^n_{N^e}(\boldsymbol{\mu}), \mathbf{W}^{n+1}(\boldsymbol{\mu})) \in \mathbb{R}^{m_I}$.

Relying on (50), instead of (44) we thus obtain the following hyper-reduced monodomain equation: given $\boldsymbol{\mu} \in \mathscr{P}$ and $\mathbf{V}^0_{N^e}(\boldsymbol{\mu})$, for each $n = 0, \ldots, n_t - 1$ solve

$$\left[\frac{C_m}{\Delta t}\mathbf{Z}_e^T\mathbf{M}_L\mathbf{Z}_e + \sum_{i=1}^{m_K}\theta_K^i(\boldsymbol{\mu})\mathbf{Z}_e^T\mathbf{K}_i\mathbf{Z}_e\right]\mathbf{V}^{n+1}_{N^e}(\boldsymbol{\mu}) = \frac{C_m}{\Delta t}\mathbf{Z}_e^T\mathbf{M}_L\mathbf{Z}_e\mathbf{V}^n_{N^e}(\boldsymbol{\mu})$$
$$+\mathbf{Z}_e^T\mathbf{M}_L\boldsymbol{\Phi}_I(\mathbf{P}^T\boldsymbol{\Phi}_I)^{-1}\mathbf{P}_I^T\tilde{\mathbf{I}}(\mathbf{Z}_e\mathbf{V}^n_{N^e}(\boldsymbol{\mu}), \mathbf{W}^{n+1}(\boldsymbol{\mu})). \tag{51}$$

The construction of the reduced basis $\mathbf{Z}_e$ and the snapshots selection strategy will be addressed in Sect. 5.3.

5.2 The Full Electrophysiology Problem

In this section we aim at developing an efficient ROM for the full electrical problem (42), which involves a PDE (monodomain equation) and a system of ODEs (ionic model). Replacing the second equation of (42) with the ROM (51) derived in the previous section, we obtain the following problem: given $\boldsymbol{\mu} \in \mathscr{D}$ and $\mathbf{V}^0_{N^e}(\boldsymbol{\mu})$ and $\mathbf{W}^0(\boldsymbol{\mu})$, for each $n = 0, \ldots, n_t - 1$ solve

$$\begin{cases} \mathbf{W}^{n+1}(\boldsymbol{\mu}) = \mathbf{W}^n(\boldsymbol{\mu}) + \Delta t\tilde{\mathbf{S}}(\mathbf{V}^n(\boldsymbol{\mu}), \mathbf{W}^n(\boldsymbol{\mu})) \\ \left[\frac{C_m}{\Delta t}\mathbf{Z}_e^T\mathbf{M}_L\mathbf{Z}_e + \sum_{i=1}^{m_K}\theta_K^i(\boldsymbol{\mu})\mathbf{Z}_e^T\mathbf{K}_i\mathbf{Z}_e\right]\mathbf{V}^{n+1}_{N^e}(\boldsymbol{\mu}) = \frac{C_m}{\Delta t}\mathbf{Z}_e^T\mathbf{M}_L\mathbf{Z}_e\mathbf{V}^n_{N^e}(\boldsymbol{\mu}) \\ \qquad + \mathbf{Z}_e^T\mathbf{M}_L\boldsymbol{\Phi}_I(\mathbf{P}_I^T\boldsymbol{\Phi}_I)^{-1}\mathbf{P}_I^T\tilde{\mathbf{I}}(\mathbf{Z}_e\mathbf{V}^n_{N^e}(\boldsymbol{\mu}), \mathbf{W}^{n+1}(\boldsymbol{\mu})). \end{cases} \tag{52}$$

Since the ODEs still depend on the high-fidelity approximation of the potential $\mathbf{V}^n$ at each time t_n, $n = 1, \ldots, n_t$, we should in principle compute the matrix-vector product $\mathbf{V}^n = \mathbf{Z}_e\mathbf{V}^n_{N^e}$ at each time t_n to evaluate $\tilde{\mathbf{S}}(\mathbf{V}^n(\boldsymbol{\mu}), \mathbf{W}^n(\boldsymbol{\mu}))$ and solve the ionic model. To avoid this operation, we can again rely on a DEIM approximation. In fact, we can express the restriction of $\tilde{\mathbf{I}}(\mathbf{V}, \mathbf{W})$ to the subset of DEIM indices $\wp_I$ as

$$\begin{aligned} \mathbf{P}_I^T\tilde{\mathbf{I}}(\mathbf{Z}_e\mathbf{V}^n_{N^e}(\boldsymbol{\mu}), \mathbf{W}^{n+1}(\boldsymbol{\mu})) &= \tilde{\mathbf{I}}(\mathbf{Z}_e\mathbf{V}^n_{N^e}(\boldsymbol{\mu}), \mathbf{W}^{n+1}(\boldsymbol{\mu}))|_{\wp_I} \\ &= \tilde{\mathbf{I}}(\mathbf{Z}_e\mathbf{V}^n_{N^e}(\boldsymbol{\mu})|_{\wp_I}, \mathbf{W}^{n+1}(\boldsymbol{\mu})|_{\wp_I}). \end{aligned} \tag{53}$$

In particular, we observe that in order to solve $(52)_2$, only the restriction of the ionic variables vector $\mathbf{W}^{n+1}(\boldsymbol{\mu})$ to the indices $\wp_I$ is required. Hence, (52) becomes:

given $\boldsymbol{\mu} \in \mathscr{D}$, $\mathbf{V}^0_{N^e}(\boldsymbol{\mu})$ and $\mathbf{W}^0(\boldsymbol{\mu})$, for each $n = 0, \dots, n_t - 1$ solve

$$\begin{cases} \mathbf{W}^{n+1}(\boldsymbol{\mu})|_{\wp_I} = \mathbf{W}^n(\boldsymbol{\mu})|_{\wp_I} + \Delta t \tilde{\mathbf{S}}(\mathbf{V}^n(\boldsymbol{\mu}), \mathbf{W}^n(\boldsymbol{\mu}))|_{\wp_I} \\ \left[\dfrac{C_m}{\Delta t} \mathbf{Z}_e^T \mathbf{M}_L \mathbf{Z}_e + \displaystyle\sum_{i=1}^{m_K} \theta_K^i(\boldsymbol{\mu}) \mathbf{Z}_e^T \mathbf{K}_i \mathbf{Z}_e \right] \mathbf{V}_{N^e}^{n+1}(\boldsymbol{\mu}) = \dfrac{C_m}{\Delta t} \mathbf{Z}_e^T \mathbf{M}_L \mathbf{Z}_e \mathbf{V}_{N^e}^n(\boldsymbol{\mu}) \\ \qquad + \mathbf{Z}_e^T \mathbf{M}_L \boldsymbol{\Phi}_I (\mathbf{P}^T \boldsymbol{\Phi}_I)^{-1} \tilde{\mathbf{I}}(\mathbf{Z}_e \mathbf{V}_{N^e}^n(\boldsymbol{\mu})|_{\wp_I}, \mathbf{W}^{n+1}(\boldsymbol{\mu})|_{\wp_I}). \end{cases} \tag{54}$$

As done in (53) for the ionic current, we write

$$\tilde{\mathbf{S}}(\mathbf{V}^n(\boldsymbol{\mu}), \mathbf{W}^n(\boldsymbol{\mu}))|_{\wp_I} = \tilde{\mathbf{S}}(\mathbf{Z}_e \mathbf{V}_{N^e}^n(\boldsymbol{\mu})|_{\wp_I}, \mathbf{W}^n(\boldsymbol{\mu})|_{\wp_I})$$

so that, exploiting the approximation $\mathbf{V}^n(\boldsymbol{\mu}) \approx \mathbf{Z}_e \mathbf{V}_{N^e}^n(\boldsymbol{\mu})$, $n = 0, \dots, n_t - 1$, the hyper-reduced order model for the electrical problem reads as follows: given $\boldsymbol{\mu} \in \mathscr{D}$, $\mathbf{V}^0_{N^e}(\boldsymbol{\mu})$ and $\mathbf{W}^0(\boldsymbol{\mu})$, for each $n = 0, \dots, n_t - 1$ solve

$$\begin{cases} \mathbf{W}^{n+1}(\boldsymbol{\mu})|_{\wp_I} = \mathbf{W}^n(\boldsymbol{\mu})|_{\wp_I} + \Delta t \tilde{\mathbf{S}}(\mathbf{Z}_e \mathbf{V}_{N^e}^n(\boldsymbol{\mu})|_{\wp_I}, \mathbf{W}^n(\boldsymbol{\mu})|_{\wp_I}) \\ \left[\dfrac{C_m}{\Delta t} \mathbf{Z}_e^T \mathbf{M}_L \mathbf{Z}_e + \displaystyle\sum_{i=1}^{m_K} \theta_K^i(\boldsymbol{\mu}) \mathbf{Z}_e^T \mathbf{K}_i \mathbf{Z}_e \right] \mathbf{V}_{N^e}^{n+1}(\boldsymbol{\mu}) = \dfrac{C_m}{\Delta t} \mathbf{Z}_e^T \mathbf{M}_L \mathbf{Z}_e \mathbf{V}_{N^e}^n(\boldsymbol{\mu}) \\ \qquad + \mathbf{Z}_e^T \mathbf{M}_L \boldsymbol{\Phi}_I (\mathbf{P}^T \boldsymbol{\Phi}_I)^{-1} \tilde{\mathbf{I}}(\mathbf{Z}_e \mathbf{V}_{N^e}^n(\boldsymbol{\mu})|_{\wp_I}, \mathbf{W}^{n+1}(\boldsymbol{\mu})|_{\wp_I}). \end{cases} \tag{55}$$

Hence, instead of solving the ODEs $(54)_1$ for each degree of freedom of $\mathscr{T}_{ep}$, we compute their solution only at m_I nodes, with $m_I \ll N_h^e$. Indeed, both the ionic model and the monodomain equation only depend on the restriction of the potential to the indices $\wp_I$; in particular, by recalling that

$$\mathbf{Z}_e \mathbf{V}_{N^e}^n(\boldsymbol{\mu})|_{\wp_I} = \mathbf{P}_I^T \mathbf{Z}_e \mathbf{V}_{N^e}^n(\boldsymbol{\mu}),$$

we notice that the matrix $\mathbf{P}_I^T \mathbf{Z}_e \in \mathbb{R}^{m_I \times N^e}$ can be precomputed offline. Solving the ODEs system at each node of the mesh would have been required if the activation equation underwent a similar reduction procedure and a ROM for the coupled electromechanical problem was built; see [13] for more on this subject.

5.3 Snapshots Selection Strategy

To build $\mathbf{Z}_e$ and $\boldsymbol{\Phi}_I$ we rely on a POD-POD strategy, namely we perform POD with respect to both the time and the parameter vector. In particular:

- for each parameter value $\boldsymbol{\mu}_1, \dots, \boldsymbol{\mu}_{n_s}$ randomly chosen in $\mathscr{P}$, we solve the high-fidelity problem (42) and perform a POD in time to compress the snapshots

$\mathbf{V}^n(\boldsymbol{\mu}_i)$, $n = 1, \ldots, n_t$, thus obtaining a basis $\mathbf{Z}_i$; finally, we obtain the POD basis $\mathbf{Z}_e$ performing a POD on the n_s bases $\mathbf{Z}_1, \ldots, \mathbf{Z}_{n_t}$;

- regarding the nonlinear term, we collect the snapshots $\tilde{\mathbf{I}}(\mathbf{Z}\mathbf{V}^n_{Ne}(\boldsymbol{\mu}_i), \mathbf{W}^{n+1}(\boldsymbol{\mu}_i))$, $n = 1, \ldots, n_t$, $i = 1, \ldots, n_s$ by solving the following problem: given $\mathbf{V}^0_{Ne}(\boldsymbol{\mu}_i)$ and $\mathbf{W}^0(\boldsymbol{\mu}_i)$, $i = 1, \ldots, n_s$, for each $n = 0, \ldots, n_t - 1$ compute

$$\begin{cases} \mathbf{W}^{n+1}(\boldsymbol{\mu}) = \mathbf{W}^n(\boldsymbol{\mu}) + \Delta t \tilde{\mathbf{S}}(\mathbf{V}^n(\boldsymbol{\mu}), \mathbf{W}^n(\boldsymbol{\mu})) \\ \left[\dfrac{C_m}{\Delta t}\mathbf{Z}^T\mathbf{M}_L\mathbf{Z} + \displaystyle\sum_{i=1}^{m_K} \theta^i_K(\boldsymbol{\mu})\mathbf{Z}^T\mathbf{K}_i\mathbf{Z}\right]\mathbf{V}^{n+1}_{Ne}(\boldsymbol{\mu}) = \dfrac{C_m}{\Delta t}\mathbf{Z}^T\mathbf{M}_L\mathbf{Z}\mathbf{V}^n_{Ne}(\boldsymbol{\mu}) \\ \qquad\qquad + \mathbf{Z}^T\mathbf{M}_L\tilde{\mathbf{I}}(\mathbf{Z}\mathbf{V}^n_{Ne}(\boldsymbol{\mu}), \mathbf{W}^{n+1}(\boldsymbol{\mu})) \end{cases} \tag{56}$$

  and then perform the same POD-POD procedure with respect to time and parameters as in the previous case; see Algorithm 4.

Here we rely on *global* reduced bases; considering local bases obtained by partitioning the snapshot set either with respect to time or the parameters, or, alternatively, produced by a clustering procedure in the physical space, represents a further improvement, currently under investigation; see [59] for further details.

Algorithm 4 ROM construction (electrophysiology)

INPUT: n_s parameter values $\{\boldsymbol{\mu}_1, \ldots, \boldsymbol{\mu}_{n_s}\}$, $\mathbf{V}^0(\boldsymbol{\mu}_i)$, $\mathbf{W}^0(\boldsymbol{\mu}_i)$, $i = 1, \ldots, n_s$
OUTPUT: $\mathbf{Z}_e$, $\boldsymbol{\Phi}_I$, $\boldsymbol{\Phi}_K$

for $i = 1, \ldots, n_s$ **do**
  $\mathbf{S}_K \leftarrow [\mathbf{S}_K \;\; vec(\mathbf{K}(\boldsymbol{\mu}_i))]$
  for $n = 0, \ldots, N_t - 1$ **do**
    Solve problem (42) for $\boldsymbol{\mu}_i$
    $\mathbf{S}_V \leftarrow [\mathbf{S}_V \;\; \mathbf{V}^{n+1}(\boldsymbol{\mu}_i)]$
  end for
  $\mathbf{Z}_i = \text{POD}(\mathbf{S}_V, \varepsilon)$
end for
$\mathbf{Z}_e = \text{POD}(\mathbf{Z}_1, \ldots, \mathbf{Z}_{n_s}; \varepsilon)$
$\boldsymbol{\Phi}_K = \text{POD}(\mathbf{S}_K; \varepsilon_{MDEIM})$, $\mathbf{K}_l = vec^{-1}(\boldsymbol{\Phi}_K(l, :))$, $l = 1, \ldots, m_K$
for $i = 1, \ldots, n_s$ **do**
  for $n = 0, \ldots, N_t - 1$ **do**
    Solve problem (56) for $\boldsymbol{\mu}_i$
    $\mathbf{S}_I \leftarrow [\mathbf{S}_I \;\; \tilde{\mathbf{I}}(\mathbf{Z}\mathbf{V}^n_{Ne}(\boldsymbol{\mu}), \mathbf{W}^{n+1}(\boldsymbol{\mu}))]$
  end for
  $\boldsymbol{\Phi}_i = \text{POD}(\mathbf{S}_I; \varepsilon_{DEIM})$
end for
$\boldsymbol{\Phi}_I = \text{POD}(\boldsymbol{\Phi}_1, \ldots, \boldsymbol{\Phi}_{n_s}; \varepsilon_{DEIM})$

6 RB Methods for Cardiac Mechanics

We now show how to take advantage of the techniques presented in Sect. 4 in order to build a ROM for the mechanical problem. This latter is a *time-driven* problem, since its solution depends on time only through its coupling with the electrophysiology model, which is intrinsically unsteady. Relying on the RB method provides a twofold advantage, since the problem has to be solved not only for different parameters, but also at several time instants.

The parametrized version of problem (28) reads: given $\mathbf{U}^{(0)}(t_n;\boldsymbol{\mu}) = \mathbf{U}(t_{n-1};\boldsymbol{\mu})$, for $n = 1,\dots,N_t$, $N_t = T/\Delta t_m$, for $k \geq 1$ solve:

$$\begin{cases} \mathbf{J}(\mathbf{U}^{(k-1)}(t_n,\boldsymbol{\mu});\mathbf{G}_f^m(\boldsymbol{\mu}))\delta\mathbf{U}(t_n;\boldsymbol{\mu}) = -\mathbf{R}(\mathbf{U}^{(k-1)}(t_n;\boldsymbol{\mu});\mathbf{G}_f^l(\boldsymbol{\mu})), \\ \mathbf{U}^{(k)}(t_n;\boldsymbol{\mu}) = \mathbf{U}^{(k-1)}(t_n;\boldsymbol{\mu}) + \delta\mathbf{U}(t_n;\boldsymbol{\mu}), \end{cases} \tag{57}$$

until $||\mathbf{R}(\mathbf{U}^{(k-1)}(t_n;\boldsymbol{\mu});\mathbf{G}_f^l(\boldsymbol{\mu})||_2 < \varepsilon$, where k is the index of the Newton iterations, t_n denotes the time instant which the mechanical problem is computed at, and $l = (n+1)D - 1$, with $D = \Delta t_m/\Delta t_e$. Here we also have made the dependence on the (discrete representation of the) fibers shortening $\mathbf{G}_f$; recall that the time stepping for this latter variable follows the one of the electrical problem.

For the mechanical problem, time is instead considered as an additional parameter, although with peculiar features. Indeed, as we are not interested in solving problem (57) for generic values of t selected online, we use the same time step Δt_m in both the offline and online stages; in particular, online we solve the reduced problem associated with (57) only in the time instants of the form $t_n = n\,\Delta t_m$ used during the offline stage for the snapshots computation.

6.1 *Jacobian Approximation by the Broyden Method*

When dealing with time dependent (and/or large scale) problems, (M)DEIM might be not efficient enough to guarantee a rapid approximation of the Jacobian matrix during the online phase. In particular, the classical DEIM technique (see Sect. 4.3) provides moderate computational speed-up when dealing with nonlinear problems which require a large number of terms m_R. On the other hand, MDEIM may require an overwhelming amount of CPU time and memory to store the snapshots $\mathbf{J}(\mathbf{U}^{(k-1)}(t_n;\boldsymbol{\mu}_i);\mathbf{G}_f^l(\boldsymbol{\mu}_i))$, and to build the basis $\boldsymbol{\Phi}_J$ during the offline phase. For these reasons, a possible alternative to approximate the reduced Jacobian matrix (38) is to rely on a purely algebraic technique such as the *Broyden method* [16], developed to effectively approximate the Jacobian matrix when its analytic form is unknown, or it is too expensive to compute. This approach has been applied for the reduction of nonlinear structural problems in [71] and further improved in [13].

This yields the following hyper-reduced order model for the mechanical problem: for each t_n, $n = 1, \ldots, N_t$, given $\mathbf{u}_{N,m}^{(0)}(t_n;\boldsymbol{\mu})$, for $k \geq 1$ solve:

$$\begin{cases} \mathbf{J}_N^B(\mathbf{u}_{N,m}^{(k)}(t_n;\boldsymbol{\mu});\mathbf{G}_f^l(\boldsymbol{\mu}))\delta\mathbf{u}_{N,m}^{(k)}(t_n;\boldsymbol{\mu}) = -\mathbf{R}_{N,m}(\mathbf{Z}\mathbf{u}_{N,m}^{(k)}(t_n;\boldsymbol{\mu});\mathbf{G}_f^l(\boldsymbol{\mu})),\\ \mathbf{u}_{N,m}^{(k+1)}(t_n;\boldsymbol{\mu}) = \mathbf{u}_{N,m}^{(k)}(t_n;\boldsymbol{\mu}) + \delta\mathbf{u}_{N,m}^{(k)}(t_n;\boldsymbol{\mu}), \end{cases} \tag{58}$$

and iterate until $||\mathbf{R}_{N,m}(\mathbf{Z}\mathbf{u}_{N,m}^{(k)}(t_n;\boldsymbol{\mu});\mathbf{G}_f^l(\boldsymbol{\mu}))||_2 < \varepsilon_{RB}$; as in (57), here $l = (n+1)D - 1$. We underline that the high-fidelity approximation of the fibers shortening variable $\mathbf{G}_f$ is considered, that is, we do not deal in this work with the reduction of the coupled electro-mechanical problem; rather, we focus on the efficient reduction of the mechanical problem only.

At each Newton step, the Broyden approximation of the reduced Jacobian matrix is computed as

$$\mathbf{J}_N^B(\mathbf{u}_{N,m}^{(k)}(t_n;\boldsymbol{\mu})) = \mathbf{J}_N^B(\mathbf{u}_{N,m}^{(k-1)}(t_n;\boldsymbol{\mu})) + \frac{\Delta\bar{\mathbf{R}}_{N,m} - \mathbf{J}_N^B(\mathbf{u}_{N,m}^{(k-1)}(t_n;\boldsymbol{\mu}))\Delta\bar{\mathbf{u}}_{N,m}}{\Delta\bar{\mathbf{u}}_{N,m}^T\Delta\bar{\mathbf{u}}_{N,m}}\Delta\bar{\mathbf{u}}_{N,m}^T, \tag{59}$$

where

$$\Delta\bar{\mathbf{R}}_{N,m} = \mathbf{R}_{N,m}(\mathbf{Z}\mathbf{u}_{N,m}^{(k)}(t_n;\boldsymbol{\mu});t_n,\boldsymbol{\mu}) - \mathbf{R}_{N,m}(\mathbf{Z}\mathbf{u}_{N,m}^{(k-1)}(t_n;\boldsymbol{\mu});t_n\boldsymbol{\mu})$$

and

$$\Delta\bar{\mathbf{u}}_{N,m} = \mathbf{u}_{N,m}^{(k)}(t_n;\boldsymbol{\mu}) - \mathbf{u}_{N,m}^{(k-1)}(t_n;\boldsymbol{\mu}).$$

The method is based on a rank-one update of the Jacobian matrix $\mathbf{J}_N^B(\mathbf{u}_{N,m}^{(k)}(t_n;\boldsymbol{\mu}))$; indeed, the second term of the right hand side of (59) is a rank one matrix since every column is a scalar multiple of $\Delta\bar{\mathbf{u}}_{N,m}^T$, and only require simple operations between vectors of dimension N. Therefore, at the online stage each Newton step only requires a residual assembling, since the update of the reduced Jacobian matrix and the solution of the low dimensional linear system are extremely fast. On the other hand, the initialization of the Jacobian matrix $\mathbf{J}_N^B(\mathbf{u}_{N,m}^{(0)}(t_n;\boldsymbol{\mu});\gamma_f(t_n,\boldsymbol{\mu}))$ at step $k = 0$ represents a critical aspect of the Broyden technique. To provide a suitable approximation of the exact Jacobian matrix, we consider a finite difference approximation of the form

$$[\mathbf{J}_N^B(\mathbf{u}_{N,m}^{(0)}(t_n;\boldsymbol{\mu}))]_{i,j} = \frac{[\mathbf{R}_{N,m}(\mathbf{Z}(\mathbf{u}_{N,m}^{(0)}(t_n;\boldsymbol{\mu}) + \eta\mathbf{e}_i))]_j - [\mathbf{R}_{N,m}(\mathbf{Z}\mathbf{u}_{N,m}^{(0)}(t_n;\boldsymbol{\mu}))]_j}{\eta},$$

where $\mathbf{e}_i$ is the i-th column of the identity matrix and $\eta \in \mathbb{R}$ is a coefficient to be properly chosen (here we take $\eta = 10^{-5}$). η shall be small enough to guarantee an accurate approximation of the derivatives, however an excessively small value may lead to undesired cancellation errors. Note that the initialization requires to assemble online N times the residual vector—an affordable operation if N is relatively small. Instead, if $N > m_R$, relying on DEIM to initialize the Jacobian matrix is more convenient than the finite difference approximation. In general, problems characterized by a large dimension of the reduced space N can be efficiently reduced exploiting the MDEIM technique, provided a small number m_J of terms can be selected to accurately approximate the Jacobian matrix.

We close this section by pointing out that, during the online stage, only the reduced basis $\mathbf{Z}$ and a DEIM basis $\boldsymbol{\Phi}_R$ for the residual approximation are required. To build this latter, we rely on a strategy similar to the one described in Sect. 4.3.1, introducing the following intermediate problem similar to (41): for each t_n, $n = 0, \ldots, N_t$, given $\mathbf{u}_N^{(0)}(t_n; \boldsymbol{\mu}) \in \mathbb{R}^N$, for $k \geq 1$ solve:

$$\begin{cases} \tilde{\mathbf{J}}_N^B(\mathbf{u}_N^{(k)}(t_n; \boldsymbol{\mu}); \mathbf{G}_f^l(\boldsymbol{\mu}))\delta\mathbf{u}_N^{(k)}(t_n; \boldsymbol{\mu}) = -\mathbf{Z}^T\mathbf{R}(\mathbf{Z}\mathbf{u}_N^{(k)}(t_n; \boldsymbol{\mu}); \mathbf{G}_f^l(\boldsymbol{\mu})), \\ \mathbf{u}_N^{(k+1)}(t_n; \boldsymbol{\mu}) = \mathbf{u}_N^{(k)}(t_n; \boldsymbol{\mu}) + \delta\mathbf{u}_N^{(k)}(t_n; \boldsymbol{\mu}), \end{cases} \tag{60}$$

and iterate until $||\mathbf{Z}^T\mathbf{R}(\mathbf{Z}\mathbf{u}_N^{(k)}(t_n; \boldsymbol{\mu}); \mathbf{G}_f^l(\boldsymbol{\mu}))||_2 < \varepsilon_{RB}$.

Precisely, we first solve the high-fidelity problem (28) n_s times to compute the reduced basis $\mathbf{Z}$; then, we solve (60) to store the residual snapshots $\mathbf{R}(\mathbf{Z}\mathbf{u}_N^{(k)}(t_n; \boldsymbol{\mu}_i))$ needed to build the DEIM basis $\boldsymbol{\Phi}_R$. Problem $(60)_1$ is an inexact Newton step, for which the Jacobian matrix $\tilde{\mathbf{J}}_N^B$ is approximated through the Broyden technique, however relying on the projection of the high-fidelity residual onto the RB space—rather than on the reduced residual, as in (59). Solving (60) is significantly faster than solving (28) since the only operation to perform is the assembling of the high-fidelity residual.The full snapshots selection procedure is summarized in the next section (see Algorithm 5).

6.2 Local-in-Time Reduced Basis

If the $\boldsymbol{\mu}$-dependent solution of a non stationary problem significantly varies throughout the time interval $[0, T]$, describing the whole set of solutions for $\boldsymbol{\mu} \in \mathscr{P}$ may require a large number of RB functions and, in particular, a large number of terms m_R when considering nonlinear problems where the residual vector is approximated by DEIM. To overcome this issue, we build *local-in-time* bases of smaller dimension, instead than a unique (larger) global basis on $[0, T]$, thus yielding a strong reduction in terms of online computational time, still retaining the same cost during the offline stage. In particular, we split the full time interval $[0, T]$

Algorithm 5 ROM construction (mechanics)

INPUT: n_s combinations of parameters $\{\boldsymbol{\mu}_1, \ldots, \boldsymbol{\mu}_{n_s}\}$, n_b time subintervals $\{I_1, \ldots, I_{n_b}\}$
OUTPUT: $\mathbf{Z}^s$, $\boldsymbol{\Phi}^s_R$, $\boldsymbol{\Phi}^s_J$ (only for MDEIM), $s = 1, \ldots, n_b$

for $i = 1, \ldots, n_s$ **do**
  for $s = 1, \ldots, n_b$ **do**
    for $t_n \in \tau_s$ **do**
      Solve problem (57) for t_n, $\boldsymbol{\mu}_i$
      At each Newton step k: $\mathbf{S}_U \leftarrow [\mathbf{S}_U \quad \mathbf{U}^{(k)}(t_n; \boldsymbol{\mu}_i)]$
      At each Newton step k: $\mathbf{S}_J \leftarrow [\mathbf{S}_J \quad vec(\mathbf{J}(\mathbf{U}^{(k)}(t_n; \boldsymbol{\mu}_i)))]$ (only for MDEIM)
    end for
    $\mathbf{Z}^{i,s} = \text{POD}(\mathbf{S}_U; \varepsilon)$; $\boldsymbol{\Phi}^s \leftarrow [\boldsymbol{\Phi}^s \quad \mathbf{Z}^{i,s}]$; delete $[\mathbf{S}_U]$
    $\boldsymbol{\Phi}^{i,s}_J = \text{POD}(\mathbf{S}_J; \varepsilon)$; $\boldsymbol{\Phi}^s_J \leftarrow [\boldsymbol{\Phi}^s_J \quad \boldsymbol{\Phi}^{i,s}_J]$ (only for MDEIM); delete $[\mathbf{S}_J]$
  end for
end for
for $s = 1, \ldots, n_b$ **do**
  $\mathbf{Z}^s = \text{POD}(\boldsymbol{\Phi}^s; \varepsilon)$
  $\boldsymbol{\Phi}^s_J \leftarrow \text{POD}(\boldsymbol{\Phi}^s_J; \varepsilon)$ (only for MDEIM)
end for
for $i = 1, \ldots, n_s$ **do**
  for $s = 1, \ldots, n_b$ **do**
    for $t_n \in \tau_s$ **do**
      Solve problem (41) (for MDEIM) or (60) (for Broyden) for t_n, $\boldsymbol{\mu}_i$
      At each Newton step k: $\mathbf{S}_R \leftarrow [\mathbf{S}_R \quad \mathbf{R}(\mathbf{U}^{(k)}(t_n; \boldsymbol{\mu}_i))]$
    end for
    $\boldsymbol{\Phi}^{i,s}_R = \text{POD}(\mathbf{S}_R; \varepsilon)$, $\boldsymbol{\Phi}^s_R \leftarrow [\boldsymbol{\Phi}^s_R \quad \boldsymbol{\Phi}^{i,s}_R]$; delete $[\mathbf{S}_R]$
  end for
end for
for $s = 1, \ldots, n_b$ **do**
  $\boldsymbol{\Phi}^s_R \leftarrow \text{POD}(\boldsymbol{\Phi}^s_R; \varepsilon)$
end for

into n_b subintervals

$$I_1 = [0, T_1],\ I_2 = [T_1, T_2],\ \ldots,\ I_{n_b} = [T_{n_b-1}, T]$$

and we build different bases on each subinterval. In particular, we compute n_b set of bases $\{\mathbf{Z}^s, \boldsymbol{\Phi}^s_R\}$, $s = 1, \ldots, n_b$ (and $\boldsymbol{\Phi}^s_J$ if using MDEIM), and each basis associated to the s-th interval is obtained by performing the POD only on the set of snapshots related to $I_s = [T_{s-1}, T_s]$.

We remark that the mechanical problem is solved in correspondence of evenly spaced time instants $\{t_n\}_{n=0}^{N_t}$, whereas the new partition in subintervals I_s is not necessarily uniform. This means that, denoting by $\tau_s = \{t_n \mid t_n \in I_s\}$, the number of elements of τ_s can be different from the number of elements of τ_r when $s \neq r$; in particular, larger subintervals I_s are taken if the solution features a small variability in time, while narrow subintervals where the solution rapidly changes. The choice of number and width of the different subintervals is not trivial and multiple options can be considered. Here, we opt for an heuristic choice of the subintervals I_s, $s = 1, \ldots, n_b$; a more-in-depth investigation is required in this respect. More general strategies to obtain multiple, local bases have been proposed in the literature; see e.g. [5, 85] for the use of proper clustering algorithms. They require to compute and store the solution snapshots, when varying time and parameters, and then to apply a clustering algorithm in order to build different bases, one for each cluster. While guaranteeing a satisfactory accuracy, snapshots storage can be quite large. Promising results in the case of cardiac electrophysiology can be found in [59].

The local-in-time basis approach combined with a POD-POD strategy allows to avoid the storage of all snapshots. Following this approach, for each $\boldsymbol{\mu}_i$, $i = 1, \ldots, n_s$, we perform a full time-driven simulation and we store separately the snapshots $\mathbf{U}^{(k)}(t_n; \boldsymbol{\mu}_i)$ associated with the subintervals I_s, $s = 1, \ldots, n_b$. Then, for each subinterval I_s, we apply the POD algorithm to the local snapshots $\mathbf{U}^{(k)}(t_n; \boldsymbol{\mu}_i)$, $t_n \in I_s$, obtaining the POD bases $\mathbf{Z}^{i,s}$. Once we have performed the high-fidelity simulations for all $\boldsymbol{\mu}_i$ and we have computed all the POD bases, we build our final bases for each subinterval I_s as:

$$\mathbf{Z}^s = POD(\mathbf{Z}^{i,s}, \varepsilon).$$

A schematic representation of the method is reported in Fig. 3. The same procedure is used also to build $\boldsymbol{\Phi}_R^s$ and, if required, $\boldsymbol{\Phi}_J^s$. In this way, different snapshots can be computed simultaneously; at the end of each simulation, we can only retain the POD bases $\{\mathbf{Z}^{i,s}, \boldsymbol{\Phi}_R^{is}, \boldsymbol{\Phi}_J^{is}\}$, thus saving significant memory resources.

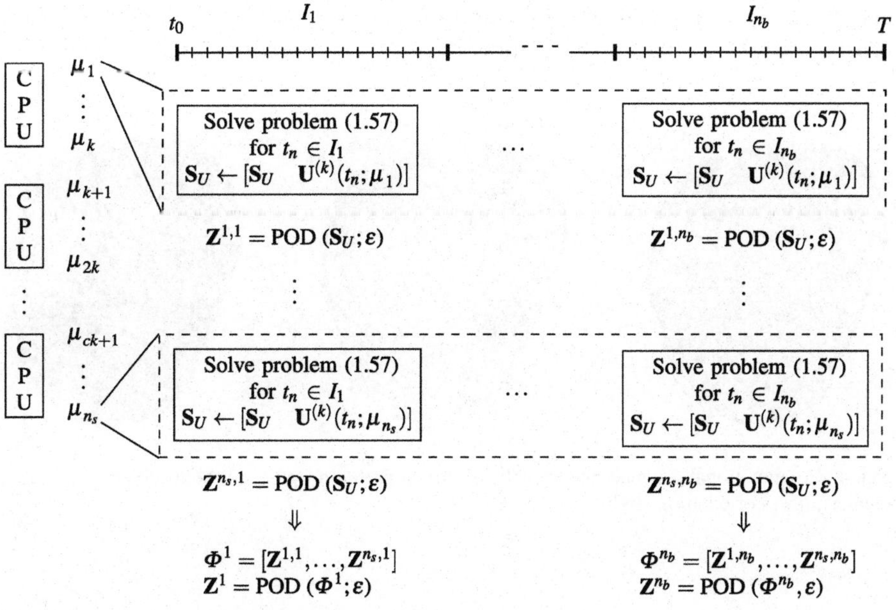

Fig. 3 Flowchart of the POD-POD procedure combined with local-in-time bases to compute the bases $\mathbf{Z}^s$, $s = 1, \ldots, n_b$, of the reduced space

The procedure described above has to be integrated with the generation of snapshots (and bases) to approximate residual vectors and Jacobian matrices; in particular, we split in two different stages the construction of $\{\mathbf{Z}^s, \boldsymbol{\Phi}^s_J\}$ and $\boldsymbol{\Phi}^s_R$. The full procedure is described in Algorithm 5 in the case either MDEIM or Broyden approximations are considered.

7 Numerical Results

We apply the strategies described in the previous sections for the reduction of an electrical and a mechanical problem, both solved on a patient-specific left ventricle geometry, by focusing on the systolic phase of the heart beat. This geometry was generated from medical images using the semi-automatic segmentation method proposed in [34]. Fibers and sheets have been computed according to the algorithm proposed in [74]; this procedure is based on the assumption that sheets are lying along the radial direction $\mathbf{s}_0$ and requires the solution of a Laplace problem over the ventricular domain to compute the sheets direction.

We rely on two different computational meshes: $\mathscr{T}_{ep}$ with 248,216 elements and $N_h^e = 45{,}817$ dofs for the electrical problem and $\mathscr{T}_m$ with 31,027 elements and $N_h = 18{,}567$ dofs for the mechanics. In Fig. 4 we show the two meshes, as well as the fibers distribution computed using the algorithm proposed in [74] with maximum fibers orientation on the epicardium and endocardium of $\theta_{epi} = -60°$ and $\theta_{endo} = 60°$, respectively. For both problems we consider as high-fidelity full-order model (FOM) the approximation obtained with linear (P1) finite elements.

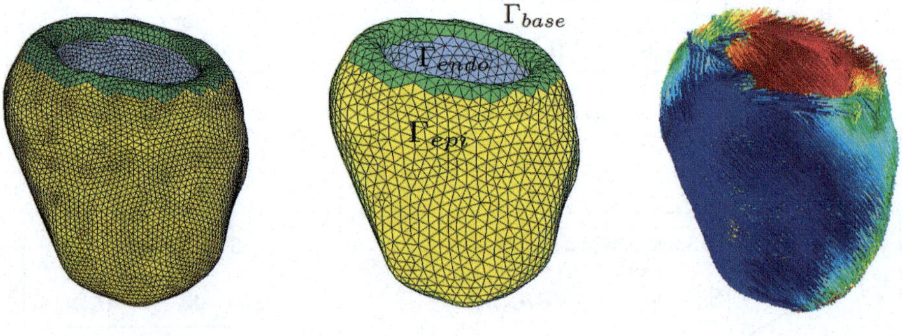

Fig. 4 Computational grids adopted for the electrical problem (left) and the mechanical problem (center); fibers orientation (right)

7.1 Test Case 1: Electrophysiology

In the case of the electrophysiology problem, the parameters that we consider are those representing the electrical conductivities:

- $\sigma_f \in [30\,k\Omega^{-1}\,cm^{-1}, 80\,k\Omega^{-1}\,cm^{-1}]$;
- $\sigma_s = \sigma_n \in [10\,k\Omega^{-1}\,cm^{-1}, 30\,k\Omega^{-1}\,cm^{-1}]$.

in the fibers direction (σ_f) and in the plane orthogonal to the former ($\sigma_s = \sigma_n$), see Eq. (6). For the time discretization of the monodomain equation and the ionic model we employ a time step $\Delta t_e = 0.02\,ms$. The depolarization wave is initialized on a layer in the bottom part of the endocardium; this choice is motivated by the fact that the Purkinje fibers terminations are mainly located near the apex of the ventricle.

By employing the POD-Galerkin method with DEIM/MDEIM approximations for the nonlinear ionic terms/parametrized diffusion matrix described in Sect. 5 we reduce the two-way coupled electrical problem (42), which involves a PDE (monodomain equation) and a system of ODEs (ionic model). The proposed ROM yields a speed-up of more than one order of magnitude, since it takes about 0.01 s for each time step, while the FOM takes 0.12 s. The RB solutions have been computed using $N_e = 217$ basis functions for the solution and $m_I = 363$ DEIM terms to approximate the ionic currents, as specified in Table 2. Provided a sufficient number of basis functions N_e, m_I is considered, the front propagation captured by the ROM is similar to the one obtained with the FOM. However, depending on the chosen parametrization, this achievement might be more difficult to obtain; for instance, if parameters describe local variations of tissue properties (such as, for instance, the presence of a scar) the behavior of the solution is much more involved, thus requiring a larger number of basis functions to reach a good accuracy.

A comparison between the high-fidelity and the reduced solutions is reported in Fig. 5, where we observe that the two models provide very similar results from a qualitative point of view. The average relative error between the ROM and the FOM is shown in Fig. 6 for three different values of m_I and is about 5%. The error has been computed over a test set of 50 randomly chosen parameters different than the values of the training sample used to compute the snapshots.

Moreover, we report in Fig. 7 the activation maps obtained for different parameter values. The electrical signal first activates the central area of the epicardium, then it spreads towards the apex and finally reaches the base. As expected, the duration

Table 2 Test case 1: numerical data

POD tolerances	$10^{-4}/5 \cdot 10^{-1}$	RB dofs	217
DEIM tolerances	$10^{-4}/5 \cdot 10^{-1}$	MDEIM tolerance	10^{-10}
DEIM terms	363	MDEIM terms	5
FE time ionic model	0.022	RB time ionic model	0.001 s
FE time monodomain	0.09 s	RB time monodomain	0.0085 s
FE time	0.12 s	RB time	0.01 s

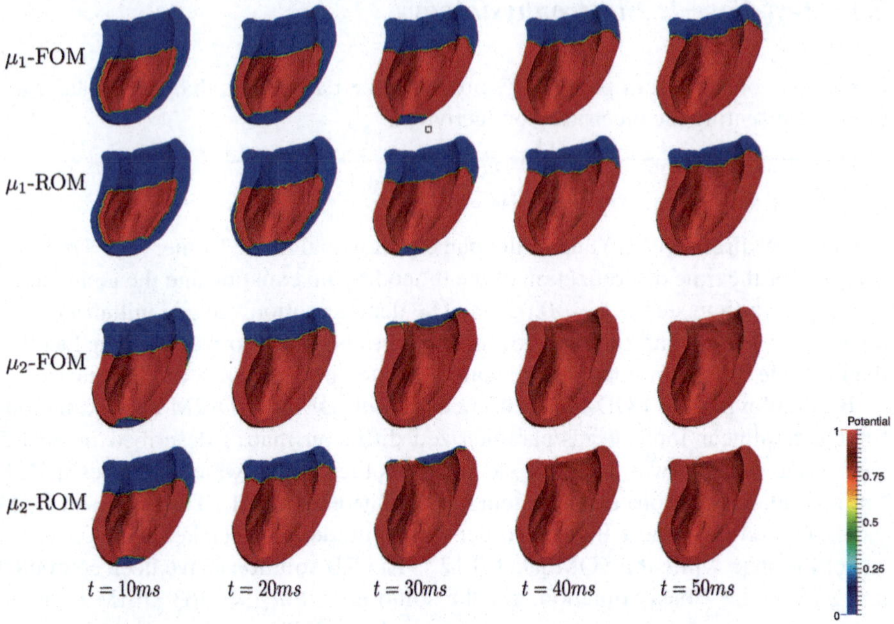

Fig. 5 Test case 1: FOM and ROM solutions computed at different time instants for $\boldsymbol{\mu}_1 = [30, 10]$ and $\boldsymbol{\mu}_2 = [80, 30]$

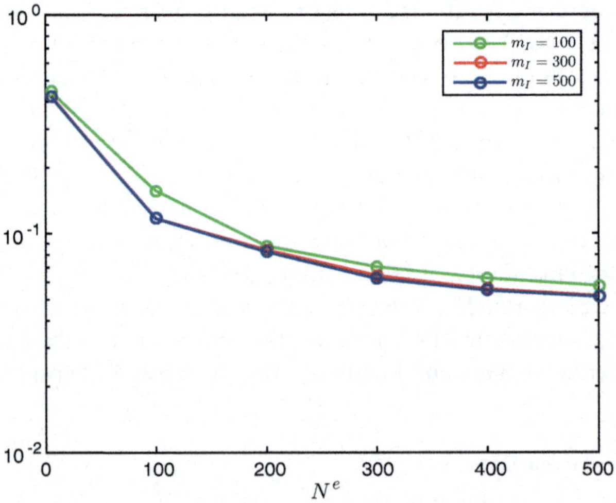

Fig. 6 Test case 1: average $L^2(0, T; H^1(\Omega_0))$ relative error between FOM and ROM solutions

of the depolarization phase is longer if electrical conductivities are smaller. The depolarization time as a function of σ_f is reported in Fig. 8.

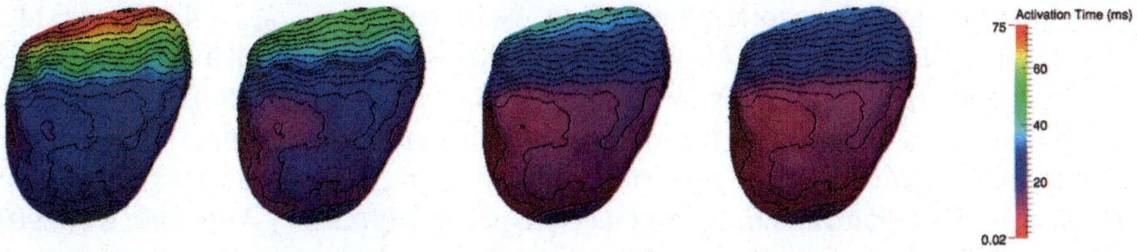

Fig. 7 Test case 1: activation maps for $\boldsymbol{\mu}_1 = [30, 10]$, $\boldsymbol{\mu}_2 = [45, 10]$, $\boldsymbol{\mu}_3 = [65, 30]$ and $\boldsymbol{\mu}_4 = [80, 30]$

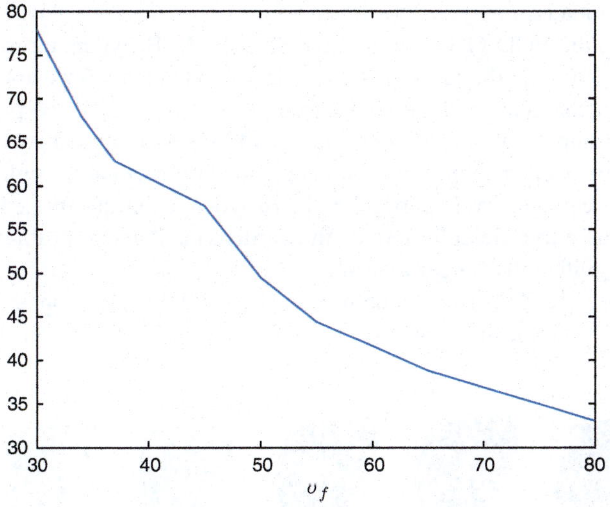

Fig. 8 Test case 1: whole depolarization times (ms) when varying σ_f, considering $\sigma_s = \sigma_n = 20$

7.2 Test Case 2: Mechanics

We now turn to the reduction of the mechanical problem, recalling that for the case at hand, the electromechanical coupling is not included in the ROM—that is, for each parameter value queried online we rely on the FOM approximation of the electrophysiology model and, in particular, of the fibers' shortening variable.

We consider as parameters the electrical conductivities (similarly to test case 1) and the orientation of the fibers:

- $\sigma_f \in [30\, k\Omega^{-1}\, cm^{-1}, 80\, k\Omega^{-1}\, cm^{-1}]$;
- $\sigma_s = \sigma_n \in [10\, k\Omega^{-1}\, cm^{-1}, 30\, k\Omega^{-1}\, cm^{-1}]$;
- $\theta_{max} \in [30°, 80°]$;
- $t \in [0, 100ms]$.

Time can be seen as an additional parameter, since the problem is quasi-static, and needs to be solved at different time steps. In particular, we split the [0, 100 ms], corresponding to the systolic phase of the heart beat, into $n_b = 4$ subintervals: [0, 30 ms], [30 ms, 60 ms], [60 ms, 90 ms], [90 ms, 100 ms]. Such a partition is introduced heuristically, taking into account the fact that the solution changes very rapidly during the last part of the systole; for this reason the last interval is smaller than the others. We consider a time step $\Delta t_e = 0.02$ ms for the time discretization of the electrical problem, and time instants equispaced with $\Delta_m = 3$ ms for $t \in [0, 90)$ and $\Delta_m = 0.5$ ms for $t \in [90, 100]$ for the mechanical problem. This latter is then solved, for different parameter values, at these time instants.

In Figs. 9 and 10 we show the displacement of the myocardium obtained with the FOM and the ROM on the whole ventricle and on a longitudinal section for two different parameters.

Employing the POD-Galerkin method with DEIM/Broyden approximations for the residual vector and the Jacobian matrix described in Sect. 6 we obtain also in the case of the mechanical problem even more promising results: the proposed ROM yields a speed-up of about 20, since the ROM takes about 20 s while the FOM requires 7 min for each solution of the mechanical problem at a single time instant, on a single-core processor. In particular, the Broyden technique turns out to be really appropriate since to efficiently reduce the cardiac mechanical problem we have to choose m_R significantly bigger than N.

The proposed ROM correctly captures the high-fidelity solution, as it can be seen also from the behavior of the average relative error shown in Fig. 11; the difference

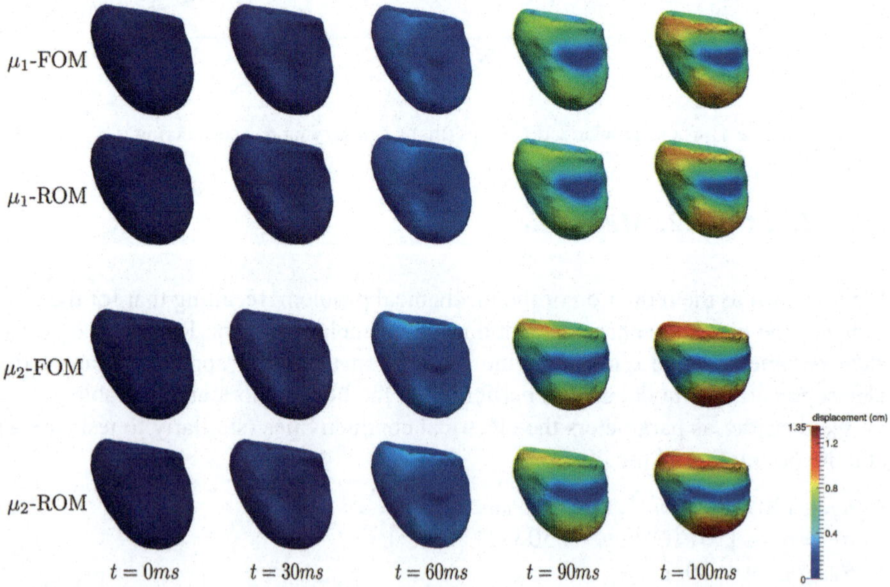

Fig. 9 Test case 2: displacement at different time instants, $\boldsymbol{\mu}_1 = [60, 10, 78°]$ (first two rows), $\boldsymbol{\mu}_2 = [80, 30, 34°]$ (last two rows), obtained with the FOM and the ROM

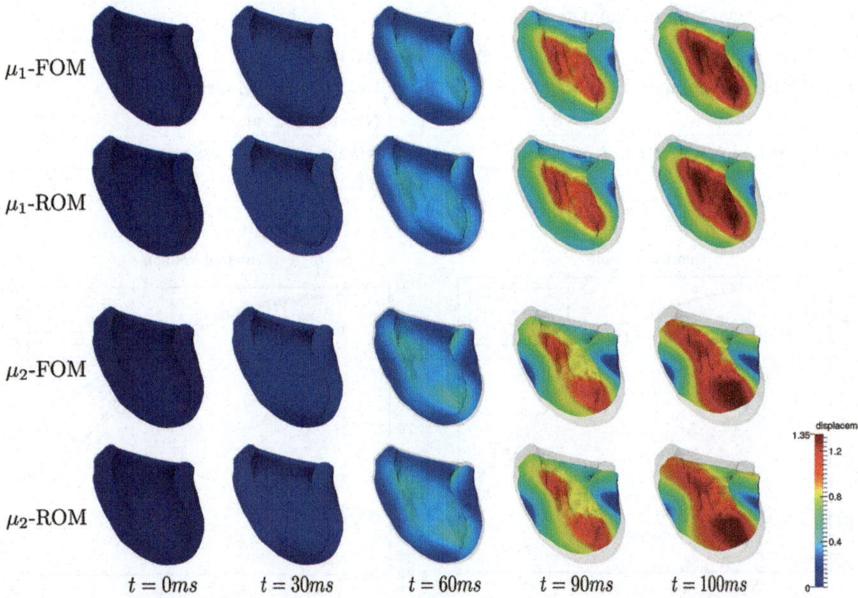

Fig. 10 Test case 2: section of the ventricle at different time instants, $\boldsymbol{\mu}_1 = [60, 10, 78°]$ (first two rows), $\boldsymbol{\mu}_2 = [80, 30, 34°]$ (last two rows), obtained with the FOM and the ROM

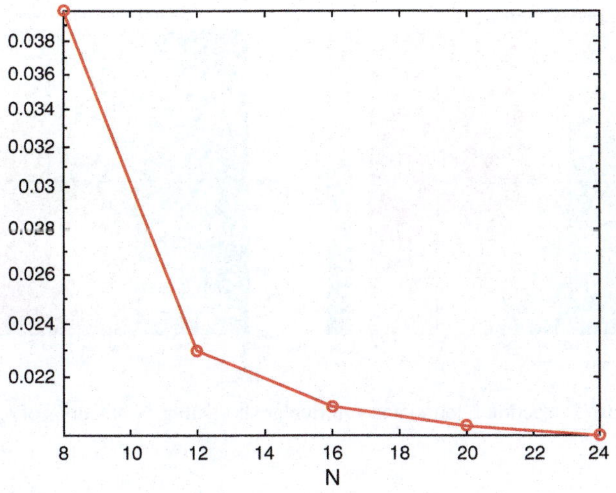

Fig. 11 Test case 2: average $L^2(0, T; H^1(\Omega_0))$ relative error

between the FOM and the ROM solutions is about 2%. Numerical data associated to this test case are reported in Table 3.

We show in Fig. 12 the variation of the volume inside the ventricular cavity for different values of θ_{max} and σ_f. In particular, the cases $\sigma_f = 60$, $\sigma_s = \sigma_n = 20$

Table 3 Test case 2: numerical data

POD tolerances	$10^{-3} - 0.05$	RB dofs	22
Residual DEIM tolerances	$10^{-5} - 10^{-2}$	Residual DEIM terms	72
Newton iterations	8	Newton tolerance	10^{-7}
FE time residual assembling	$4\,s$	RB time residual assembling	$0.52\,s$
FE time	$7\,min$	RB time	$20\,s$

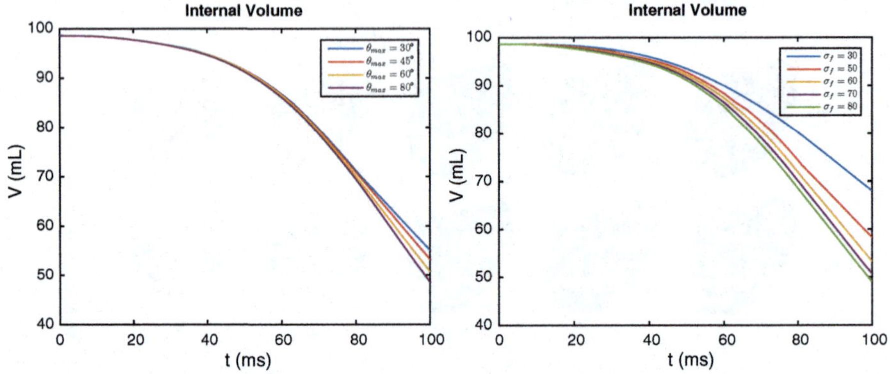

Fig. 12 Test case 2: internal volume variation when varying θ_{max} (left) and σ_f (right)

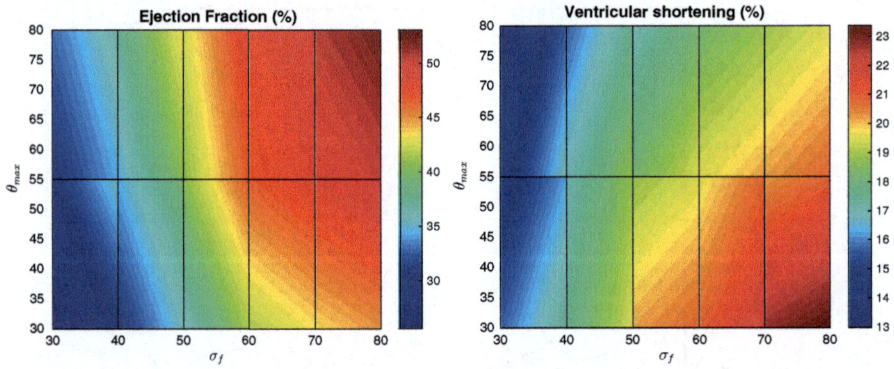

Fig. 13 Test case 2: ejection fraction and ventricular shortening as functions of σ_f and θ_{max}

(left) and $\theta_{max} = 60°$, $\sigma_s = \sigma_n = 20$ (right) have been considered. The associated ejection fraction has been reported in Fig. 13, where also the ventricular shortening is shown. These analyses have been carried out by varying θ_{max} and σ_f since σ_s and σ_n have a moderate effect on the solution. The quantity of blood ejected by the ventricle is larger when θ_{max} and σ_f assume large values. On the contrary, shortening is higher when considering small values of θ_{max}. This behavior appears counterintuitive since we may expect that higher shortening corresponds to higher

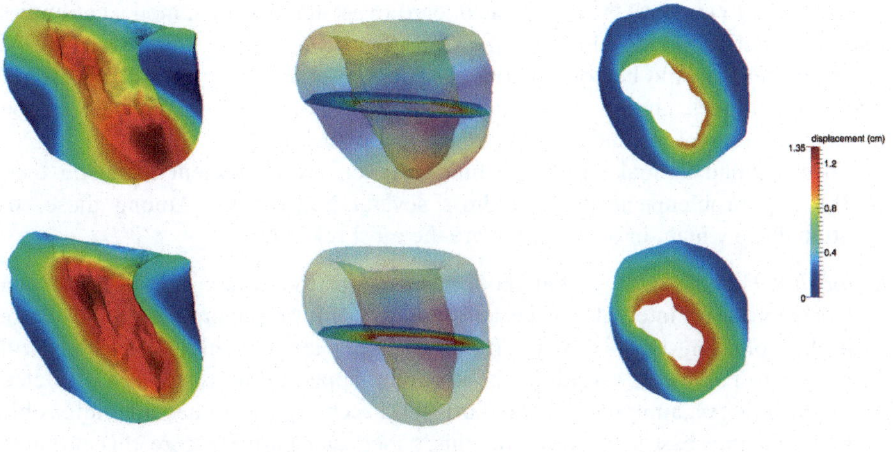

Fig. 14 Displacement of the myocardium for $\theta_{max} = 30°$ (up) and $\theta_{max} = 80°$

ejection fraction. To explain this phenomenon, we report in Fig. 14 the displacement of the muscle for two different values of θ_{max}: even if for $\theta_{max} = 80°$ the shortening is smaller, the cavity is more shrunk due to a larger wall thickening.

The model is able to reproduce the wall thickening and the ventricular shortening of the heart contraction. In particular, we obtain a ventricular shortening ranging from 13% to 23%, coherent with physiological values. The ejection fraction, usually measured by an echocardiogram, varies between 50% and 55% in healthy patients.

8 Challenges and Perspectives

The proper integration of several techniques to perform both solution space reduction (Galerkin-POD method) and system approximation (DEIM/MDEIM or, alternatively to this latter, Broyden approximation) has enabled the application of the reduced basis method for parametrized PDEs to problems arising in cardiac electromechanics. Cardiac electrophysiology and mechanics problems pose several challenges to ROM techniques, because of their complex, nonlinear, multiscale (both in space and time) nature; moreover, parameter dependence might be extremely involved, for instance when aiming at describing subject-specific clinical data and exploring inter-patient variability. On the other hand, ROM techniques provide a unique opportunity to solve relevant problems related with data-model integration, such as model calibration, uncertainty characterization and propagation, parameter identification and inverse problems, data assimilation.

All these problems are of primary importance in order to translate mathematical models into clinical care. Quantitative insights coming from the repeated solution of electrophysiology corresponding to different values of material parameter could provide better understanding of heart (mis)functionality; parameter identification has potential to improve the diagnosis of cardiovascular diseases; model calibration may be beneficial to develop therapies tailored to the subject characteristics. All these problems would be computationally unaffordable when relied only on high-fidelity techniques, aiming at considering variability of geometries, a wide number of scenarios to explore, and several parameters to deal with.

The roadmap to make ROM techniques even more efficient to tackle these challenging problems needs to address several bottlenecks. Among these, we mention those which, in our opinion, are the most relevant:

1. *local ROMs*. Using a global reduced basis for the whole parameter set and the whole time interval can be an extremely limiting approach. For instance, if the solution shows moving fronts and this latter is highly sensible with respect to parameter variations (as it might happen in the case of cardiac electrophysiology), applying the standard ROM techniques can become unfeasible. Local-in-time bases, as shown in this paper, can partially cure this problem, however more general and robust techniques to build *local ROMs* are required. As shown in [59, 60], a k-means clustering in the state space of the snapshots for both the solution and the nonlinear term can be a viable strategy to overcome this bottleneck;
2. *time behavior*. So far, POD has been applied also with respect to the time independent variable, to compress information carried by a set of solution snapshots computed at different time steps, no matter which is the time behavior characterizing the problem. In general, however, reduction of parametrized PDEs becomes much more difficult when passing from elliptic (or dissipative time-dependent) to hyperbolic problems (for instance pure transport equations). More ad-hoc strategies, able for instance to detect time invariances, or to track traveling waves, should be considered in the case of cardiac electrophysiology, because of the sharpness of the front and its extremely rapid dynamics;
3. *coupled and multiphysics problems*. Designing efficient ROMs (either monolithic or segregated) to couple different problems, such as cardiac electrophysiology and mechanics, is still an open issue, and almost untouched in the field of cardiovascular applications; preliminary investigations on the electromechanical case can be found in [13]. In this latter work it is also shown a possible strategy to realize a fully coupled electromechanical problem by including the mechano-electrical feedback only during the online stage, and relying on different ROMs for the two subproblems; Similarly, in this work the choice of using the monodomain model is performed. Such a model is adequate in physiological conditions, however the richer—but more complicated—bidomain model is required for treating some pathological conditions. The methods described in the previous sections can also apply to the bidomain model, however entailing an extra burden from a computational standpoint, because of the presence of two fields to be computed, and the different overall nature of the PDEs system.
4. *multiscale problems*. The multiscale nature of the electromechanical model should be properly taken into account in a ROM aiming at describing multiple behaviors at different scales. Among the mentioned open problems, this is by

far the most involved one, because of the intrinsic difficulty of the models at the cellular scale. Several extensions of the RB method to multiscale problems showing highly oscillating coefficients have been considered in the last decade, however only focusing on elliptic problems [1, 2, 40, 57]. When applied to these problems, the RB method enables to speed up a large number of similar computations on the fine local mesh for each new realization of the coefficient; its application to complex nonlinear materials, however, is still completely open.

5. *multi-fidelity problems*. In view of exploiting reduced order modeling techniques to solve uncertainty quantification problems, the use of models characterized by different fidelities is also foreseen. Following, e.g., the approach relying on multi-fidelity sampling and a Bayesian formulation proposed in [12], information from an approximate, low-fidelity model can be rigorously exploited and incorporated to perform output evaluations, estimates on their variability and, ultimately, parameter studies. In this respect, the eikonal model and a recently proposed reaction-eikonal source model [56], this latter offering the computational advantages of the eikonal model while preserving the full biophysical details of a computationally costly reaction-diffusion model, could be considered as an extremely cheap, yet detailed, low-fidelity model.

The investigation and development of reliable and efficient reduced-order modeling techniques is a very active field of numerical analysis and scientific computing; with no doubt, cardiovascular applications represent one of the most challenging and significant environments.

References

1. Abdulle, A., Bai, Y.: Reduced basis finite element heterogeneous multiscale method for high-order discretizations of elliptic homogenization problems. J. Comput. Phys. **231**(21), 7014–7036 (2012)
2. Abdulle, A., Budáč, O.: A reduced basis finite element heterogeneous multiscale method for stokes flow in porous media. Comput. Methods Appl. Mech. Eng. **307**, 1–31 (2016)
3. Ambrosi, D., Arioli, G., Nobile, F., Quarteroni, A.: Electromechanical coupling in cardiac dynamics: the active strain approach. SIAM J. Appl. Math. **71**(2), 605–621 (2011). https://doi.org/10.1137/100788379
4. Ambrosi, D., Pezzuto, S.: Active stress vs. active strain in mechanobiology: constitutive issues. J. Elast. **107**(2), 199–212 (2012)
5. Amsallem, D., Zahr, M., Farhat, C.: Nonlinear model order reduction based on local reduced-order bases. Int. J. Numer. Meth. Eng. **92**(10), 891–916 (2012)
6. Ashikaga, H., Coppola, B., Yamazaki, K., Villarreal, F., Omens, J., Covell, J.: Changes in regional myocardial volume during the cardiac cycle: implications for transmural blood flow and cardiac structure. Am. J. Physiol. Heart. Circ. Physiol. **295**(2), H610–H618 (2008)
7. Ballarin, F., Faggiano, E., Ippolito, S., Manzoni, A., Quarteroni, A., Rozza, G., Scrofani, R.: Fast simulations of patient-specific haemodynamics of coronary artery bypass grafts based on a POD–Galerkin method and a vascular shape parametrization. J. Comput. Phys. **315**, 609–628 (2016)

8. Ballarin, F., Faggiano, E., Manzoni, A., Quarteroni, A., Rozza, G., Ippolito, S., Antona, C., Scrofani, R.: Numerical modeling of hemodynamics scenarios of patient-specific coronary artery bypass grafts. Biomech. Model. Mechanobiol. **16**(4), 1373–1399 (2017)
9. Balzani, D., Deparis, S., Fausten, S., Forti, D., Heinlein, A., Klawonn, A., Quarteroni, A., Rheinbach, O., Schroder, J.: Aspects of Arterial Wall Simulations: Nonlinear Anisotropic Material Models and Fluid Structure Interaction. Dekan der Fak. für Mathematik und Informatik (2014)
10. Barrault, M., Maday, Y., Nguyen, N., Patera, A.: An empirical interpolation method: application to efficient reduced-basis discretization of partial differential equations. C. R. Math. **339**(9), 667–672 (2004)
11. Baumann, M.: Nonlinear model order reduction using pod/deim for optimal control of Burgers' equation. Ph.D. thesis, TU Delft, Delft University of Technology (2013)
12. Biehler, J., Gee, M., Wall, W.: Towards efficient uncertainty quantification in complex and large scale biomechanical problems based on a Bayesian multi fidelity scheme. Biomech. Model. Mechanobiol. **14**(3), 489–513 (2015)
13. Bonomi, D.: Reduced-order models for the parametrized cardiac electromechanical problem. Ph.D. thesis, Politecnico di Milano (2017)
14. Bonomi, D., Manzoni, A., Quarteroni, A.: A matrix DEIM technique for model reduction of nonlinear parametrized problems in cardiac mechanics. Comput. Methods Appl. Mech. Eng. **324**, 300–326 (2017)
15. Boulakia, M., Schenone, E., Gerbeau, J.: Reduced-order modeling for cardiac electrophysiology. Application to parameter identification. Int. J. Numer. Meth. Biomed. Eng. **28**(6–7), 727–744 (2012)
16. Broyden, C.: A class of methods for solving nonlinear simultaneous equations. Math. Comput. **19**(92), 577–593 (1965)
17. Bueno-Orovio, A., Cherry, E., Fenton, F.: Minimal model for human ventricular action potentials in tissue. J. Theor. Biol. **253**(3), 544–560 (2008). https://doi.org/10.1016/j.jtbi.2008.03.029
18. Carlberg, K., Farhat, C., Cortial, J., Amsallem, D.: The GNAT method for nonlinear model reduction: effective implementation and application to computational fluid dynamics and turbulent flows. J. Comput. Phys. **242**, 623–647 (2013)
19. Carlberg, K., Tuminaro, R., Boggs, P.: Preserving Lagrangian structure in nonlinear model reduction with application to structural dynamics. SIAM J. Sci. Comput. **37**(2), B153–B184 (2015)
20. Chapelle, D., Gariah, A., Sainte-Marie, J.: Galerkin approximation with proper orthogonal decomposition: new error estimates and illustrative examples. ESAIM: Math. Model. Numer. Anal. **46**(4), 731–757 (2012)
21. Chaturantabut, S., Sorensen, D.: Nonlinear Model Reduction via Discrete Empirical Interpolation. SIAM J. Sci. Comput. **32**(5), 2737–2764 (2010). https://doi.org/10.1137/090766498
22. Chaturantabut, S., Sorensen, D.: Application of POD and DEIM on dimension reduction of non-linear miscible viscous fingering in porous media. Math. Comp. Model. Dyn. **17**(4), 337–353 (2011)
23. Cherubini, C., Filippi, S., Nardinocchi, P., Teresi, L.: An electromechanical model of cardiac tissue: constitutive issues and electrophysiological effects. Prog. Biophys. Mol. Biol. **97**(2–3), 562–573 (2008)
24. Clayton, R., Panfilov, A.: A guide to modelling cardiac electrical activity in anatomically detailed ventricles. Prog. Biophys. Mol. Biol. **96**(1), 19–43 (2008). https://doi.org/10.1016/j.pbiomolbio.2007.07.004
25. Colciago, C., Deparis, S., Quarteroni, A.: Comparisons between reduced order models and full 3d models for fluid–structure interaction problems in haemodynamics. J. Comput. Appl. Math. **265**, 120–138 (2014)

26. Colciago, C.M., Deparis, S., Forti, D.: Fluid-structure interaction for vascular flows: from supercomputers to laptops. In: Frei, S., Holm, B., Richter, T., Wick, T., Yang, H. (eds.) Fluid-Structure Interaction: Modeling, Adaptive Discretisations and Solvers. Radon Series on Computational and Applied Mathematics, vol. 20. De Gruyter, Berlin (2017)
27. Colli Franzone, P., Pavarino, L.F., Taccardi, B.: Simulating patterns of excitation, repolarization and action potential duration with cardiac bidomain and monodomain models. Math. Biosci. **197**(1), 35–66 (2005)
28. Colli Franzone, P., Pavarino, L.F., Scacchi, S.: Mathematical Cardiac Electrophysiology. Modeling, Simulation and Applications Series, vol. 13. Springer, Milano (2014)
29. Colli Franzone, P., Pavarino, L.F., Scacchi, S.: Parallel multilevel solvers for the cardiac electromechanical coupling. Appl. Numer. Math. **95**, 140–153 (2015)
30. Corrado, C., Lassoued, J., Mahjoub, M., Zemzemi, N.: Stability analysis of the POD reduced order method for solving the bidomain model in cardiac electrophysiology. Math. Biosci. **272**, 81–91 (2016)
31. Dal, H., Goktepe, S., Kaliske, M., Kuhl, E.: A fully implicit finite element method for bidomain models of cardiac electromechanics. Comput. Methods Appl. Mech. Eng. **253**, 323–336 (2013)
32. Deparis, S., Forti, D., Quarteroni, A.: A rescaled localized radial basis function interpolation on non-cartesian and nonconforming grids. SIAM J. Sci. Comput. **36**(6), A2745–A2762 (2014)
33. Eriksson, T., Prassl, A., Plank, G., Holzapfel, G.: Influence of myocardial fiber/sheet orientations on left ventricular mechanical contraction. Math. Mech. Solids **18**, 592–606 (2013)
34. Fedele, M., Faggiano, E., Barbarotta, L., Cremonesi, F., Formaggia, L., Perotto, S.: Semi-automatic three-dimensional vessel segmentation using a connected component localization of the region-scalable fitting energy. In: 2015 9th International Symposium on Image and Signal Processing and Analysis (ISPA), pp. 72–77. IEEE, Piscataway, NJ (2015)
35. Gerbeau, J., Lombardi, D., Schenone, E.: Reduced order model in cardiac electrophysiology with approximated lax pairs. Adv. Comput. Math. **41**(5), 1103–1130 (2015)
36. Gerbi, A., Dede', L., Quarteroni, A.: A monolithic algorithm for the simulation of cardiac electromechanics in the human left ventricle. Tech. rep., MOX - Politecnico di Milano (2017). Report 51/2017
37. Göktepe, S., Kuhl, E.: Electromechanics of the heart: a unified approach to the strongly coupled excitation–contraction problem. Comput. Mech. **45**(2–3), 227–243 (2010)
38. Heidenreich, E., Ferrero, J., Doblare, M., Rodriguez, J.: Adaptive macro finite elements for the numerical solution of monodomain equations in cardiac electrophysiology. Ann. Biomed. Eng. **38**(7), 2331–2345 (2010)
39. Helfenstein, J., Jabareen, M., Mazza, E., Govindjee, S.: On non-physical response in models for fiber-reinforced hyperelastic materials. Int. J. Solids Struct. **47**(16), 2056–2061 (2010)
40. Hesthaven, J.S., Zhang, S., Zhu, X.: Reduced basis multiscale finite element methods for elliptic problems. Multiscale Model. Simul. **13**(1), 316–337 (2015)
41. Hesthaven, J., Rozza, G., Stamm, B.: Certified Reduced Basis Methods for Parametrized Partial Differential Equations. SpringerBriefs Mathematics. Springer, Cham (2016)
42. Hodgkin, A., Huxley, A.: A quantitative description of membrane current and its application to conduction and excitation in nerve. J. Physiol. **117**(4), 500–544 (1952)
43. Holzapfel, G., Ogden, R.: Constitutive modelling of passive myocardium: a structurally based framework for material characterization. Philos. Trans. A Math. Phys. Eng. Sci. **367**(1902), 3445–3475 (2009). https://doi.org/10.1098/rsta.2009.0091
44. Keldermann, R., Nash, M., Panfilov, A.: Modeling cardiac mechano-electrical feedback using reaction-diffusion-mechanics systems. Physica D **238**(11), 1000–1007 (2009)
45. Krysl, P., Lall, S., Marsden, J.: Dimensional model reduction in non-linear finite element dynamics of solids and structures. Int. J. Numer. Meth. Eng. **51**(4), 479–504 (2001)
46. Kuzmin, D.: A Guide to Numerical Methods for Transport Equations. University Erlangen-Nuremberg, Erlangen (2010)
47. Lassila, T., Manzoni, A., Quarteroni, A., Rozza, G.: Boundary control and shape optimization for the robust design of bypass anastomoses under uncertainty. ESAIM-Math. Model. Numer. **47**(4), 1107–1131 (2013)

48. Lassila, T., Manzoni, A., Quarteroni, A., Rozza, G.: A reduced computational and geometrical framework for inverse problems in haemodynamics. Int. J. Numer. Methods Biomed. Eng. **29**(7), 741–776 (2013)
49. Maday, Y., Nguyen, N.C., Patera, A.T., Pau, G.S.H.: A general multipurpose interpolation procedure: the magic points. Commun. Pure Appl. Anal. **8**(1), 383–404 (2009)
50. Manzoni, A., Quarteroni, A., Rozza, G.: Shape optimization for viscous flows by reduced basis methods and free-form deformation. Int. J. Numer. Meth. Fluids **70**(5), 646–670 (2012)
51. Nardinocchi, P., Teresi, L.: On the active response of soft living tissues. J. Elast. **88**(1), 27–39 (2007)
52. Nash, M., Hunter, P.: Computational mechanics of the heart. J. Elast. **61**, 113–141 (2001). https://doi.org/10.1023/A:1011084330767
53. Nash, M., Panfilov, A.: Electromechanical model of excitable tissue to study reentrant cardiac arrhythmias. Prog. Biophys. Mol. Biol. **85**(2–3), 501–522 (2004). https://doi.org/10.1016/j.pbiomolbio.2004.01.016
54. Negri, F.: Efficient reduction techniques for the simulation and optimization of parametrized systems: Analysis and applications. Ph.D. thesis, Ecole Polytechnique Fédérale de Lausanne (2016)
55. Negri, F., Manzoni, A., Amsallem, D.: Efficient model reduction of parametrized systems by matrix discrete empirical interpolation. J. Comput. Phys. **303**, 431–454 (2015)
56. Neic, A., Campos, F., Prassl, A., Niederer, S., Bishop, M., Vigmond, E., Plank, G.: Efficient computation of electrograms and ECGs in human whole heart simulations using a reaction-eikonal model. J. Comput. Phys. **346**, 191–211 (2017)
57. Nguyen, N.: A multiscale reduced-basis method for parametrized elliptic partial differential equations with multiple scales. J. Comput. Phys. **227**(23), 9807–9822 (2008)
58. Noble, D., Garny, A., Noble, P.: How the hodgkin-huxley equations inspired the cardiac physiome project. J. Physiol. **590**(11), 2613–28 (2012)
59. Pagani, S.: Reduced-order models for inverse problems and uncertainty quantification in cardiac electrophysiology. Ph.D. thesis, Politecnico di Milano (2017)
60. Pagani, S., Manzoni, A., Quarteroni, A.: Numerical approximation of parametrized problems in cardiac electrophysiology by a local reduced basis method. Comput. Methods Appl. Mech. Eng. **340**, 530–558 (2018)
61. Pathmanathan, P., Whiteley, J.: A numerical method for cardiac mechanoelectric simulations. Ann. Biomed. Eng. **37**(5), 860–873 (2009)
62. Pathmanathan, P., Chapman, S., Gavaghan, D., Whiteley, J.: Cardiac electromechanics: the effect of contraction model on the mathematical problem and accuracy of the numerical scheme. J. Mech. Appl. Math. **63**, 375–399 (2010)
63. Pathmanathan, P., Mirams, G., Southern, J., Whiteley, J.: The significant effect of the choice of ionic current integration method in cardiac electro-physiological simulations. Int. J. Num. Meth. Biomed. Eng. **27**(1), 1751–1770 (2011). https://doi.org/10.1002/cnm. http://onlinelibrary.wiley.com/doi/10.1002/cnm.1494/full
64. Peherstorfer, B., Butnaru, D., Willcox, K., Bungartz, H.: Localized discrete empirical interpolation method. SIAM J. Sci. Comput. **36**(1), A168–A192 (2014)
65. Pezzuto, S.: Mechanics of the heart – constitutive issues and numerical experiments. Ph.D. thesis, Politecnico di Milano (2013)
66. Potse, M., Dubé, B., Vinet, A., Cardinal, R.: A comparison of monodomain and bidomain propagation models for the human heart. Conf. Proc. IEEE Eng. Med. Biol. Soc. **53**(12), 3895–3898 (2006). https://doi.org/10.1109/IEMBS.2006.259484
67. Quarteroni, A., Valli, A.: Numerical Approximation of Partial Differential Equations, vol. 23. Springer Science and Business Media, Berlin (2008)
68. Quarteroni, A., Manzoni, A., Negri, F.: Reduced Basis Methods for Partial Differential Equations: An Introduction. Unitext, vol. 92. Springer, Cham (2016)
69. Quarteroni, A., Lassila, T., Rossi, S., Ruiz-Baier, R.: Integrated heart – coupling multiscale and multiphysics models for the simulation of the cardiac function. Comput. Methods Appl. Mech. Eng. **314**, 345–407 (2017)

70. Quarteroni, A., Manzoni, A., Vergara, C.: The cardiovascular system: mathematical modeling, numerical algorithms, clinical applications. Acta Numer. **26**, 365–590 (2017)
71. Radermacher, A., Reese, S.: POD-based model reduction with empirical interpolation applied to nonlinear elasticity. Int. J. Numer. Meth. Eng. **107**(6), 477–495 (2016)
72. Rossi, S.: Anisotropic modeling of cardiac mechanical activation. Ph.D. thesis, Ecole Politechnique Federale de Lausanne (2014)
73. Rossi, S., Ruiz-Baier, R., Pavarino, L.F., Quarteroni, A.: Orthotropic active strain models for the numerical simulation of cardiac biomechanics. Int. J. Numer. Meth. Biomed. Eng. **28**(6–7), 761–788 (2012)
74. Rossi, S., Lassila, T., Ruiz-Baier, R., Sequeira, A., Quarteroni, A.: Thermodynamically consistent orthotropic activation model capturing ventricular systolic wall thickening in cardiac electromechanics. Eur. J. Mech. A/Sol. **48** (2013). https://doi.org/10.1016/j.euromechsol.2013.10.009
75. Ruiz-Baier, R., Gizzi, A., Rossi, S., Cherubini, C., Laadhari, A., Filippi, S., Quarteroni, A.: Mathematical modelling of active contraction in isolated cardiomyocytes. Math. Med. Biol. **31**(3), 259–283 (2014)
76. Sainte-Marie, J., Chapelle, D., Cimrman, R., Sorine, M.: Modeling and estimation of the cardiac electromechanical activity. Comput. Struct. **84**(28), 1743–1759 (2006)
77. Sansour, C.: On the physical assumptions underlying the volumetric-isochoric split and the case of anisotropy. Eur. J. Mech. A/Sol. **27**(1), 28–39 (2008). https://doi.org/10.1016/j.euromechsol.2007.04.001
78. Smith, N., Nickerson, D., Crampin, E., Hunter, P.: Multiscale computational modelling of the heart. Acta Numer. **13**, 371–431 (2004). https://doi.org/10.1017/S0962492904000200
79. Strobeck, J., Sonnenblick, E.: Myocardial contractile properties and ventricular performance. In: The Heart and Cardiovascular System: Scientific Foundations, pp. 31–49. Raven Press, New York (1986)
80. Sundnes, J., Wall, S., Osnes, H., Thorvaldsen, T., McCulloch, A.: Improved discretisation and linearisation of active tension in strongly coupled cardiac electro-mechanics simulations. Comput. Meth. Biomech. Biomed. Eng. **17**(6), 604–615 (2014)
81. Taber, L., Perucchio, R.: Modeling heart development. J. Elast. **61**(1–3), 165–197 (2000)
82. Trayanova, N., Eason, J., Aguel, F.: Computer simulations of cardiac defibrillation: a look inside the heart. Comput. Vis. Sci. **4**(4), 259–270 (2002)
83. Tung, L.: A bi-domain model for describing ischemic myocardial DC potentials. Ph.D. thesis, Massachusetts Institute of Technology (1978)
84. Wang, Y., Haynor, D., Kim, Y.: An investigation of the importance of myocardial anisotropy in finite element modeling of the heart: methodology and application to the estimation of defibrillation efficacy. IEEE Trans. Biomed. Eng. **48**(12), 1377–1389 (2001)
85. Washabaugh, K., Amsallem, D., Zahr, M., Farhat, C.: Nonlinear model reduction for cfd problems using local reduced-order bases. In: 42nd AIAA Fluid Dynamics Conference and Exhibit, Fluid Dynamics and Co-located Conferences, AIAA Paper, vol. 2686, pp. 1–16 (2012)
86. Whiteley, J., Bishop, M., Gavaghan, D.: Soft tissue modelling of cardiac fibres for use in coupled mechano-electric simulations. Bull. Math. Biol. **69**(7), 2199–2225 (2007)
87. Wirtz, D., Sorensen, D., Haasdonk, B.: A posteriori error estimation for DEIM reduced nonlinear dynamical systems. SIAM J. Sci. Comput. **36**(2), A311–A338 (2014)
88. Yang, H., Veneziani, A.: Efficient estimation of cardiac conductivities via pod-deim model order reduction. Appl. Numer. Math. **115**, 180–199 (2017)

Andrea Manzoni is Assistant Professor of Numerical Analysis at Politecnico di Milano since 2017. After earning the PhD in Mathematics at the EPFL (Swiss Federal Institute of Technology), Lausanne (Switzerland) in 2012, he was postdoctoral researcher and research associate at SISSA (International School for Advanced Studies), Trieste (Italy) and EPFL. He works on the development of reduced order modeling techniques for the rapid and reliable approximation of parameter-dependent PDEs; his research activity also focuses on uncertainty quantification,

PDE-constrained optimization, computational and statistical inverse problems. His interests in the field of mathematical and numerical modeling of the heart are related with electrophysiology, mechanical problems, and haemodynamics.

Diana Bonomi got her Bachelor and Master Degrees in Mathematical Engineering at Politecnico di Milano, then her Ph.D. in Mathematical Models and Methods in Engineering in 2017, still at Politecnico di Milano, with a subject on reduced-order models for parametrized cardiac electromechanical problems. This work is partially based on her PhD research thesis.

Alfio Quarteroni is Full Professor of Numerical Analysis at the Politecnico of Milan (Italy), since 1989. He has been director of the Chair of Modelling and Scientific Computing at the EPFL (Swiss Federal Institute of Technology), Lausanne (Switzerland) from 1998 to 2017. He is member of the Italian Academy of Science (Accademia Nazionale dei Lincei), the European Academy of Science, and the Academia Europaea. He is the author of 22 books, editor of 5 books, author of more than 300 papers. He is the recipient of the ERC Advanced Grant "MATHCARD" in 2008, of two ERC PoC (Proof of Concept) grants: "Math2Ward" in 2012 and "Math4AAARisk" in 2015, and of another ERC Advanced Grant "iHEART" in 2017, all about the mathematical modelling of the cardiovascular system.

Aortic Endovascular Surgery

Michele Conti, Simone Morganti, Alice Finotello, Rodrigo M. Romarowski, Alessandro Reali, and Ferdinando Auricchio

Abstract The continuous technological improvements of medical instruments and devices make minimally-invasive approaches a real and valid alternative to standard open surgery in more and more cases. Recent developments in cardiovascular surgery, in particular, have led to the success of thoracic endovascular repair (TEVAR) and transcatheter aortic valve implantation (TAVI). If, on the one hand, minimally-invasive interventions induce shorter hospital stays, faster recovery, and thus reduced costs, on the other hand, since, for obvious reasons, the direct control of the operator on the procedure is much more limited, operation planning and decision-making steps cover a crucial importance. In this context, computational tools have demonstrated to play a remarkable role, providing the surgeon with predictive information regarding the potential optimality of the treatment strategy. In the present chapter, we aim at describing recent developments of TEVAR and TAVI modeling, from both the structural and fluid-dynamic point of view.

M. Conti · A. Reali · F. Auricchio
Department of Civil Engineering and Architecture, University of Pavia, Pavia, Italy
e-mail: michele.conti@unipv.it; alessandro.reali@unipv.it; auricchio@unipv.it

S. Morganti (✉)
Department of Electrical, Computer, and Biomedical Engineering, University of Pavia, Pavia, Italy
e-mail: simone.morganti@unipv.it

A. Finotello
Department of Experimental Medicine, University of Genoa, Genoa, Italy
e-mail: alice.finotello@edu.unige.it

R. M. Romarowski
3D Simulation Lab, IRCCS Policlinico San Donato, San Donato MI, Italy
e-mail: Rodrigo.Romarowski@grupposandonato.it

D. Boffi et al. (eds.), *Mathematical and Numerical Modeling of the Cardiovascular System and Applications*, SEMA SIMAI Springer Series 16,
https://doi.org/10.1007/978-3-319-96649-6_7

1 Introduction

Cardiovascular diseases still represent the major cause of death in Western countries [21]. Among the many different cardiovascular pathologies, the structural deterioration and impairment of arteries (e.g., aortic aneurysm or carotid artery stenosis), as well as of heart valves (e.g., aortic valve stenosis or mitral valve prolapse) certainly plays a crucial role.

In the recent years, for the treatment of such a class of pathologies, minimally-invasive endovascular approaches have met an increasing interest and stimulated continued developments. In fact, endovascular procedures, consisting in the insertion of miniature devices through a blood vessel for the treatment of the specific vascular disease, present many advantages over open surgery. Benefits of minimally-invasive procedures include: (1) shorter hospital stay, being much less traumatic than standard open procedures, and, consequently, (2) reduced costs for institutions, (3) fewer complications traditionally associated with open surgery, and (4) faster recovery and return to normal physical activity. For these reasons, currently, in many cases and whenever possible, surgeons opt for endovascular treatments, especially for high-risk patients.

However, the less invasiveness of endovascular treatments makes the direct control of the procedure more difficult for the surgeon, since only radiological images (usually angiographic images) can guide him when performing the operation. Moreover, like all medical procedures, also endovascular surgery is not without risks. In particular, possible risks that can occur include: (endo- or paravalvular) leaks, device movement and migration, device excessive deformation preventing optimal performance, etc. All these possible complications make the procedure heavily dependent on the correct choice of the device type, size, and positioning strategy; which is the main reason why physicians are more and more recognizing the importance of pre-surgical planning for these interventions.

In this context, in order to predict the complications that may occur and, at the same time, in order to optimize the entire surgical strategy, a very promising approach is given by virtual computer-based simulations taking into account the specific details of the anatomy and characteristics of arteries and valves, as they can greatly differ from one individual to another. There is little doubt that patient-specific simulations will help minimally-invasive procedures expand their recommended patient population.

In the present chapter we focus on two of the most routinely performed endovascular procedures, i.e., thoracic endovascular repair (TEVAR) and transcatheter aortic valve implantation (TAVI), with the aim of presenting computational strategies and tools mainly based on finite element to obtain predictions that may support the decision-making process and procedure planning. In particular, after a brief description of the state-of-the art, we will present both structural and fluid-dynamics results obtained by simulating TEVAR and TAVI in patient-specific cases.

2 Structural Simulation of TEVAR

TEVAR is a consolidated procedure to treat thoracic aortic diseases such as aneurysms and dissections, especially in those patients who are unsuitable for standard open surgery [12]. The procedure is performed through a catheter-guided deployment of one or more stent-grafts, which are metallic tubular structures covered by a polymeric skirt. Such devices are normally used to diverge the arterial blood flow within pathologic vascular anatomies; for instance, stent-grafts are used to reduce the pressurization of aneurysmatic sac or to exclude the perfusion of false lumen in a dissected artery. The rapid evolution of self-expandable stent-grafts in the last decade motivates the use of structural Finite Element Analysis (FEA) also in this emerging sector.

One of the pioneer investigations in this area was presented in 2008 by Kleinstreuer et al. [18] who discussed a FEA of tubular, diamond-shaped stent grafts under representative cyclic loading for abdominal aortic aneurysm (AAA) repair. In particular, the authors studied the mechanical behavior and fatigue performance of different materials resembling commercially available stent-graft systems, evaluating and comparing the effects of the different stages of the device usage (crimping, deployment, and cyclic pressure loading) on stent-graft fatigue life, radial force, and wall compliance through the numerical simulations.

In 2012, De Bock and colleagues [7] experimentally validated the use of structural FEA to virtually deploy a bifurcated stent graft (Medtronic Talent) in a patient-specific model of an abdominal aorta. The authors modeled the entire deployment procedure, with the stent graft being crimped and bent according to the vessel geometry, and subsequently released. The numerical procedure was validated comparing the simulation outcomes with the in-vitro data regarding the placement of the device in a silicone mock aneurysm, scan by high resolution Computed Tomography (CT) scan.

In the same year, Demanget et al. [8] simulated numerically the bending of two manufactured stent-grafts (Aorfix by Lombard Medical and Zenith by Cook Medical Europe) using FEA. The authors studied the overall behavior of the stent-grafts by assessing stent spacing variation and cross-section deformation. The study is motivated by the potential relationship between the clinical complications and the insufficient stent-graft flexibility, especially when devices are deployed in tortuous arteries.

The same authors validated the numerical procedure in a subsequent study [9] where the two commercially available stent-grafts were subjected to severe bending tests and their 3D geometries in undeformed and bent configurations were imaged from X-ray microtomography to set up stent-graft numerical models, subjected to the boundary conditions measured experimentally. The computational framework was further used to numerically assess the flexibility and mechanical stresses undergone by stents and fabric of currently marketed stent-graft limbs (Aorfix, Anaconda, Endurant, Excluder, Talent, Zenith Flex, Zenith LP, and Zenith Spiral-Z) [10].

In 2013, Auricchio and colleagues [2] described the use of a custom-made stent-graft to perform a fully endovascular repair of an asymptomatic ascending aortic pseudoaneurysm in a patient who was a poor candidate for open surgery. The authors also discussed the possible contribution of a dedicated medical images analysis and patient-specific simulation as support to procedural planning. In particular, the authors have compared the simulation prediction based on pre-operative images with post-operative outcomes. The agreement between the computer-based analysis and reality demonstrated by this study further encourages the use of FEA-based simulations not only as a tool for device designers but also as a procedural planning tool for the physicians.

Perrin and colleagues [22] investigated the expansion of five marketed stent-graft iliac limbs to evaluate quantitatively their mechanical performances. The simulation of the deployment was modeled in a simplified manner according to the following steps: stent-graft crimping and insertion in the delivery sheath, removal of the sheath and stent-graft deployment in the aneurysm, application of arterial pressure. Their results showed that in the most curved aneurysm and for some devices kinks and apposition defects are present especially at the outer curve of the artery.

Analogously, the same research group [23] proposed a numerical methodology to predict stent-graft final deployed shapes after surgery, applied on three clinical cases, using preoperative scans to generate patient-specific vessel models. The numerical results were compared to the actual deployed geometry of the stent-grafts after surgery that was extracted from postoperative scans. They observed relevant matching between simulated and actual deployed stent-graft geometries, especially for proximal and distal stents outside the aneurysm sac which are particularly important for practitioners. Such a good agreement between numerical results and clinical cases makes finite element simulation very promising for preoperative planning of endovascular repair; in fact, part of such technology is the ground of PrediSurge company.

The aforementioned studies clearly show that, nowadays, structural simulation of endografting is a consolidated technology. Certainly the time of proof-of-concept case studies is gone; the simulations are ready for the validation with perspective studies and to benchmarked by the clinical market. In such a scenario it is essential to promote the clinical applications of such tools. With this aim, in the following we briefly provide an example about the use of structural FEA to plan the surgery, accounting for patient-specific anatomical features, which are particular important in borderline patients where the pre-operative decisions based on static images are not straightforward.

Two patients with thoracic aortic dilations underwent contrast enhanced CT scans to visualize the diseased morphology. Patient 1 (P1) presented a saccular aneurysm having a maximum diameter of 55 mm just below the aortic arch, whereas Patient 2 (P2) had a maximum diameter of 40 mm in the same position. Both CT scans were segmented with the Vascular Modeling Toolkit (VMTK) open source library in order to create a 3D surface of the vessel wall.

Based on TEVAR guidelines to evaluate the length of the proximal landing zone and the radius of curvature of the arch, TEVAR was performed in P1 with the deployment of one stent-graft Medtronic Valiant 28-24-150 (Medtronic, Santa Rosa, CA, USA), consisting of 8 NiTiNOL rings covering a polyester skirt to limit blood flow into the sac plus a proximal uncovered ring to improve stability without occluding the left common carotid artery. Due to the borderline morphology of P2 arch, characterized by a short proximal aneurysm neck and high arch angulation, decision to undergo TEVAR was not taken immediately. Surgeons proposed to virtually deploy a stent-graft according to aortic characteristics, in this case a Medtronic Valiant 26-26-100.

Virtual stent-graft models were created for both patients to define meshes suitable for structural FEA using Abaqus/Explicit v. 6.16 (Simulia, Dassault Systèmes, Providence, RI, USA). Meshes consisted on a group of nodes connected by two sets of elements: one composed by linear brick elements (C3D8R) representing the struts and one of triangular membrane elements (M3D3) representing the fabric coverage. Struts were made of Nitinol whereas the fabric was made of woven polyester adopting the material properties reported in [18]. Aorta was considered rigid and surface meshes of triangular elements (R3D3) were also then imported in Abaqus.

Set-up of the input file for the structural FEA was supported by an in-house Python script allowing the user to select the given stent-graft model and the proximal landing point within the diseased vessel. Numerical analysis of stent-graft deployment in the aortic model is a non-linear problem involving large deformations and contact; Abaqus/Explicit was used as finite element solver. The steps of the deployment for the case of patient P1 are depicted in Fig. 1: first the stent-graft is crimped by a catheter and curved from a straight position to the vessel centerline; in this step, a frictionless contact interaction is enforced between the catheter and the struts of the prosthesis. Once the stent is in place, the widening of the catheter surface allows for the stent-graft to re-expand and thus simulate its deployment. To do so, a contact pair between the stent-graft struts and the luminal surface of the artery is activated.

Results of the deployment in P2 are shown in Fig. 2. To evaluate the quality of the apposition to the vessels wall, the point-wise distances between the surface of the

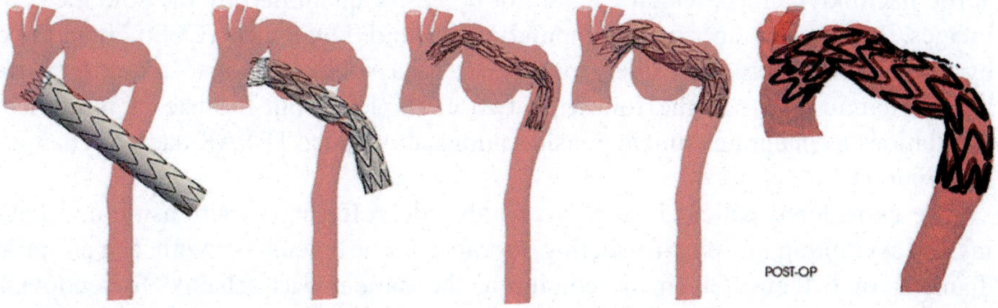

Fig. 1 On the left, steps of the endograft simulation in Patient 1 model; on the right, 3D reconstruction from actual post-operative CT scan, showing the endograft configuration after the implant, which matches the prediction of numerical analysis

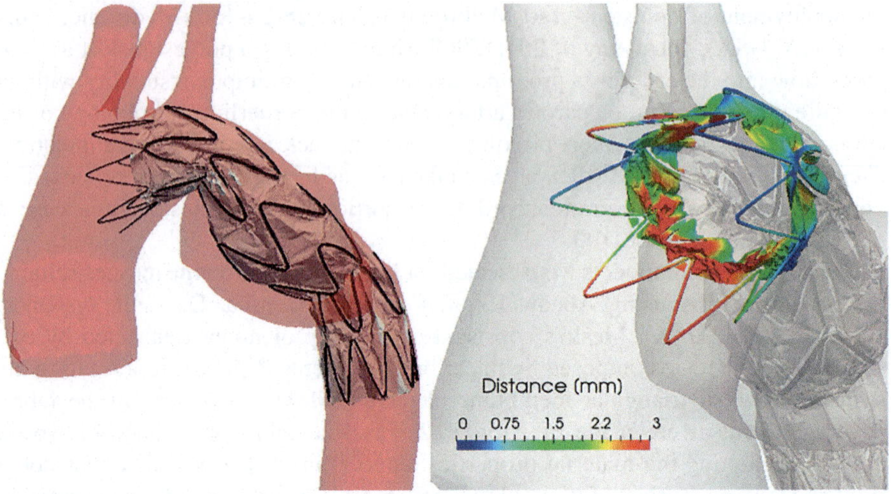

Fig. 2 Final configuration of Patient 2 device together with a map of the distances between the stent-graft and the vessel. A significant bird-beak—incomplete endograft apposition to the aortic wall—is shown

deployed stent-graft and the aorta were calculated; P2 showed a significant proximal bird-beak and a diameter mismatch in the distal landing zone, leading to endograft folding.

Based on our results, the impossibility to undergo TEVAR in P2 was confirmed. Considering the simulations and the clinical condition of the patient, vascular surgeons opted for an alternative treatment.

3 Fluid-Dynamics Simulation of TEVAR

The implant of endovascular prosthesis has an obvious impact on the post-operative aortic haemodynamics, which cannot not be easily quantified by the sole medical images. With such an aim, Computational Fluid Dynamics (CFD), integrated by clinical data, has extensively proved to be a valuable tool to quantify aorta hemodynamics [14]. In the following two examples about the use of numerical simulations to integrate clinical considerations about post-TEVAR haemodynamics are reported.

The considered patient is a 51-year-old male suffering hypertension and having an asymptomatic post-dissecting thoracic aortic aneurysm with a maximum diameter of 6.1 cm. Given his condition, the patient was selected for endovascular exclusion of the aneurysm using two cTAG stent-grafts (conformable TAG, W.L. GORE & Associates. Inc., Flagstaff, Arizona, USA), sized 34-34-20 and 28-28-15 mm, with the proximal landing zone covering the origin of the left subclavian artery. At 6 months follow-up, imaging showed successful exclusion of the aneurysm without endoleak but the post-operative CT scan showed a bird-beak configuration of the proximal prosthesis. Further collection of clinical data and medical images has been performed also 1 year after the intervention. In particular, magnetic resonance imaging (MRI) was performed using a 1.5-T scan (Magnetom Sonata Maestro Class, Siemens, Erlangen, Germany) to acquire phase-contrast (PC) sequences able to provide in-vivo measures of the flow at predefined aortic cross-sections. Such a measurements have been integrated with CFD to compute the actual haemodynamic conditions of the patient after the treatment. Although the main clinical concern was firstly directed to the endoluminal protrusion of the prosthesis, the CFD simulations have demonstrated that there are two other important areas where the local hemodynamics is impaired and a disturbed blood flow is present: the first one is the ostium of the left subclavian artery, which is partially closed by the graft; the second one is the stenosis of the distal thoracic aorta. Besides the clinical relevance of these specific findings, this study highlights how CFD analyses allow observing important flow effects resulting from the specific features of patient vessel geometries. Such results demonstrate the potential impact of computational biomechanics not only on the basic knowledge of physiopathology, but also on the clinical practice, thanks to a quantitative extraction of knowledge made possible by merging medical data and mathematical models [3] (see Fig. 3).

The same clinical case has been analyzed to evaluate how the aortic haemodynamics evolves in time [15]; also in this case the integration of numerical simulations and clinical data has been performed, pre-op, post-op, and 1-year follow up data. We focus on both geometrical features like curvature, torsion

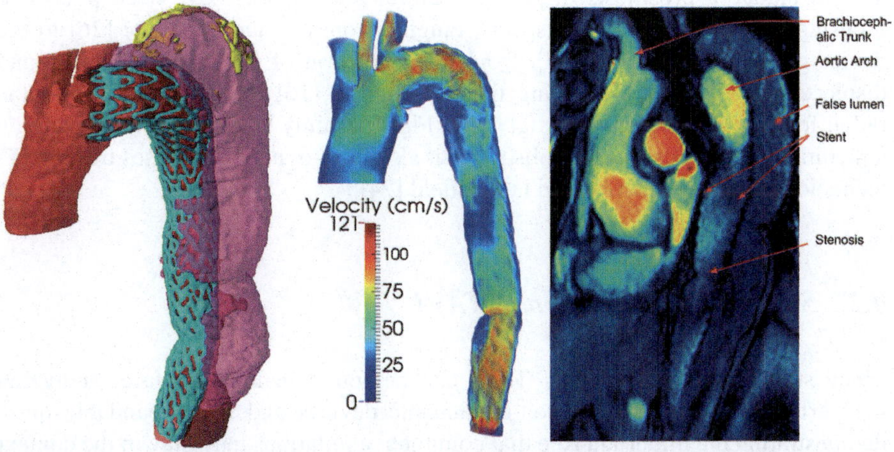

Fig. 3 Illustrative representation of the patient-specific post-operative haemodynamics, predicted by Computational Fluid Dynamics. The analysis is based on dedicated medical image elaboration and can be compared to in-vivo measures, performed through phase-contrast MRI. The images are adapted from Auricchio et al. [3]

and area variations, as well as near-wall and intravascular flow-related quantities (i.e., wall shear stress-based descriptors and helicity). Comparison of the different morphologies indicates a partial restoration of normal flow in the region of interest, even though low wall shear stresses are still present with the associated risks. Overall, this study demonstrates the efficacy of quantitative computational tools in understanding the long-term impact of TEVAR.

4 TAVI

4.1 Background and State of the Art

TAVI procedure is a quite innovative approach for the treatment of severe aortic stenosis. Instead of standard open-heart surgery, a catheter carrying a compressed prosthetic heart valve is placed usually in the femoral artery (or alternatively through the subclavian artery or even directly through the apex of the heart) and guided into the left ventricle to let the device be positioned over the diseased aortic valve.

This new approach for the treatment of symptomatic patients with severe aortic stenosis has been shown to be feasible and safe in patients at very high or prohibitive surgical risk and it has developed rapidly since the first-in-man implantation performed by Cribier in 2002 [6].

In the last decade, also numerical models and simulations aiming at predicting the behavior of implanted prosthetic device have significantly grown. From the first publication by Dwyer et al. [11] aiming at characterizing the blood ejection force able to induce a prosthesis migration, several other works have been published focusing on different specific aspects, ranging from the aortic rupture [26] to the device deployment [2, 4, 5, 17], from the extraction of postoperative stress and displacements by post-processing medical images [16] to the evaluation of the radial force [25]. An example is given in Fig. 4. Mainly contributions focusing on structural aspects have been published, but also fluid-dynamic analyses using CFD technology have been performed in this field [24].

4.2 Structural Simulation of TAVI

Many structural simulations of TAVI can be found in the literature, analyzing the performance of both balloon-expandable devices and self-expandable ones, demonstrating the important role that computer simulations may play in the context of predictive medicine.

We here mention and detail two examples of application of the finite element technology in real clinical cases: one taking into consideration a balloon-expandable device (i.e., the Edwards Sapien valve), and one taking into consideration a

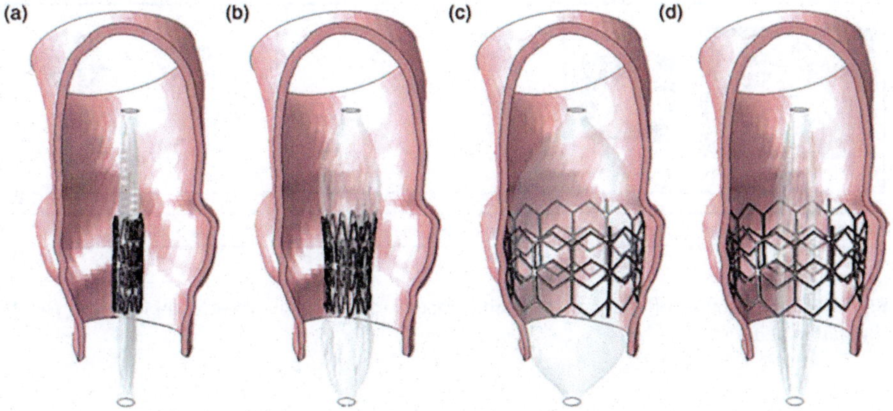

Fig. 4 Different frames of balloon expansion and stent apposition: (**a**) initial configuration; (**b**) the balloon starts to deploy the stent; (**c**) the balloon is fully expanded and the stent fully deployed; (**d**) final configuration after balloon deflation. The figure is adapted from [2]

self-expandable one (i.e., the Medtronic Corevalve valve). We finally draw some considerations of modeling the structural problem of TAVI using finite elements.

4.2.1 The Balloon-Expandable Edwards Sapien Device: A Clinical Application of FEA

Two patients who underwent TAVI with very different procedure outcomes were retrospectively selected. In this chapter section, we aim at demonstrating the capability of computational tools and, in particular, of FEA of reproducing the device performance as measured by physicians in the postoperative course. This can consequently prove the predictive potential of computer-based simulations in terms of valve performance and possible complications.

The simulation procedure is quite complex and can be summarized in seven principal steps [2]: (1) creation of stent model, (2) simulation of stent crimping, (3) creation of the patient-specific aortic valve model, (4) simulation of balloon inflation and stent expansion, (5) creation of prosthetic valve model, (6) simulation of valve mapping onto the deployed stent and (7) simulation of valve closure.

Both patients were treated with an Edwards Sapien XT valve, size 26. An accurate geometrical model of the device is reconstructed based on a high-resolution micro-CT scan (Skyscan 1172 with a resolution of 0.17 micron) of a real device sample. The obtained stent model is then meshed using about 85,000 solid elements with reduced integration. A Von Mises plasticity model with an isotropic hardening is adopted for the stent, which is made of a cobaltchromium alloy. In particular, the following parameters are used: 233 GPa, 0.35, 414 MPa, 933 MPa and 44.5% in terms of Young modulus, Poissons coefficient, yield stress, ultimate stress, and deformation at break, respectively [20].

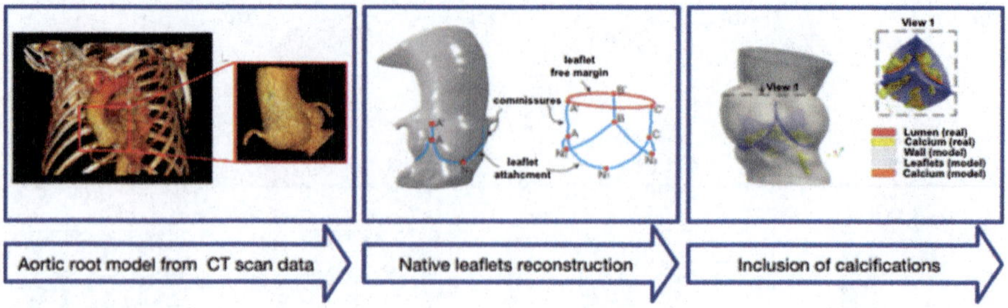

Fig. 5 Reconstruction of the patient-specific model of the aortic valve complex used for the simulation of TAVI

The native aortic valve complex model is obtained by processing preoperative CT medical images (see Fig. 5). In particular, the aortic root wall geometry is reconstructed using standard approaches well known for blood vessel reconstruction. We assume hyperelastic isotropic properties of the aortic tissue, modeled using a nearly-incompressible reduced sixth-order polynomial formulation. Parameters are tabled in [19]. The leaflets are constructed in open-configuration by coupling information coming from the CT scan (related to the leaflet attachment lines) and measurements of the free margin from ultrasound. Calcifications are also extracted from CT data and modeled assuming elastic behavior (Young modulus equal to 10 MPa and Poissons ratio equal to 0.35).

After simulating the crimping of the device, we perform the simulation of device implantation by mimicking the effects of a balloon, as happens in clinical practice. Once the stent is implanted, we simulate the prosthetic leaflets mapping on the stent and valve closure. All the numerical analyses are non-linear problems involving large deformations and contact. For this reason, Abaqus/Explicit solver is used to perform large deformation analyses; in particular, quasi-static procedures are used again assuming that inertia forces do not change the solution. Kinetic energy is monitored to ensure that the ratio of kinetic energy to internal energy remains less than 10%.

Stress distribution is characterized by concentrated spots of higher stress values induced by the contact between the stent and the aortic wall (see Fig. 6). On one side, higher stress values can be related to higher force of adherence between stent and aortic wall; on the other side, high stress patterns concentrated in the annular region can indicate a major risk of aortic rupture, which is a possible TAVI complication leading to cardiac tamponade and subsequent fatal events.

Paravalvular leakage is one of the most frequent complications which may occur after TAVI due to incomplete adherence of the prosthetic stent to the aortic wall. For the considered patients, we can quantitatively evaluate the area of the paravalvular orifices (see Fig. 7), which can be assumed to be proportional to the amount of retrograde paravalvular blood flow. Interestingly, the obtained results agree with postoperative medical data, which showed a greater regurgitation in the

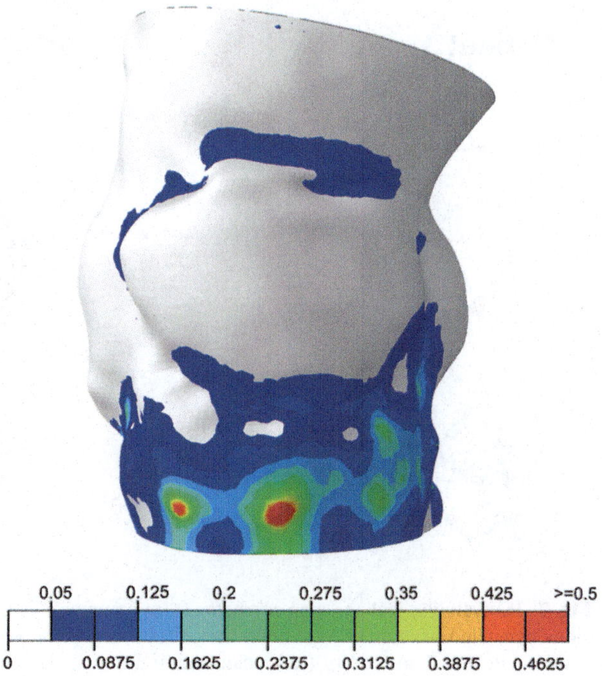

Fig. 6 Von Mises stress distribution on the aortic root (and, in particular, on the left ventricular outflow tract) after prosthesis implantation

same patient with bigger (virtually reproduced) paravalular orifices. This may be seen as a qualitative validation of the developed computational procedure.

4.2.2 The Self-Expandable Medtronic Corevalve Device: A Clinical Application of FEA

A similar framework has been developed for the simulation of the most common self-expandable device, namely the Medtronic Corevalve prosthesis. In this chapter section we aim at demonstrating the ability of computational tools in providing to the medical operator predictive information about the optimal intervention strategy. To this end, we focus on a single real clinical case of Corevalve implantation. The native aortic valve and the calcifications are reconstructed from CT images and included in the finite element model, while the prosthetic device is generated from a micro-CT scan, thus allowing very realistic simulation outcomes. Our ultimate goal is to understand and quantitatively evaluate the effects of the stent implantation angle and depth in terms of clinically relevant parameters, i.e., post-implant asymmetries, stress/strain patterns, paravalvular leakage, and valve coaptation.

The investigated case-study concerns a 75 year-old male patient treated with a Corevalve 26 device to cure severe aortic stenosis. The reconstruction procedure of the patient-specific geometry (aortic root wall, native leaflets, and calcifications) is very similar to that adopted in the previous case (i.e., the Edwards Sapien case).

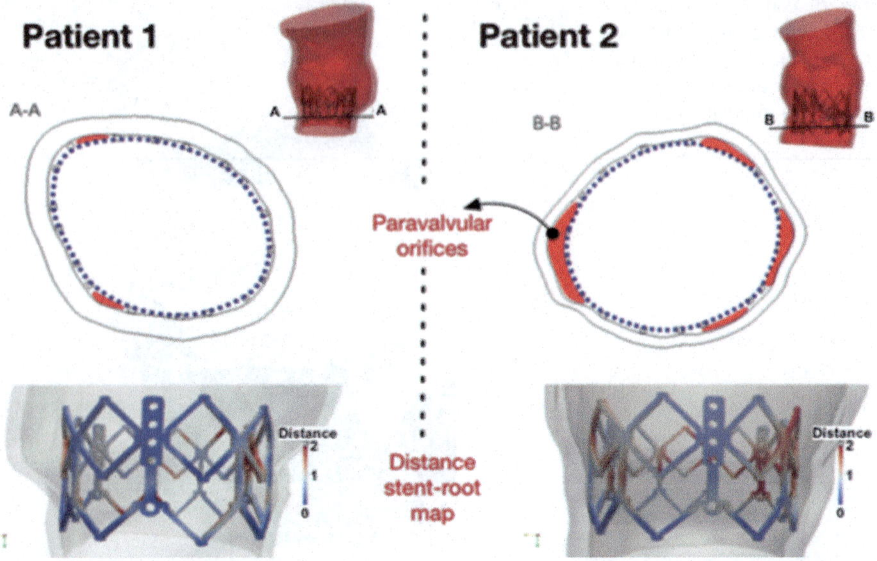

Fig. 7 Paravalvular leakage measurements from computational analyses: both the orifice dimensions generated after stent apposition and distance maps showing the quality of the adherence between stent and aortic root (red means maximum distance equal to 2 mm, blue means perfect adherence) can be measured

The main difference with the case described in the previous chapter section is the prosthetic model, having a very different geometry that we are able to accurately model again by scanning the device using a micro CT machine and very different material properties, being made of NiTiNOL, whose superelastic effect is modeled using the same material parameters as reported in [1]. After native aortic valve modeling, the simulation framework requires that (1) the Corevalve stent is crimped to reach the device dimension during catheter insertion; (2) then it is expanded by gradually removing the radial constraint imposed by the catheter and, finally, (3) the prosthetic leaflets are mapped inside the stent and closed by applying a uniform pressure equivalent to a physiologic diastolic ventriculo-aortic pressure gradient.

From the simulation of the different implantation strategies (reported in Fig. 8), it is possible to evaluate (and quantitatively measure) the different post-operative performance, make a comparison among them, and identify the potentially optimal implantation option.

To this end, (1) stent-root interaction pressure and area can measured to extract information about the device anchoring to the aortic wall, (2) the stress pattern on the aortic root can give indications about the triggering of possible inflammatory processes due to excessive stress, (3) paravalular leakage and, (4) prosthetic leaflets coaptation measures can be extracted as predictors of procedure success or failure. As an example, in Fig. 9 we provide the measure of stent eccentricity after implantation. Clearly, the more the stent is deformed from the ideal (original)

ID	description	d [mm]
T0	high	1
T1	medium	4
T2	low	8

ID	description	Φ [deg]
T1_R-10	left	-15°
T1	vertical	0°
T1_R+10	right	+15°

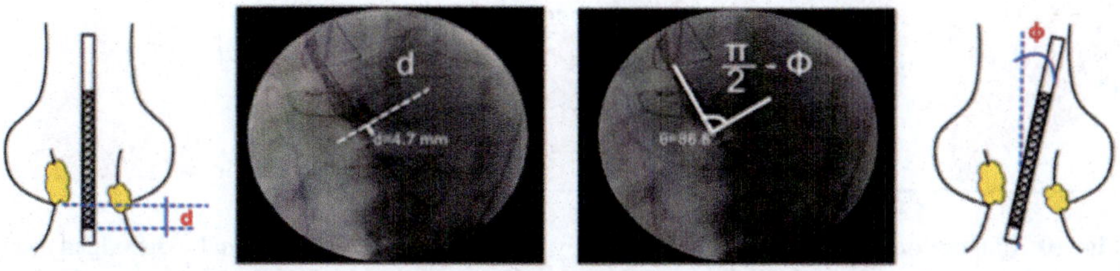

Fig. 8 Different positioning strategies may be investigated using computational tools. In particular, the device depth (d) and inclination angle (θ) are angiographic measurable parameters during the intervention that can be analyzed using FEA. Three different options for each parameter are taken into consideration

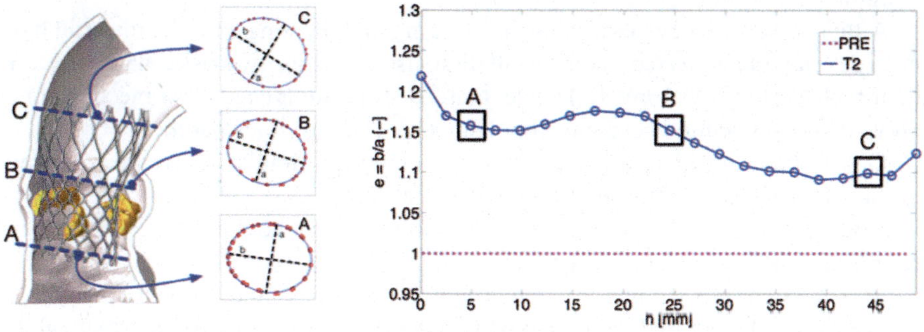

Fig. 9 An example of post-processing of the simulation results. The eccentricity of the implanted stent can be measured to get a measure of its deformation after expansion. This information can be also related to prosthesis performance since a distorted stent can induce non-proper functioning of the bioprosthetic leaflets

geometry (eccentricity = 1 everywhere) the more the prosthesis performance are compromised.

4.3 Fluid-Dynamics Simulation of TAVI

Among the risks of TAVI, stroke induced by dislodged debris (as depicted in Fig. 10) is the most detrimental; a (clinically evident) post-TAVI stroke rate up to 9% has been observed. Even more prevalent but more difficult to assess is the occurrence of sub-clinical brain insults due to micro-emboli that are released during the procedure and migrate to the brain: clinically silent but morphologically detectable perfusion deficits occur in 84% of patients undergoing a TAVI procedure. Even if the potential

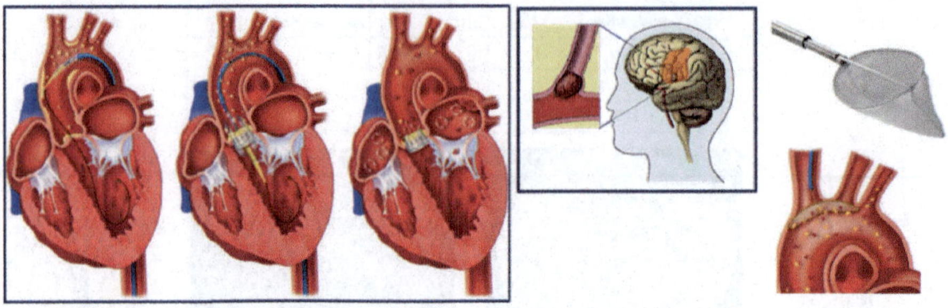

Fig. 10 Illustration of the embolisms induced by TAVI procedure. Emboli can be dislodged along the whole procedure, from the delivery stage up to the prosthesis deployment. Right: examples of commercial embolic filters. Adapted from [13]

impact of these lesions to the neurocognitive function in the long term remains unknown, these numbers are worrisome.

Although several devices aiming at capturing or deflecting embolic material have been developed and tested, their installation also holds its own risks, and therefore, their real-world use is limited. Hence, there is still a high need for a mechanism to prevent peri-procedural cerebral embolism, especially of micro-emboli.

Given such considerations, we have developed a simulation framework integrating experimental (in-vitro) and numerical simulation (in-silico) approaches, intended to support the design of novel embo-protective solutions. Preliminary results are reported in the following.

The in-vitro set-up consists of a hydraulic simulation of the aorta where particles can be released, captured, and counted to quantify how the emboli distribute along the supra-aortic branches under different hemodynamic conditions. The set-up is composed of a fluid circuit (Fig. 11) with an aorta model, a centrifugal pump, and a reservoir; an injection system for releasing particles; two filters for capturing particles; a data acquisition system to record flow and pressure patterns. In the aorta model, the supra-aortic branches are distally merged in order to enable a flow quantification in two major directions: cranial (CRA), i.e., towards the cerebral circulation; caudal (CAU), i.e., towards the descending aorta. For our preliminary experiments, the CRA/CAU flow distribution ratio was kept constant at 35/65. A bolus with particles with an average diameter of 0.7 mm, (Amberlite, Sigma-Aldrich), was injected at the level of the aortic valve (root) by a syringe pump over 4 s. After travelling through the aorta model, the particles were captured in either the cranial or caudal filter and counted. The distribution (split) of particles was repeatedly assessed for four different combinations of flow rate (range 3–5 l/min) and release location (left or right). The standard deviation for the repeated measures ($5\times$ per setting) was $<6\%$, suggesting that the release and catch method is sufficiently consistent to use as a test platform for comparing various emboli-prevention devices.

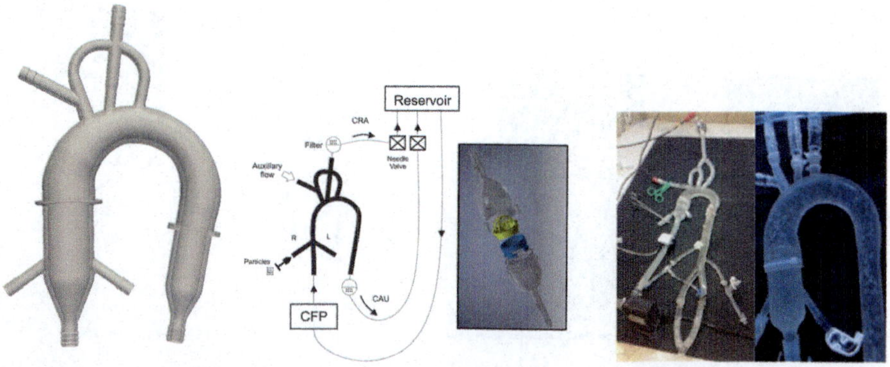

(a) Mock aortic model created using 3D printing and average geometrical data retrieved from literature review.

(b) Left: schematic representation of the mock circulatory loop; Right: assembly of the plug and play filter elements

(c) Actual set-up and enhanced contrast imaging of the released particles

Fig. 11 From the 3D printed mock aortic model (**a**) to the mock circulatory loop (**b**) and analysis set-up (**c**)

Computer Simulations The experiments have been corroborated by CFD simulations, widely used for similar purposes. In particular, the aortic model used for the in-vitro experiments has been reproduced by CAD software. CFD simulations have been carried out on this model by solving the incompressible unsteady Navier-Stokes equations in the region of interest. Three types of boundaries exist: the model entry, the two exits, and the vessel walls. The latter have prescribed null velocity as boundary condition, assuming a non-compliant wall. On the inflow section we prescribed a constant flow rate of 4 L/min. On the exit sections, the flow distribution conditions adopted for the experiments were prescribed: 35% of the outflow in the cranial direction and 65% in the caudal direction. The simulation results were then post-processed to perform particle tracking, i.e., plot the path of a set of particles and assess the particle distribution (CRA/CAU).

First Embodeviation Approach A first concept of an active embo-deviator was tested in both the computer model and the in vitro model. It represents the insertion of a catheter in the right subclavian artery (RSA) through which a bolus of saline is ejected at the time of aortic valve deployment. This has the advantage that perfusion to the brain is maintained, and that potentially the flow in the aorta can be disturbed such that particles move to the less critical descending aorta (caudal) and avoid the brain. Simulations and experiments were performed for various bolus flow rates and results are depicted in Fig. 12. Both platforms confirm the same result, namely that a bolus flow of 0.5 l/min completely prevents emboli from entering the brain via the right carotid artery and reduces the embolic load in the left carotid artery.

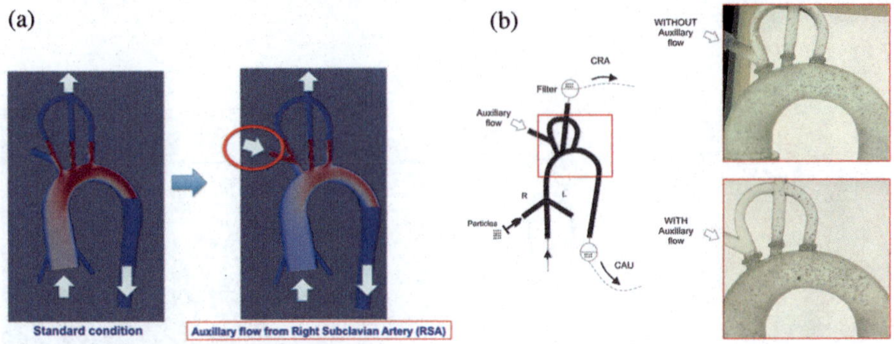

Fig. 12 CFD simulations and in vitro tests on the same mock aortic model. (**a**) CFD simulation of standard condition when auxiliary flow is not activated (Left panel) and when right subclavian artery is perfused shielding the brachiocephalic trunk. The images color refers to the computed flow velocity: red high flow velocity region, blue: low flow velocity region. (**b**) On the left, schematic representation of the mock circulatory loop with red box highlighting the region depicted in the right side of the figure showing the case when particles are released without auxiliary flow from right subclavian artery (RSA) (top) and when the auxiliary flow is present (bottom)

Acknowledgements MC acknowledges the support of ESC Research Grant 2016 and Prof. S. Demertzis (MD), Dr. E. Ferrari (MD) and Dr. S. Vandenberghe—Cardiocentro Ticino for the activity regarding TAVI embolism.

References

1. Auricchio, F., Conti, M., Morganti, S., et al.: Shape-memory alloys: from constitutive modeling to finite element analysis of stent deployment. Comput. Model Eng. Sci. **57**, 225–243 (2010)
2. Auricchio, F., Conti, M., Marconi, S., et al.: Patient-specific aortic endografting simulation: from diagnosis to prediction. Comput. Biol. Med. **43**(4), 386–394 (2013)
3. Auricchio, F., Conti, M., Lefieux, A., et al.: Patient-specific analysis of post-operative aortic hemodynamics: a focus on thoracic endovascular repair (TEVAR). Comput. Mech. **54**(4), 943–953 (2014)
4. Auricchio, F., Conti, M., Morganti, S., et al.: Simulation of transcatheter aortic valve implantation: a patient-specific finite element approach. Comput. Methods Biomech. Biomed. Eng. **17**(12), 1347–1357 (2014)
5. Capelli, C., Bosi, G.M., Cerri, E., et al.: Patient-specific simulations of transcatheter aortic valve stent implantation. Med. Biol. Eng. Comput. **50**(2), 183–192 (2012)
6. Cribier, A., Eltchaninoff, H., Bash, A., et al.: Percutaneous transcatheter implantation of an aortic valve prosthesis for calcific aortic stenosis: first human case description. Circulation **106**, 3006–3008 (2002)
7. De Bock, S., Iannaccone, F., De Santis, G., et al.: Virtual evaluation of stent graft deployment: a validated modeling and simulation study. J. Mech. Behav. Biomed. Mater. **13**, 129–139 (2012)
8. Demanget, N., Avril, S., Badel, P., et al.: Computational comparison of the bending behavior of aortic stent-grafts. J. Mech. Behav. Biomed. Mater. **5**(1), 272–282 (2012)
9. Demanget, N., Latil, P., Orgeas, L., et al.: Severe bending of two aortic stent-grafts: an experimental and numerical mechanical analysis. Ann. Biomed. Eng. **40**(12), 2674–2686 (2012)

10. Demanget, N., Duprey, A., Badel, P., et al.: Finite element analysis of the mechanical performances of 8 marketed aortic stent-grafts. J. Endovasc. Ther. **20**(4), 523–535 (2013)
11. Dwyer, H.A., Matthews, P.B., Azadan, A., et al.: Computational fluid dynamics simulation of transcatheter aortic valve degeneration. Interact. Cardiovasc. Thorac. Surg. **9**(2), 301–308 (2009)
12. Erbel, R., Aboyans, V., Boileau, C., et al.: ESC guidelines on the diagnosis and treatment of aortic diseases. Eur. Heart J. **35**, 2873–2926 (2014)
13. Fanning, J.P., Walters, D.L., Platts, D.G., et al.: Characterization of neurological injury in transcatheter aortic valve implantation. Circulation **129**(4), 504–515 (2014)
14. Figueroa, C.A., Taylor, C.A., Chiou, A.J., et al.: Magnitude and direction of pulsatile displacement forces acting on thoracic aortic endografts. J. Endovasc. Ther. **16**(3), 350–358 (2009)
15. Gallo, D., Lefieux, A., Morganti, S., et al.: A patient-specific follow up study of the impact of thoracic endovascular repair (TEVAR) on aortic anatomy and on post-operative hemodynamics. Comput. Fluids **141**, 54–61 (2016)
16. Gessat, M., Hopf, R., Polloket, T., et al.: Image-based mechanical analysis of stent deformation: concept and exemplary implementation for aortic valve stents. IEEE Trans. Biomed. Eng. **61**(1), 4–15 (2014)
17. Gunning, P.S., Vaughan, T.J., McNamara, L.M.: Simulation of self expanding transcatheter aortic valve in a realistic aortic root: implications of deployment geometry on leaflet deformation. Ann. Biomed. Eng. **42**(9), 1989–2001 (2014)
18. Kleinstreuer, C., Li, Z., Basciano, C.A., et al.: Computational mechanics of Nitinol stent grafts. J. Biomech. **41**(11), 2370–2378 (2008)
19. Morganti, S., Conti, M., Aiello, M., et al.: Simulation of transcatheter aortic valve implantation through patient-specific finite element analysis: two clinical cases. J. Biomech. **47**(11), 2547–2555 (2014)
20. Morlacchi, S., Colleoni, S.G., Cardenes, R., et al.: Patient-specific simulations of stenting procedures in coronary bifurcations: two clinical cases. Med. Eng. Phys. **35**, 1272–1281 (2013)
21. Mozaffarian, D., Benjamin, E.J., Go, A.S., et al.: Executive summary: heart disease and stroke statistics-2016: update: a report from the American Heart Association. Circulation **133**, 447–454 (2016)
22. Perrin, D., Demanget, N., Badel, P., et al.: Deployment of stent grafts in curved aneurysmal arteries: toward a predictive numerical tool. Int. J. Numer. Methods Biomed. Eng. **31**(1), e02698 (2015)
23. Perrin, D., Badel, P., Orgeas, L., et al.: Patient-specific numerical simulation of stent-graft deployment: validation on three clinical cases. J. Biomech. **48**(10), 1868–1875 (2015)
24. Sirois, E., Wang, Q., Sun, W.: Fluid simulation of a transcatheter aortic valve deployment into a patient-specific aortic root. Cardiovasc. Eng. Technol. **2**(3), 186–195 (2011)
25. Tzamtzis, S., Viquerat, J., Yapet, J., et al.: Numerical analysis of the radial force produced by the Medtronic-CoreValve and Edwards-SAPIEN after transcatheter aortic valve implantation (TAVI). Med. Eng. Phys. **35**(1), 125–130 (2013)
26. Wang, Q., Kodali, S., Primiano, C., et al.: Simulations of transcatheter aortic valve implantation: implications for aortic root rupture. Biomech. Model Mechanobiol. **14**(1), 29–38 (2014)

Michele Conti is assistant professor of Industrial Bioengineering at the University of Pavia, Italy. He has coauthored over 40 publications on Peer Reviewed Journals. His current main research interest is cardiovascular biomechanics combining in-silico and in-vitro analysis.

Simone Morganti is assistant professor of Mechanics of Solids at the University of Pavia, Italy. He has coauthored over 30 publications on Peer Reviewed Journals. His current research interests range from the modeling of transcatheter aortic valve implantation to the simulation of the additive manufacturing process.

Alice Finotello is currently a PhD student in the Department of Experimental Medicine (University of Genoa) and she is also a research member of the CompMech Group (University of Pavia). She received her MSc in Bioengineering from University of Pavia in 2015. Her research interests include medical image processing and computational modelling.

Rodrigo M. Romarowski is a researcher in the group of Vascular Surgery at the Policlinico San Donato research hospital in Milan, Italy. His research interests include medical image processing, computational fluid dynamics as well as the translational aspect of cardiovascular research. He closely collaborates with the group of Computational Mechanics and Advanced Materials of the University of Pavia to develop virtual tools for endovascular treatment in the aorta.

Alessandro Reali is Full Professor of Mechanics of Solids and Structures at the University of Pavia and a renowned expert of Computational Mechanics (with broad research interests including among others isogeometric analysis, finite elements, strong-form methods). He is an ISI Highly Cited Researcher and an ERC grantee, as well as the recipient of many awards (including, e.g., the Bruno Finzi Prize, the IACM Fellows Award, the TUM-IAS Hans Fischer Fellowship, the IACM John Argyris Award, the AIMETA Junior Prize, the ECCOMAS O.C. Zienkiewicz Award).

Ferdinando Auricchio is currently the Head of the Department of Civil Engineering and Architecture (University of Pavia), a Full Professor of Mechanics of Solids at the University of Pavia, and a Research Associate at IMATI-CNR (Pavia, Italy). He received his PhD in Civil Engineering at the University of California at Berkeley in 1995. He is also vice-president of the European Community of Computational Methods in Applied Sciences (ECCOMAS), and Member of the Additive Manufacturing Thematic Group Steering Committee (Lombardy Association for Intelligent Industry). His research interests include biomechanics, 3D printing, material constitutive modeling.

Combined Parameter and Model Reduction of Cardiovascular Problems by Means of Active Subspaces and POD-Galerkin Methods

Marco Tezzele, Francesco Ballarin, and Gianluigi Rozza

Abstract In this chapter we introduce a combined parameter and model reduction methodology and present its application to the efficient numerical estimation of a pressure drop in a set of deformed carotids. The aim is to simulate a wide range of possible occlusions after the bifurcation of the carotid. A parametric description of the admissible deformations, based on radial basis functions interpolation, is introduced. Since the parameter space may be very large, the first step in the combined reduction technique is to look for active subspaces in order to reduce the parameter space dimension. Then, we rely on model order reduction methods over the lower dimensional parameter subspace, based on a POD-Galerkin approach, to further reduce the required computational effort and enhance computational efficiency.

1 Introduction

Numerical simulations of biomedical problems is a topic of large interest nowadays, especially for what concerns the application of shape optimization techniques aimed at the improvement of long-term outcomes of clinical interventions [4, 49]. Several challenging aspects can be identified in such a task, especially when seeking a personalized (patient-specific) treatment [15, 35, 65]: model construction and segmentation, numerical solution of the underlying fluid dynamics equations, assimilation of clinical data (e.g. for boundary conditions), choice of the cost functional and medical indices to be optimized [1, 2], as well as model deformation during the optimization procedure. The latter is a topic of remarkable interest, since it is well known that local geometrical features may severely affect the computational fluid dynamics (CFD) simulation and thus the results of the optimization [30].

M. Tezzele · F. Ballarin · G. Rozza (✉)
Mathematics Area, mathLab, SISSA, Trieste, Italy
e-mail: marco.tezzele@sissa.it; francesco.ballarin@sissa.it; gianluigi.rozza@sissa.it

D. Boffi et al. (eds.), *Mathematical and Numerical Modeling of the Cardiovascular System and Applications*, SEMA SIMAI Springer Series 16,
https://doi.org/10.1007/978-3-319-96649-6_8

A challenge in the applicability of optimization procedures in the clinical environment is the large computational time that each procedure would entail. Indeed, such optimal control problems are usually tackled by means of iterative solvers that require several expensive CFD simulations for different geometrical configurations [37]. To this end, several authors have proposed to employ computational reduction techniques based on reduced order (or surrogate) models. We refer to [7, 10, 28, 36, 48, 50] for a few representative applications, as well as e.g. to [21, 38] for introductory textbooks on the underlying methodology. These methods rely on the definition of a parameter space, which is related to the set of admissible deformations that can be considered during the iterative optimization procedure, and its exploration to retain the most relevant features of the CFD solution on the parameter space. It is however well known that reduced order methods suffer from the so-called "curse of dimensionality" if the parameter space is high-dimensional. Although there exists techniques to account for high-dimensional parameter spaces in the reduced order modelling framework, based e.g. on sparse grids [20, 64] or on a proper weighting of the parameter space [19, 63], the approach that we propose in this manuscript is different and aims at reducing the high-dimensional parameter space as well, while preserving a very broad set of admissible deformations.

We have recently dealt with several techniques concerning efficient shape parametrization techniques in the framework of reduced order modelling. A first possible choice is related to the shape morphing method itself. For our goals it suffices to classify them in two groups: *general purpose* or *problem specific*. The design of a *problem specific* shape parametrization technique should aim at reducing the high-dimensional parameter space; for instance, the centerlines-based approach proposed in [7] is able to reduce the parameter space accounting only for deformations in a cylindrical coordinates frame of reference. In contrast, in this chapter we show a parameter space reduction technique for *general purpose* shape morphing methods. Among possible *general purpose* methods we mention *Free Form Deformation* (FFD) [45, 59], *Radial Basis Functions* (RBF) interpolation [16, 48, 53] or *Inverse Distance Weighting* (IDW) interpolation [33, 60, 66]. Broadly speaking, the aforementioned methods require the displacement of some control points to induce a deformation on the domain, and we identify the parameters as the displacements of the control points. In these context, our earlier approaches to parameter space reduction have relied on screening procedures based on Morris' randomized one-at-a-time design [5, 52], modal analysis [33] or semi-automatic reduction of the number of control points [9]. Each of these approaches results in lower dimensional parameter spaces by retaining an "optimal" subset of the possible control points. Unfortunately, when requiring very low dimensional parameter spaces (order of unity) the resulting set of admissible deformations may be considerably shrunk. Not only, another goal to achieve for complex shape parametrizations is also an efficient positioning of control points, requiring versatility and "full" capability in the geometric representation. To this end, in this manuscript we propose to exploit an active subspaces (AS) method [22], building on our previous experience

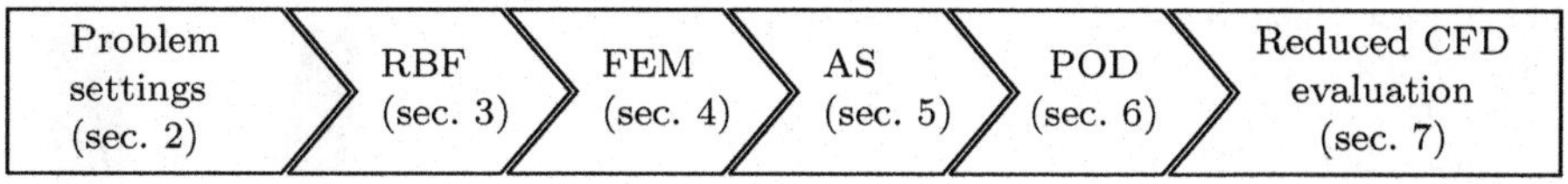

Fig. 1 Outline of the chapter

on a shape optimization of a naval engineering problem in [62]. AS has been employed in many real world problems; among others, we mention aerodynamic shape optimization [47], integrated hydrologic model [42], the parameter reduction for the HyShot II scramjet model [24], a satellite system model [40]. The main difference between our previous approaches and the AS property is that our former approaches were constraining the search of a lower dimensional parameter subspace to be parallel to a subset of the axes of the parameter space, while the AS method will automatically identify the "optimal" lower dimensional subspace without any such constraint, taking a linear combination of all the original parameters. To show an example of the resulting methodology we will consider a cardiovascular test case related to the computation of the pressure drop of a series of deformed carotids.

The outline of the chapter is presented in Fig. 1. The formulation of the problem of estimation of the pressure drop across a stenosed carotid artery is summarized in Sect. 2. Shape morphing by means of RBF interpolation is then introduced in Sect. 3. The high fidelity method, based on finite elements, is briefly summarized in Sect. 4. The computed values of the quantity of interest will be used to train the AS reduction of the parameter space in Sect. 5. The same high fidelity solver will be used in Sect. 6 to train a Proper Orthogonal Decomposition (POD)-Galerkin method on the lower dimensional parameter subspace. This combination will further enhance computational efficiency for the procedure. Numerical results and error analyses of the whole pipeline will be presented in Sect. 7. Conclusions and future perspectives follow in Sect. 8.

2 A Model Cardiovascular Problem: Pressure Drop Estimation Across a Stenosis

In this section we introduce the problem of the estimation of the pressure drop across two parametrized stenoses in a carotid bifurcation.

Let $\Omega \subset \mathbb{R}^n$, $n = 3$, be a domain (see Fig. 2), obtained from the INRIA 3D Meshes Research Database [41], that describes an idealized carotid bifurcation. We will call Ω the *reference* domain; for practical reasons this domain happens to correspond to the healthy case (no stenoses), even though this assumption is not fundamental for the remainder of the paper.

Let $\mathbb{D} \subset \mathbb{R}^m$, be the set of parameters, that we assume to be a box in $\mathbb{R}^m$, for $m \in \mathbb{N}$. Moreover, let $\mathcal{M}(\boldsymbol{x}; \boldsymbol{\mu}) : \mathbb{R}^n \to \mathbb{R}^n$, with $\boldsymbol{\mu} \in \mathbb{D}$, be a shape morphing

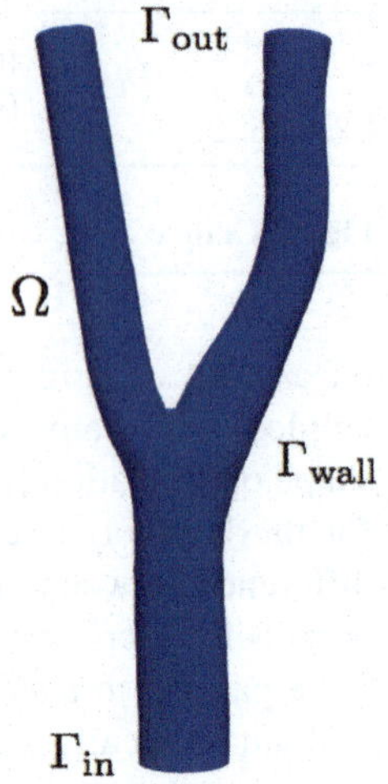

Fig. 2 Representation of the reference domain Ω, inlet section Γ_{in}, rigid walls Γ_{wall} and outlet section Γ_{out}

that maps the reference domain Ω into the deformed domain $\Omega(\boldsymbol{\mu})$ as follows:

$$\Omega(\boldsymbol{\mu}) = \mathcal{M}(\Omega; \boldsymbol{\mu}).$$

We refer to Sect. 3 for the actual definition of m and $\mathcal{M}$ for the case at hand.

Let us consider the following steady Navier-Stokes equations: for any $\boldsymbol{\mu} \in \mathbb{D}$, find $(\boldsymbol{u}(\boldsymbol{\mu}), p(\boldsymbol{\mu})) \in H^1(\Omega(\boldsymbol{\mu}); \mathbb{R}^n) \times L^2(\Omega(\boldsymbol{\mu}))$ such that:

$$\begin{cases} -\nu \Delta \boldsymbol{u}(\boldsymbol{\mu}) + \boldsymbol{u}(\boldsymbol{\mu}) \cdot \nabla \boldsymbol{u}(\boldsymbol{\mu}) + \nabla p(\boldsymbol{\mu}) = \boldsymbol{0} & \text{in } \Omega(\boldsymbol{\mu}), \\ \operatorname{div} \boldsymbol{u}(\boldsymbol{\mu}) = 0 & \text{in } \Omega(\boldsymbol{\mu}), \\ \boldsymbol{u}(\boldsymbol{\mu}) = \boldsymbol{u}_{\text{in}} & \text{on } \Gamma_{\text{in}}, \\ \boldsymbol{u}(\boldsymbol{\mu}) = \boldsymbol{0}, & \text{on } \Gamma_{\text{wall}}(\boldsymbol{\mu}), \\ \nu \dfrac{\partial \boldsymbol{u}(\boldsymbol{\mu})}{\partial \boldsymbol{n}} - p(\boldsymbol{\mu}) \boldsymbol{n} = \boldsymbol{0}, & \text{on } \Gamma_{\text{out}}, \end{cases} \tag{1}$$

where Γ_{in} is the inlet section, Γ_{out} the outlet section and $\Gamma_{\text{wall}}(\boldsymbol{\mu})$ are (rigid) walls of the carotid artery. Since our interest is to vary the degree of stenosis immediately after the bifurcation point (see Fig. 3), we assume that Γ_{in} and Γ_{out} are far away from the bifurcation and are not affected by $\boldsymbol{\mu}$. Here $\boldsymbol{u}(\boldsymbol{\mu})$ represents the unknown velocity, while $p(\boldsymbol{\mu})$ the unknown pressure. Moreover, the inlet velocity $\boldsymbol{u}_{\text{in}}$ is a parabolic profile and the viscosity ν is chosen such that the resulting Reynolds number is equal to 400, corresponding to the average Reynolds number over a cardiac cycle [67].

As quantity of interest we would have liked to consider the pressure drop across the stenoses. However, due to the lack of physiological boundary conditions that prescribe the pressure at the inlet, as well as non-physiological homogeneous Neumann boundary conditions that prescribe zero pressure at the outlet, this

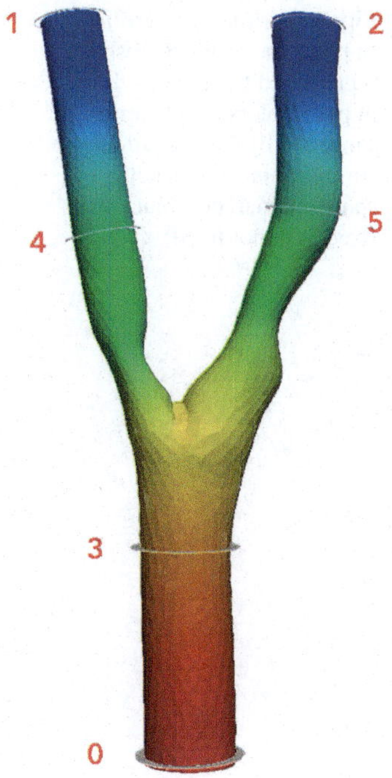

Fig. 3 Sections $S_0, \ldots, S_5$ employed in the relative pressure drop computation

quantity of interest would be severely affected by a variation of the degree of stenosis. To mitigate this, we resort to a relative pressure drop dividing by the inlet pressure (i.e., the pressure drop between the inlet and the outlet). To be more precise, denote with $P_i(\boldsymbol{\mu})$ the average of the pressure on the section S_i, $i = 0, \ldots, 5$, as shown in Fig. 3. Sections S_3 and S_4 (S_5, respectively) are used to quantify the pressure drop for the stenosis on the left (right, resp.) branch, which is then divided by the pressure drop between the inlet S_0 and the left (right, resp.) outlet S_1 (S_2, resp.). Therefore, the quantity of interest that we consider is the sum of the relative pressure drop of the two branches:

$$f(\boldsymbol{\mu}) = \frac{P_3(\boldsymbol{\mu}) - P_4(\boldsymbol{\mu})}{P_0(\boldsymbol{\mu}) - P_1(\boldsymbol{\mu})} + \frac{P_3(\boldsymbol{\mu}) - P_5(\boldsymbol{\mu})}{P_0(\boldsymbol{\mu}) - P_2(\boldsymbol{\mu})}. \tag{2}$$

3 Shape Morphing Based on Radial Basis Functions Interpolation

Radial Basis Functions (RBF) are a powerful tool for shape parametrization due to their good approximation properties [16, 48]. In this section we summarize RBF-based shape morphing following the presentation in [33]. All the algorithms have

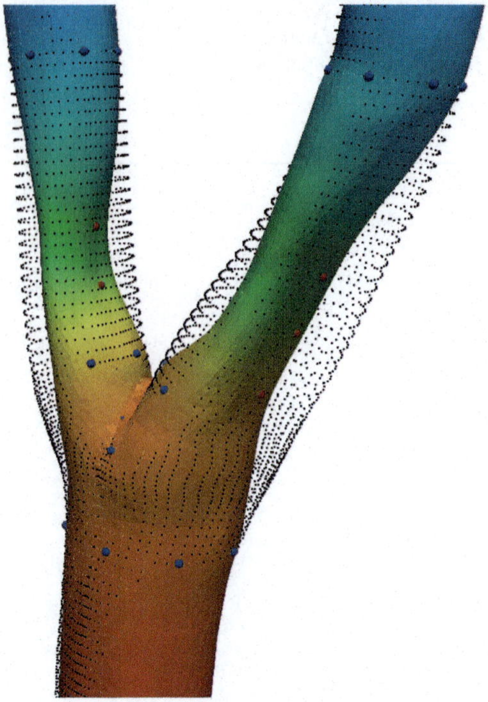

Fig. 4 Portion of a carotid. In black the original mesh represented by points, while in plain colors the deformed carotid. RBF interpolation control points are denoted by spherical markers (blue: fixed, red: not fixed)

been implemented in the open source python package PyGeM [55], which is used to perform the shape morphing in the numerical results showed in Sect. 7.

A radial basis function is any smooth real-valued function $\widetilde{\varphi} : \mathbb{R}^n \to \mathbb{R}$ such that it exists $\varphi : \mathbb{R}^+ \to \mathbb{R}$ and $\widetilde{\varphi}(\boldsymbol{x}) = \varphi(\|\boldsymbol{x}\|)$, where $\| \cdot \|$ is the Euclidean norm in $\mathbb{R}^n$.

Let us recall that parameters are denoted by $\boldsymbol{\mu} \in \mathbb{D}$. The RBF shape parametrization technique is based on the map $\mathcal{M}(\boldsymbol{x}; \boldsymbol{\mu}) : \mathbb{R}^n \to \mathbb{R}^n$, defined as follows

$$\mathcal{M}(\boldsymbol{x}; \boldsymbol{\mu}) = q(\boldsymbol{x}; \boldsymbol{\mu}) + \sum_{i=1}^{\mathcal{N}_C} \gamma_i(\boldsymbol{\mu}) \, \varphi(\|\boldsymbol{x} - \boldsymbol{x}_{C_i}\|). \tag{3}$$

where $q(\boldsymbol{x}; \boldsymbol{\mu})$ is a polynomial term to be determined, $\{\gamma_i(\boldsymbol{\mu})\}_{i=1}^{\mathcal{N}_C}$ are weights to be determined, and $\{\boldsymbol{x}_{C_i}\}_{i=1}^{\mathcal{N}_C}$ are control points selected by the user (denoted by spherical markers in Fig. 4), and $\boldsymbol{x} \in \Omega$. Among all the possible RBF for modeling shapes we select the so-called thin plate splines [32] defined as

$$\varphi(r; R) = \left(\frac{r}{R}\right)^2 \ln\left(\frac{r}{R}\right)$$

where r is the radial coordinate and $R > 0$ is a given radius.

In order to find the unknowns $q(\boldsymbol{x};\boldsymbol{\mu})$ and $\{\gamma_i(\boldsymbol{\mu})\}_{i=1}^{\mathcal{N}_C}$, let us assume that $q(\boldsymbol{x};\boldsymbol{\mu})$ is a polynomial function of degree 1, that is

$$q(\boldsymbol{x};\boldsymbol{\mu}) = \mathbf{c}(\boldsymbol{\mu}) + \mathbf{Q}(\boldsymbol{\mu})\boldsymbol{x},$$

for some unknown $\mathbf{c}(\boldsymbol{\mu}) \in \mathbb{R}^n$ and $\mathbf{Q}(\boldsymbol{\mu}) \in \mathbb{R}^{n\times n}$. Therefore Eq. (3) can be rewritten in matrix form as follows:

$$\mathcal{M}(\boldsymbol{x};\boldsymbol{\mu}) = \mathbf{c}(\boldsymbol{\mu}) + \mathbf{Q}(\boldsymbol{\mu})\boldsymbol{x} + \mathbf{G}^T(\boldsymbol{\mu})\mathbf{d}(\boldsymbol{x}), \tag{4}$$

being $\mathbf{d}(\boldsymbol{x}) = [\varphi(\|\boldsymbol{x} - \boldsymbol{x}_{C_1}\|), \dots, \varphi(\|\boldsymbol{x} - \boldsymbol{x}_{\mathcal{N}_C}\|)] \in \mathbb{R}^{\mathcal{N}_C}$ the vector constructed evaluating the radial basis function on the Euclidean distance between the control points position $\boldsymbol{x}_{C_i}$ and $\boldsymbol{x}$, and the unknown $\mathbf{G}(\boldsymbol{\mu}) = [\gamma_1(\boldsymbol{\mu}), \dots, \gamma_{\mathcal{N}_C}(\boldsymbol{\mu})] \in \mathbb{R}^{\mathcal{N}_C\times n}$. To compute the unknowns $\mathbf{c}(\boldsymbol{\mu})$, $\mathbf{Q}(\boldsymbol{\mu})$ and $\mathbf{G}(\boldsymbol{\mu})$ we enforce interpolation conditions on the set of control points, that is, given their initial position as

$$\boldsymbol{x}_C = [\boldsymbol{x}_{C_1}, \dots, \boldsymbol{x}_{C_{\mathcal{N}_C}}] \in \mathbb{R}^{\mathcal{N}_C\times n}$$

and their $\boldsymbol{\mu}$-dependent deformed positions as

$$\mathbf{y}_C(\boldsymbol{\mu}) = [\mathbf{y}_{C_1}(\boldsymbol{\mu}), \dots, \mathbf{y}_{C_{\mathcal{N}_C}}(\boldsymbol{\mu})] \in \mathbb{R}^{\mathcal{N}_C\times n}, \tag{5}$$

we enforce that

$$\mathcal{M}(\boldsymbol{x}_{C_i};\boldsymbol{\mu}) = \mathbf{y}_{C_i}(\boldsymbol{\mu}) \qquad \forall i \in \{1, \dots, \mathcal{N}_C\}. \tag{6}$$

The system is then completed by additional constraints that represent the conservation of the total force and momentum [16, 31, 53], due to the presence of the polynomial term, as follows

$$\sum_{i=1}^{\mathcal{N}_C} \gamma_i(\boldsymbol{\mu}) = 0, \tag{7}$$

$$\sum_{i=1}^{\mathcal{N}_C} \gamma_i(\boldsymbol{\mu})\boldsymbol{x}_{C_{1,i}} = \cdots = \sum_{i=1}^{\mathcal{N}_C} \gamma_i(\boldsymbol{\mu})\boldsymbol{x}_{C_{n,i}} = 0, \tag{8}$$

being $\boldsymbol{x}_{C_i} = [\boldsymbol{x}_{C_{1,i}}, \dots, \boldsymbol{x}_{C_{n,i}}]$ a vector collecting the i-th coordinates of all control points. These additional constraints, together with the presence of the polynomial term of degree one, ensures that the resulting linear system (6)–(8) has always a unique solution $(\mathbf{c}(\boldsymbol{\mu}), \mathbf{Q}(\boldsymbol{\mu}), \mathbf{G}(\boldsymbol{\mu}))$. Once the system is solved, we can deform all the points of the mesh through $\mathcal{M}(\cdot;\boldsymbol{\mu})$ to obtain the deformed configuration.

In order to exemplify (5) in our case, let us consider again Fig. 4, where control points are denoted by spherical markers. In order to enforce the deformation to be

Fig. 5 Three different deformations produced varying the $m = 10$ parameters

localized, we will move all control points close to the stenosis (colored in red) in the normal direction, thus varying the occlusion, while we will keep fixed a few control points far away from the stenosis (colored in blue). Therefore, for all fixed control points the right hand side of (6) is defined employing

$$\mathbf{y}_{C_i}(\boldsymbol{\mu}) = \boldsymbol{x}_{C_i} \qquad \forall i \in \{1, \ldots, \mathcal{N}_C^{\text{fixed}}\},$$

while for all non-fixed control points

$$\mathbf{y}_{C_i}(\boldsymbol{\mu}) = \boldsymbol{x}_{C_i} - \mu_i \mathbf{n}_i \qquad \forall i \in \{1, \ldots, m\},$$

where $\mathcal{N}_C = \mathcal{N}_C^{\text{fixed}} + m$, μ_i denotes the i-th element in $\boldsymbol{\mu}$ and $\mathbf{n}_i$ the outer unit normal to the wall evaluated at $\boldsymbol{x}_{C_i}$. In our case we have chosen $m = 10$, $\mathcal{N}_C^{\text{fixed}} = 55$ (not all control points are shown in Fig. 4), and the parameter range is $\mathbb{D} = [0, 0.3]^m$, resulting in a wide range of possible different stenosis scenarios (see Fig. 5 for an idea of the possible configurations).

4 High Fidelity Solver Based on the Finite Element Method

In this section we summarize the high fidelity solver that will be used in the training of both the active subspace (see Sect. 5) and the POD-Galerkin reduction (see Sect. 6). Let (V, Q) be an inf-sup stable finite element (FE) pair, being $V_\delta(\boldsymbol{\mu}) \subset H^1(\Omega_\delta(\boldsymbol{\mu}); \mathbb{R}^n)$ and $Q_\delta(\boldsymbol{\mu}) \subset L^2(\Omega_\delta(\boldsymbol{\mu}))$, being δ the maximum diameter size of a tetrahedralization[1] Ω_δ of Ω; Taylor-Hood $\mathbb{P}^2 - \mathbb{P}^1$ FE has been employed in the numerical experiments shown in Sect. 7.

[1]In order to simplify the exposition we will report the FE formulation on the deformed domain $\Omega(\boldsymbol{\mu})$. However, it should be noted that only the mesh Ω_δ of the reference domain Ω is generated, and deformed meshes $\Omega_\delta(\boldsymbol{\mu})$ are obtained through the mapping $\mathcal{M}(\cdot; \boldsymbol{\mu})$.

The high fidelity approximation of (1) reads: for any $\boldsymbol{\mu} \in \mathbb{D}$, find $(\boldsymbol{u}_\delta(\boldsymbol{\mu}), p_\delta(\boldsymbol{\mu})) \in V_\delta(\boldsymbol{\mu}) \times Q_\delta(\boldsymbol{\mu})$ such that $\boldsymbol{u}_\delta(\boldsymbol{\mu}) = \boldsymbol{u}_{\text{in}}$ on Γ_{in}, $\boldsymbol{u}_\delta(\boldsymbol{\mu}) = \mathbf{0}$, on $\Gamma_{\text{wall}}(\boldsymbol{\mu})$ and:

$$\begin{cases} a(\boldsymbol{u}_\delta, \boldsymbol{v}_\delta; \boldsymbol{\mu}) + b(\boldsymbol{v}_\delta, p_\delta; \boldsymbol{\mu}) + c(\boldsymbol{u}_\delta, \boldsymbol{u}_\delta, \boldsymbol{v}_\delta; \boldsymbol{\mu}) = \mathbf{0}, & \forall \boldsymbol{v}_\delta \in V_\delta(\boldsymbol{\mu}), \\ b(\boldsymbol{u}_\delta, q_\delta; \boldsymbol{\mu}) = 0, & \forall q \in Q_\delta(\boldsymbol{\mu}), \end{cases} \tag{9}$$

where

$$a(\boldsymbol{u}_\delta, \boldsymbol{v}_\delta; \boldsymbol{\mu}) = \int_{\Omega(\boldsymbol{\mu})} \nu \nabla \boldsymbol{u}_\delta : \nabla \boldsymbol{v}_\delta \, d\boldsymbol{x}, \qquad b(\boldsymbol{v}_\delta, q_\delta; \boldsymbol{\mu}) = -\int_{\Omega(\boldsymbol{\mu})} q_\delta \operatorname{div} \boldsymbol{v}_\delta \, d\boldsymbol{x}$$

are the bilinear forms associated to the diffusion and divergence operators, respectively, whereas

$$c(\boldsymbol{u}_\delta, \boldsymbol{v}_\delta, \boldsymbol{z}_\delta; \boldsymbol{\mu}) = \int_{\Omega(\boldsymbol{\mu})} (\nabla \boldsymbol{v}_\delta \, \boldsymbol{u}_\delta) \cdot \boldsymbol{z}_\delta \, d\boldsymbol{x}$$

is the trilinear form related to the nonlinear advection term. Let $\{\boldsymbol{\varphi}_i^\delta\}_{i=1}^{N_{\mathbf{u}}^\delta}$ and $\{\zeta_k^\delta\}_{k=1}^{N_p^\delta}$ the bases of $V_\delta(\boldsymbol{\mu})$ and $Q_\delta(\boldsymbol{\mu})$, respectively. Then, (9) can be equivalently rewritten as the following nonlinear system: for any $\boldsymbol{\mu} \in \mathbb{D}$, find $(\mathbf{u}(\boldsymbol{\mu}), \mathbf{p}(\boldsymbol{\mu})) \in \mathbb{R}^{N_{\mathbf{u}}^\delta} \times \mathbb{R}^{N_p^\delta}$ such that[2]:

$$\begin{bmatrix} \mathbf{A}(\boldsymbol{\mu}) + \mathbf{C}(\mathbf{u}(\boldsymbol{\mu}); \boldsymbol{\mu}) & \mathbf{B}^T(\boldsymbol{\mu}) \\ \mathbf{B}(\boldsymbol{\mu}) & \mathbf{0} \end{bmatrix} \begin{bmatrix} \mathbf{u}(\boldsymbol{\mu}) \\ \mathbf{p}(\boldsymbol{\mu}) \end{bmatrix} = \begin{bmatrix} \mathbf{f}(\boldsymbol{\mu}) \\ \mathbf{g}(\boldsymbol{\mu}) \end{bmatrix}, \tag{10}$$

where, for $1 \le i, j \le N_{\mathbf{u}}^\delta$ and $1 \le k \le N_p^\delta$:

$$\begin{aligned} (\mathbf{A}(\boldsymbol{\mu}))_{ij} = a(\boldsymbol{\varphi}_j^\delta, \boldsymbol{\varphi}_i^\delta; \boldsymbol{\mu}), \qquad (\mathbf{B}(\boldsymbol{\mu}))_{kl} = b(\boldsymbol{\varphi}_i^\delta, \zeta_k^\delta; \boldsymbol{\mu}), \\ (\mathbf{C}(\mathbf{u}(\boldsymbol{\mu}); \boldsymbol{\mu}))_{ij} = c\left(\boldsymbol{u}_\delta(\boldsymbol{\mu}), \boldsymbol{\varphi}_j^\delta, \boldsymbol{\varphi}_i^\delta; \boldsymbol{\mu}\right), \end{aligned} \tag{11}$$

and

$$\boldsymbol{u}_\delta(\boldsymbol{\mu}) = \sum_{i=1}^{N_{\mathbf{u}}^\delta} u_i(\boldsymbol{\mu}) \boldsymbol{\varphi}_i^\delta, \quad p_\delta(\boldsymbol{\mu}) = \sum_{k=1}^{N_{\mathbf{p}}^\delta} p_k(\boldsymbol{\mu}) \zeta_k^\delta,$$

For any $\boldsymbol{\mu} \in \mathbb{D}$ the nonlinear system (10) is solved via a Newton method, and the quantity of interest $f(\boldsymbol{\mu})$ is computed according to (2) in a postprocessing stage. The solution of the Navier-Stokes problem for a representative value of $\boldsymbol{\mu}$ is shown in Fig. 6.

[2]The non-homogeneous right-hand side accounts for boundary conditions via a lifting.

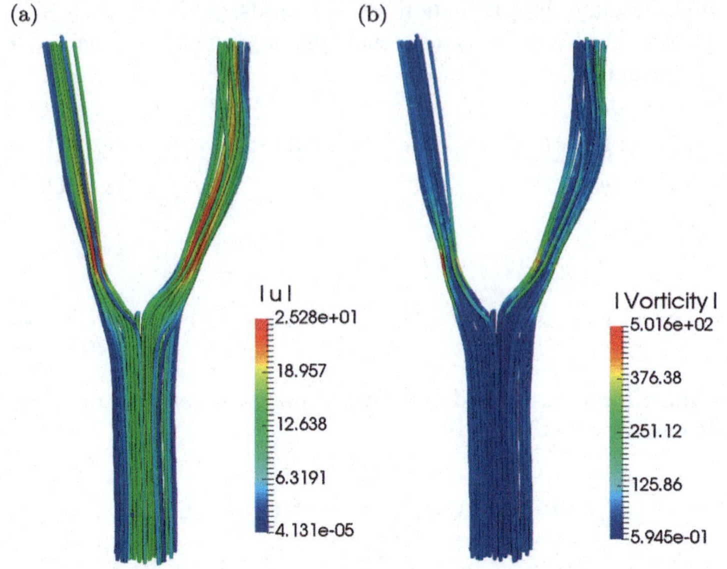

Fig. 6 Velocity and vorticity fields for a representative choice of $\boldsymbol{\mu} \in \mathbb{D}$. (**a**) Velocity field. (**b**) Vorticity

5 Parameter Space Reduction by Active Subspaces Property

The active subspaces (AS) property has been emphasized recently, see for example P. Constantine in [22]. It concerns the properties of a parametric scalar function and it is exploited for dimension reduction in parameter studies. The main idea behind the active subspaces is the following: we rescale the inputs (the parameters $\boldsymbol{\mu}$, in our case) and then rotate the inputs domain with respect to the origin in such a way lower dimension behavior of the output function (the normalized pressure drop $f(\boldsymbol{\mu})$, in our case) is revealed. The active subspaces identify a set of important directions in the space of all inputs, instead of identifying a subset of the inputs as important. The latter approach would have indeed resulted in similar limitations as in our previous works [5, 9, 33]. If the output of the simulation does not change on average along a particular direction of the parameters, then we can safely ignore that direction in the parameter study. When an active subspace is identified for the problem of interest, it is possible to perform different parameter studies such as integration, optimization, response surfaces [14], and statistical inversion [43].

Now we review the process of finding active subspaces. Let us assume[3] $f : \mathbb{R}^m \to \mathbb{R}$ is a scalar function and $\rho : \mathbb{R}^m \to \mathbb{R}^+$ a probability density function.

[3]In this section we will omit the dependence on $\boldsymbol{\mu}$. It should be understood that $f = f(\boldsymbol{\mu})$, $\rho = \rho(\boldsymbol{\mu})$, etc.

For our sake a uniform probability density will suffice, as all possible geometrical configurations can be drawn with equal probability. In particular, we assume f continuous and differentiable in the support of ρ, with continuous and square-integrable (with respect to the measure induced by ρ) derivatives. The active subspaces of the pair (f, ρ) are the eigenspaces of the covariance matrix associated to the gradients $\nabla_{\boldsymbol{\mu}} f$. To this end we define the so-called uncentered covariance matrix of the gradients of f (among others see [29] for a more deep understanding of these operators), denoted by $\boldsymbol{\Sigma}$, whose elements are the average products of partial derivatives of the simulations' input/output map, i.e.:

$$\boldsymbol{\Sigma} = \mathbb{E}\,[\nabla_{\boldsymbol{\mu}} f \, \nabla_{\boldsymbol{\mu}} f^T] = \int_{\mathbb{D}} (\nabla_{\boldsymbol{\mu}} f)(\nabla_{\boldsymbol{\mu}} f)^T \rho \, d\boldsymbol{\mu}, \tag{12}$$

where $\mathbb{E}[\cdot]$ is the expected value. To approximate the eigenpairs of this matrix it is common to use a Monte Carlo method as follows [23, 51]:

$$\boldsymbol{\Sigma} \approx \frac{1}{N_{\text{train}}^{\text{AS}}} \sum_{i=1}^{N_{\text{train}}^{\text{AS}}} \nabla_{\boldsymbol{\mu}} f_i \, \nabla_{\boldsymbol{\mu}} f_i^T, \tag{13}$$

where we draw $N_{\text{train}}^{\text{AS}}$ independent samples $\boldsymbol{\mu}^{(i)}$ from the measure ρ and where $\nabla_{\boldsymbol{\mu}} f_i = \nabla_{\boldsymbol{\mu}} f(\boldsymbol{\mu}^{(i)})$. The matrix $\boldsymbol{\Sigma}$ is symmetric and positive semidefinite, so it admits a real eigenvalue decomposition

$$\boldsymbol{\Sigma} = \mathbf{W} \boldsymbol{\Lambda} \mathbf{W}^T, \tag{14}$$

where $\mathbf{W}$ is a $m \times m$ orthogonal matrix of eigenvectors, and $\boldsymbol{\Lambda}$ is the diagonal matrix of non-negative eigenvalues arranged in descending order.

We now form a lower dimensional parameter subspace by selecting the first M eigenvectors, for some $M < m$. On average, perturbations in the first set of coordinates change f more than perturbations in the second set of coordinates. While low eigenvalues suggest that the corresponding vector is in the nullspace of the covariance matrix. Discarding these vectors we can construct an approximation of f. For the sake of notation, let us partition $\boldsymbol{\Lambda}$ and $\mathbf{W}$ as follows:

$$\boldsymbol{\Lambda} = \begin{bmatrix} \boldsymbol{\Lambda}_1 & \\ & \boldsymbol{\Lambda}_2 \end{bmatrix}, \qquad \mathbf{W} = [\mathbf{W}_1 \quad \mathbf{W}_2],$$

where $\boldsymbol{\Lambda}_1 = \text{diag}(\lambda_1, \dots, \lambda_M)$, and $\mathbf{W}_1$ contains the first M eigenvectors. The active subspace is the range of $\mathbf{W}_1$. The inactive subspace is the range of the remaining eigenvectors in $\mathbf{W}_2$. The linear combinations of the input parameters with weights from the important eigenvectors are the active variables. We approximate the behaviour of the objective function by projecting the full parameter space onto the active subspace.

Given the input parameters in the full space we can map forward to the active subspace. Respectively we have the following formulas for the active variable $\boldsymbol{\mu}_M$ and the inactive variable $\boldsymbol{\eta}$:

$$\boldsymbol{\mu}_M = \mathbf{W}_1^T \boldsymbol{\mu} \in \mathbb{R}^M, \qquad \boldsymbol{\eta} = \mathbf{W}_2^T \boldsymbol{\mu} \in \mathbb{R}^{m-M}. \tag{15}$$

That means that any point in the parameter space $\boldsymbol{\mu} \in \mathbb{R}^m$ can be expressed in terms of $\boldsymbol{\mu}_M$ and $\boldsymbol{\eta}$ as follows:

$$\boldsymbol{\mu} = \mathbf{W}\mathbf{W}^T \boldsymbol{\mu} = \mathbf{W}_1\mathbf{W}_1^T \boldsymbol{\mu} + \mathbf{W}_2\mathbf{W}_2^T \boldsymbol{\mu} = \mathbf{W}_1 \boldsymbol{\mu}_M + \mathbf{W}_2 \boldsymbol{\eta}.$$

So it is possible to rewrite f as

$$f(\boldsymbol{\mu}) = f(\mathbf{W}_1 \boldsymbol{\mu}_M + \mathbf{W}_2 \boldsymbol{\eta}),$$

and construct a surrogate quantity of interest g using only the active variables

$$f(\boldsymbol{\mu}) \approx g(\mathbf{W}_1^T \boldsymbol{\mu}) = g(\boldsymbol{\mu}_M).$$

In our pipeline, the surrogate quantity of interest g will be obtained querying a POD-Galerkin reduced order model, as described in the next section.

Active subspaces can be seen in the more general context of ridge approximation (see for example [44, 54]). It can be proved that, under certain conditions, the active subspace is a good starting point in optimal ridge approximation and it is nearly stationary as shown in [25, 39].

6 Model Order Reduction Based on a POD-Galerkin Method

In this section we now briefly summarize the POD-Galerkin method that we employ for the model order reduction of the high fidelity approximation (10) of the Navier-Stokes equations (1), based on the usual offline-online paradigm [6, 12, 38, 56]. The main novelty in this section is that the training during the offline stage (as well as the testing during the online one) will be carried out *only* over the active (parameter) subspace, and not over the full parameter space $\mathbb{D}$.

Let us denote by $\Xi_{\text{train}}^{\text{POD}} = \{\boldsymbol{\mu}_M^{(i)}\}_{i=1}^{N_{\text{train}}^{\text{POD}}} \subset \mathbb{D}$ a training set of $N_{\text{train}}^{\text{POD}}$ points chosen randomly over the active subspace, i.e. the range of $\mathbf{W}_1$. During the offline stage, we assemble the following *snapshots* matrices:

$$S_{\mathbf{u}} = [\mathbf{u}(\boldsymbol{\mu}_M^{(1)}) \mid \ldots \mid \mathbf{u}(\boldsymbol{\mu}_M^{(N_{\text{train}})})] \in \mathbb{R}^{N_{\mathbf{u}}^{\delta} \times N_{\text{train}}^{\text{POD}}},$$

$$S_p = [\mathbf{p}(\boldsymbol{\mu}_M^{(1)}) \mid \ldots \mid \mathbf{p}(\boldsymbol{\mu}_M^{(N_{\text{train}})})] \in \mathbb{R}^{N_p^{\delta} \times N_{\text{train}}^{\text{POD}}},$$

$$S_{\mathbf{s}} = [\mathbf{s}(\boldsymbol{\mu}_M^{(1)}) \mid \ldots \mid \mathbf{s}(\boldsymbol{\mu}_M^{(N_{\text{train}})})] \in \mathbb{R}^{N_{\mathbf{u}}^{\delta} \times N_{\text{train}}^{\text{POD}}},$$

where $(\mathbf{u}(\cdot), \mathbf{p}(\cdot))$ is the FE solution of (10), and the supremizer solution $\mathbf{s}(\cdot)$ is obtained by a FE approximation of the following elliptic equation: for each $\boldsymbol{\mu} \in \Xi_{\text{train}}^{\text{POD}}$, assuming (an approximation of) $p_\delta(\boldsymbol{\mu})$ to be known, find $s_\delta(\boldsymbol{\mu}) \in V_\delta(\boldsymbol{\mu})$ such that

$$(s_\delta(\boldsymbol{\mu}), \boldsymbol{v}_\delta)_{V_\delta(\boldsymbol{\mu})} = b(\boldsymbol{v}_\delta, p_\delta(\boldsymbol{\mu}); \boldsymbol{\mu}) \quad \forall \boldsymbol{v}_\delta \in V_\delta(\boldsymbol{\mu}),$$

where $(\cdot, \cdot)_{V_\delta(\boldsymbol{\mu})}$ represents the inner products in $H^1(\Omega(\boldsymbol{\mu}); \mathbb{R}^n)$. Indeed, $s_\delta(\boldsymbol{\mu})$ is such that

$$s_\delta(\boldsymbol{\mu}) = \arg \sup_{\boldsymbol{v}_\delta \neq \mathbf{0}} \frac{b(\boldsymbol{v}_\delta, p_\delta(\boldsymbol{\mu}); \boldsymbol{\mu})}{\|\boldsymbol{v}_\delta\|_{V_\delta(\boldsymbol{\mu})}},$$

and the inf-sup constant

$$\beta_\delta(\boldsymbol{\mu}) = \inf_{q_\delta \neq 0} \sup_{\boldsymbol{v}_\delta \neq \mathbf{0}} \frac{b(\boldsymbol{v}_\delta, p_\delta(\boldsymbol{\mu}); \boldsymbol{\mu})}{\|\boldsymbol{v}_\delta\|_{V_\delta(\boldsymbol{\mu})} \|q_\delta\|_{Q_\delta(\boldsymbol{\mu})}}$$

is related to the supremizer solution as follows:

$$(\beta_\delta(\boldsymbol{\mu}))^2 = \inf_{q_\delta \neq 0} \frac{\|s_\delta(\boldsymbol{\mu})\|_{V_\delta(\boldsymbol{\mu})}}{\|q_\delta\|_{Q_\delta(\boldsymbol{\mu})}}.$$

Such supremizers are employed at the reduced order level to enhance the inf-sup stability of the reduced system, which is essential to obtain an accurate approximation of the pressure. We refer to [6, 58] for further details on supremizers, as well as to [3, 17, 61] for possible alternative approaches.

A POD basis for the velocity, pressure and supremizer spaces can be obtained by considering the singular value decomposition of the following matrices [13, 57]

$$X_{\mathbf{u}}^{1/2} S_{\mathbf{u}}, \qquad X_p^{1/2} S_p, \qquad X_{\mathbf{u}}^{1/2} S_{\mathbf{s}},$$

being $X_{\mathbf{u}}$ and X_p FE matrices corresponding to the discretization of the inner products in $H^1(\Omega; \mathbb{R}^n)$ and $L^2(\Omega)$, respectively. The first $N_{\mathbf{u}}$, N_p, $N_{\mathbf{s}}$ (respectively) left singular vectors are then considered as basis functions $\{\boldsymbol{\varphi}_n\}_{n=1}^{N_{\mathbf{u}}}$, $\{\zeta_n\}_{n=1}^{N_p}$, $\{\boldsymbol{\phi}_n\}_{n=1}^{N_{\mathbf{s}}}$ of the reduced basis spaces. Therefore, the reduced spaces for velocity V_N and pressure Q_N, of cardinality $N_{\mathbf{u},\mathbf{s}} = N_{\mathbf{u}} + N_{\mathbf{s}}$ and N_p, respectively, are then obtained as

$$V_N = \text{span}(\{\boldsymbol{\varphi}_n\}_{n=1}^{N_{\mathbf{u}}}, \{\boldsymbol{\phi}_n\}_{n=1}^{N_{\mathbf{s}}}), \quad Q_N = \text{span}(\{\zeta_n\}_{n=1}^{N_p}).$$

Finally, let us introduce the corresponding basis functions matrices

$$\mathbf{Z}_{\mathbf{u},\mathbf{s}} = [\boldsymbol{\varphi}_1 | \dots | \boldsymbol{\varphi}_{N_{\mathbf{u}}} | \boldsymbol{\phi}_1 | \dots | \boldsymbol{\phi}_{N_{\mathbf{s}}}] \in \mathbb{R}^{N_{\mathbf{u}}^\delta \times N_{\mathbf{u},\mathbf{s}}},$$

$$\mathbf{Z}_p = [\zeta_1 | \dots | \zeta_{N_p}] \in \mathbb{R}^{N_p^\delta \times N_p}.$$

Let us now denote by $\Xi_{\text{test}}^{\text{POD}} = \{\boldsymbol{\mu}_M^{(j)}\}_{j=1}^{N_{\text{test}}^{\text{POD}}} \subset \mathbb{D}$ a testing set of $N_{\text{test}}^{\text{POD}}$ points chosen randomly over the active subspace. During the online stage, for any $\boldsymbol{\mu}_M \in \Xi_{\text{test}}^{\text{POD}}$, we solve the following reduced nonlinear system: find $(\mathbf{u}_N(\boldsymbol{\mu}_M), \mathbf{p}_N(\boldsymbol{\mu}_M)) \in \mathbb{R}^{N_{\mathbf{u},\mathbf{s}}} \times \mathbb{R}^{N_p}$ such that

$$\begin{bmatrix} \mathbf{A}_N(\boldsymbol{\mu}_M) + \mathbf{C}_N(\mathbf{u}_N(\boldsymbol{\mu}_M); \boldsymbol{\mu}_M) & \mathbf{B}_N^T(\boldsymbol{\mu}_M) \\ \mathbf{B}_N(\boldsymbol{\mu}_M) & 0 \end{bmatrix} \begin{bmatrix} \mathbf{u}_N(\boldsymbol{\mu}_M) \\ \mathbf{p}_N(\boldsymbol{\mu}_M) \end{bmatrix} = \begin{bmatrix} \mathbf{f}_N(\boldsymbol{\mu}_M) \\ \mathbf{g}_N(\boldsymbol{\mu}_M) \end{bmatrix},$$

where

$$\mathbf{A}_N(\boldsymbol{\mu}_M) = \mathbf{Z}_{\mathbf{u},\mathbf{s}}^T\, \mathbf{A}(\boldsymbol{\mu}_M)\, \mathbf{Z}_{\mathbf{u}},\mathbf{s}, \quad \mathbf{B}_N(\boldsymbol{\mu}_M) = \mathbf{Z}_p^T\, \mathbf{B}(\boldsymbol{\mu}_M)\, \mathbf{Z}_{\mathbf{u},\mathbf{s}},$$

$$\mathbf{C}_N(\mathbf{w}; \boldsymbol{\mu}_M) = Z_{\mathbf{u},\mathbf{s}}^T\, \mathbf{C}(\mathbf{Z}_{\mathbf{u},\mathbf{s}}\, \mathbf{w}; \boldsymbol{\mu}_M)\, \mathbf{Z}_{\mathbf{u},\mathbf{s}}.$$

In order to obtain the maximum efficiency, the reduced system should not require evaluations of quantities defined on the high fidelity mesh; a standard approach based on the empirical interpolation method [11] and its discrete variants, as well as alternatives based on gappy POD [18], could be considered. We omit any additional detail on this topic for the sake of brevity, as this is already extensively discussed in the literature cited in this section.

7 Numerical Results

In this section we present the results of the complete pipeline applied to a specific artery bifurcation. Moreover we demonstrate the improvements obtained using the pipeline with respect to the POD approach on the full parameter space.

The mesh is discretized using tetrahedral cells; a FE approximation by $\mathbb{P}^2 - \mathbb{P}^1$ elements is used, resulting in 265049 degrees of freedom. FEniCS is employed for the implementation of the high fidelity solver [46].

Let us recall that the parameter space is a $m = 10$ dimensional space. In particular the parameters are the displacements of 10 different RBF control points along the orthogonal direction with respect to the surface. The moving control points are located in the two branches just after the bifurcation in order to simulate a stenosed carotid. In Fig. 4 it is possible to observe the original undeformed carotid in black and the moving control points in red with a possible deformation. The PyGeM open source package is used to perform the deformation [55].

Since the deformations are made with respect to the reference geometry, the quality of the resulting mesh after interpolation could decrease. To address this problem we computed the aspect ratio [34] of all the tetrahedra of each deformed mesh. In Fig. 7a we plot the minimum, the maximum and the mean of such ratio. Even though the maximum values of such index are increased by the deformation process, a sensible deterioration in mesh quality affects at most 0.07% of the total number of tetrahedra. Results are reported in Fig. 7b, which summarizes the

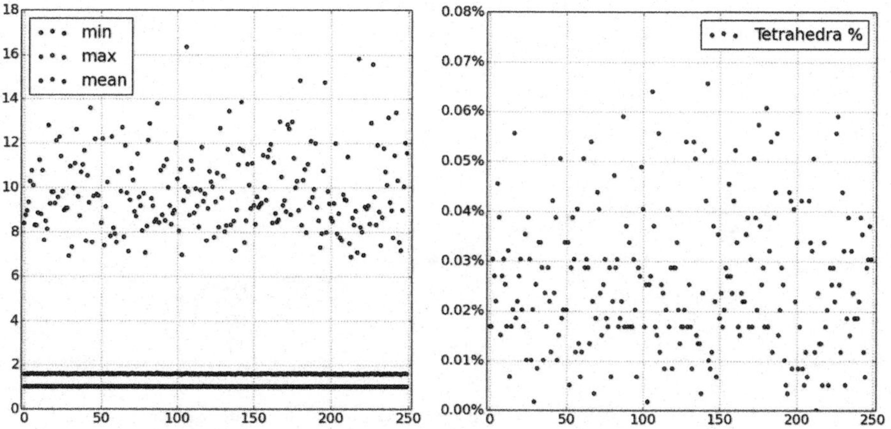

(a) Maximum, minimum, and mean value of the aspect ratio for all the mesh deformations.

(b) Percentage of tetrahedra with an aspect ratio above the maximum value of the reference geometry.

Fig. 7 Aspect ratio for each deformed mesh in the training set (**a**) and percentage of cells above the maximum aspect ratio of the reference mesh with respect to the total number of cells (**b**)

percentage of cells for which the aspect ratio is above the maximum value it had in the reference configuration. Thus, we conclude that the deformations impact on the mesh quality is negligible.

Recalling Sect. 5, we need to construct the uncentered covariance matrix $\boldsymbol{\Sigma}$ defined in (12). As shown in (13), we use a Monte Carlo method, in order to construct the matrix $\boldsymbol{\Sigma}$, using the software [26]. The number of training samples that we employ is $N_{\text{train}}^{\text{AS}} = 250$. Even though it may be challenging to explore a 10 dimensional space, heuristics reported in [22] suggest this choice of $N_{\text{train}}^{\text{AS}}$ is enough for the purposes of the active subspaces identification. In order to approximate the gradients of the pressure drop f with respect to the parameters, that is $\nabla_{\mu} f$, we use a local linear model that approximates the gradients with the best linear approximation using 17 nearest neighbors. After constructing the matrix $\boldsymbol{\Sigma}$ we calculate its real eigenvalue decomposition.

The eigenvalues of the covariance matrix in descending order are depicted in Fig. 8. The presence of gaps between the eigenvalues supports the existence of an active subspace. We can investigate the proper dimension of the active subspace using scatter plots that contain all available regression information that are called sufficient summary plots [27]. Recalling (15), Fig. 9 shows $f(\boldsymbol{\mu})$ against $\boldsymbol{\mu}_M = \mathbf{W}_1^T \boldsymbol{\mu}$, where $\mathbf{W}_1^T$ contains the first one and the first two eigenvectors, respectively. An active subspace of dimension one could suffice, but the band-width of the scatter points is quite large, so we prefer to retain more information about the output function by using a two dimensional active subspace.

To support this decision we construct a response surface varying the dimension of the active subspace and the order of the polynomial surface and we compute the

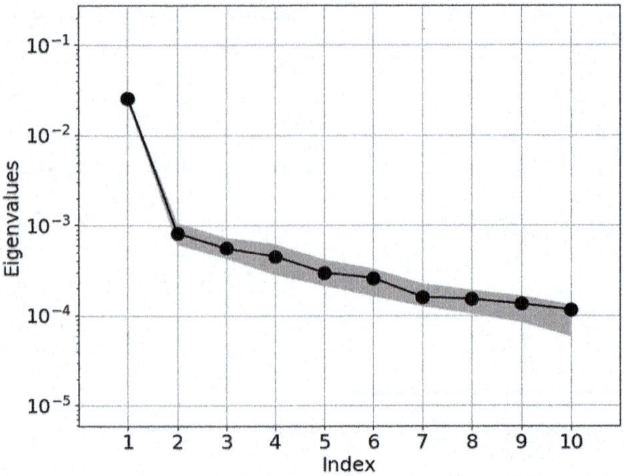

Fig. 8 Eigenvalue estimates in block circles with the bootstrap intervals (grey region). The order-of-magnitude gaps between the eigenvalues suggest confidence in the dominance of the active subspace

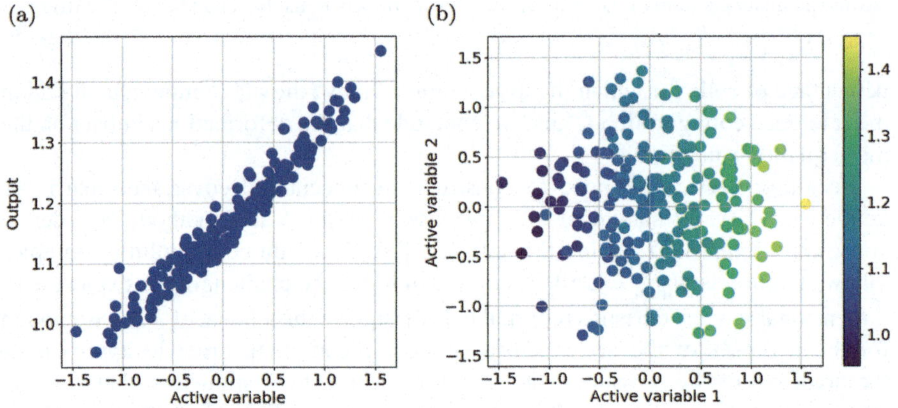

Fig. 9 Sufficient summary plots for (**a**) one and (**b**) two active variables using the training dataset

relative error with respect to a test dataset. We can see from Fig. 10 that on average the bidimensional subspace is the best choice in terms of information retention and dimension of the reduced parameter space. We underline that the choice of the active subspace dimension depends on the problem, the accuracy, and the goal you want to achieve. For the purpose of this chapter the choice we made is a very good compromise and does not affect the following results.

Once the active subspace $\mathbf{W}_1$ has been identified, we turn to the POD-Galerkin model order reduction defined in Sect. 6. The proposed combined methodology (denoted as "ROM + AS" in Figs. 11 and 12) will be compared to the standard

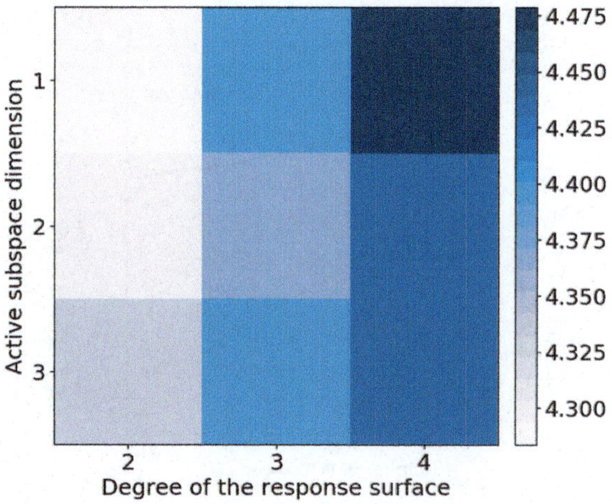

Fig. 10 Surrogate model error with respect to the active subspace dimension and the order of the response surface

POD-Galerkin approach on the full parameter space (denoted as "ROM" in Figs. 11 and 12) in order to highlight the effectiveness of our proposed method. The open source library RBniCS [8, 38] is employed to implement both methodologies. During the training phase, we select a training set of size $N_{\text{train}}^{\text{POD}}$ and compute a POD of the resulting snapshots. In the case of the combined methodology we have chosen $N_{\text{train}}^{\text{POD}} = 100$, while in the standard approach we have chosen $N_{\text{train}}^{\text{POD}} = N_{\text{train}}^{\text{AS}}$. Corresponding POD singular values are show in Fig. 11 for velocity, supremizers and pressure, as a function of the number N of selected POD modes. The results show a slower decay for the standard approach when compared to the combined one, meaning that the standard approach has to deal with a considerably larger solution manifold. If the final goal is to evaluate the quantity of interest, it is much more convenient to use the combined method, which is able to provide a much smaller solution manifold by neglecting the inactive (and so less interesting) directions. Indeed, Fig. 12 shows that the combined methodology is able to reach relative errors which are up to an order of magnitude smaller when compared to the standard one, for both velocity (Fig. 12a) and pressure (Fig. 12b) when $N = 20$. The errors are average relative errors on a testing set of cardinality $N_{\text{test}}^{\text{POD}} = 100$. A similar error analysis can be carried out for the quantity of interest, showing a trend similar to the one in Fig. 12b.

8 Conclusions and Perspectives

In this chapter we have presented a combined parameter and computational model reduction by means of active subspaces and POD-Galerkin methods, and we applied the proposed combined method on a synthetic problem related to the estimation of the pressure drop across a stenosed artery bifurcation. First, we reduced the high dimensional parameter space into a lower dimensional parameter

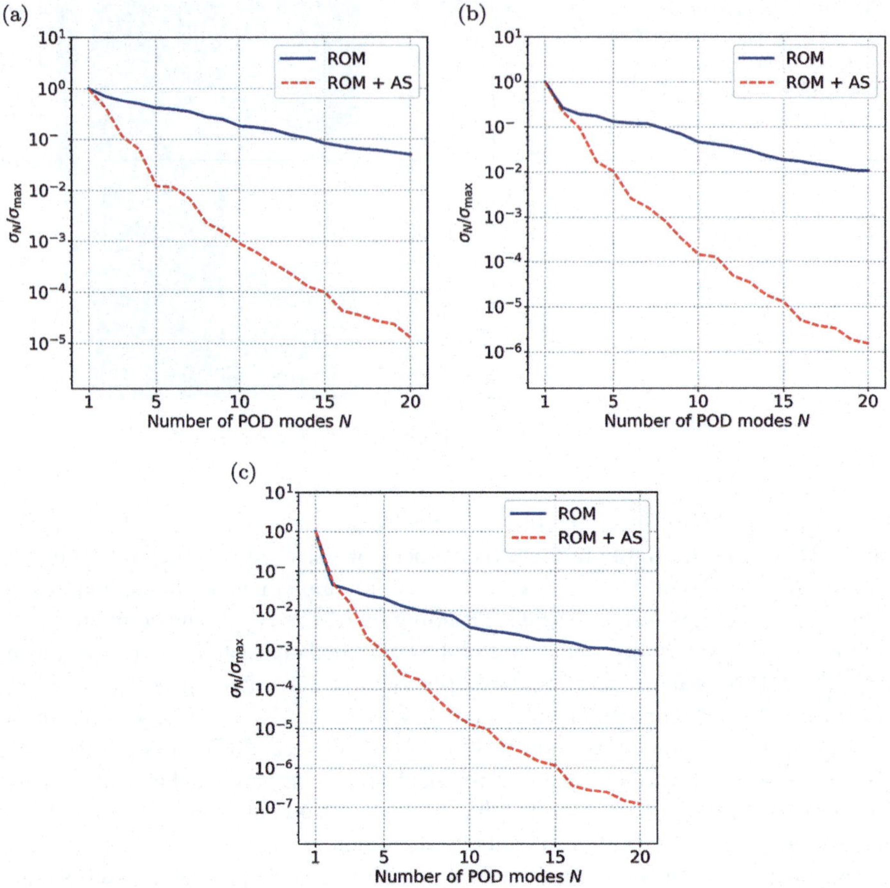

Fig. 11 POD singular values as a function of the number N of POD modes for (**a**) velocity, (**b**) supremizers, and (**c**) pressure

subspace by means of the active subspaces property. Our numerical test case, related to the deformation of two stenoses, shows an effective reduction of the dimensionality of the parameter space, from 10 control points displacements to 2 active variables. Just the active parameter subspace is then employed for a further model reduction by means of a POD-Galerkin method. When comparing the performance (in terms of errors) of the resulting reduced order model with a standard one (without active subspaces preprocessing), the proposed approach shows better results up to an order of magnitude. This is due to the fact that the standard approach has to account for several directions in the parameter space (the inactive subspaces) which only account for negligible variations in the pressure drop and thus could have been neglected. The proposed methodology could find further developments in more realistic cardiovascular problems, for what concerns

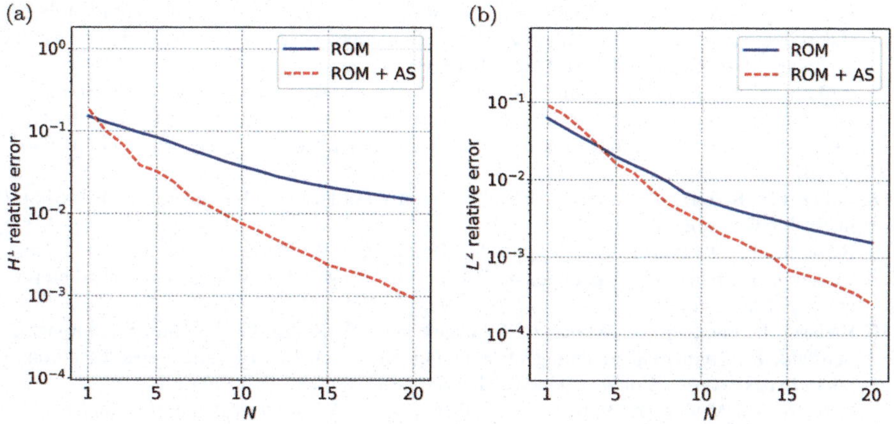

Fig. 12 Relative errors between the high fidelity solution and the reduced order one, as a function N, for (**a**) velocity and (**b**) pressure component

both the geometry (e.g. patient's personalization) and the mathematical model (e.g. unsteadiness, compliance). Moreover, several enhancements on the combination of the two approaches could be foreseen; among the possible ones, we mention a more tight coupling between the training stages of active subspaces and POD-Galerkin sampling methods based on a greedy approach in order to avoid the solution of several high fidelity problems for two (possibly large) training sets.

Acknowledgements This work was partially supported by the INDAM-GNCS 2017 project "Advanced numerical methods combined with computational reduction techniques for parameterised PDEs and applications", and by European Union Funding for Research and Innovation—Horizon 2020 Program—in the framework of European Research Council Executive Agency: H2020 ERC CoG 2015 AROMA-CFD project 681447 "Advanced Reduced Order Methods with Applications in Computational Fluid Dynamics" P.I. Gianluigi Rozza.

References

1. Agoshkov, V., Quarteroni, A., Rozza, G.: A mathematical approach in the design of arterial bypass using unsteady Stokes equations. J. Sci. Comput. **28**, 139–165 (2006)
2. Agoshkov, V., Quarteroni, A., Rozza, G.: Shape design in aorto-coronaric bypass anastomoses using perturbation theory. SIAM J. Numer. Anal. **44**(1), 367–384 (2007)
3. Ali, S., Ballarin, F., Rozza, G.: Stabilized reduced basis methods for parametrized Stokes and Navier-Stokes equations. (2018, in preparation)
4. Ambrosi, D., Quarteroni, A., Rozza, G.: Modeling of Physiological Flows. MS&A – Modeling, Simulation and Applications, vol. 5. Springer, Berlin (2012)
5. Ballarin, F., Manzoni, A., Rozza, G., Salsa, S.: Shape optimization by Free-Form Deformation: existence results and numerical solution for Stokes flows. J. Sci. Comput. **60**(3), 537–563 (2014)

6. Ballarin, F., Manzoni, A., Quarteroni, A., Rozza, G.: Supremizer stabilization of POD–Galerkin approximation of parametrized steady incompressible Navier–Stokes equations. Int. J. Numer. Methods Eng. **102**(5), 1136–1161 (2015)
7. Ballarin, F., Faggiano, E., Ippolito, S., Manzoni, A., Quarteroni, A., Rozza, G., Scrofani, R.: Fast simulations of patient-specific haemodynamics of coronary artery bypass grafts based on a POD–Galerkin method and a vascular shape parametrization. J. Comput. Phys. **315**, 609–628 (2016)
8. Ballarin, F., Sartori, A., Rozza, G.: RBniCS – reduced order modelling in fenics (2016). http://mathlab.sissa.it/rbnics
9. Ballarin, F., D'Amario, A., Perotto, S., Rozza, G.: A POD-selective inverse distance weighting method for fast parametrized shape morphing (2017, submitted). arXiv preprint arXiv:1710.09243
10. Ballarin, F., Faggiano, E., Manzoni, A., Quarteroni, A., Rozza, G., Ippolito, S., Antona, C., Scrofani, R.: Numerical modeling of hemodynamics scenarios of patient-specific coronary artery bypass grafts. Biomech. Model. Mechanobiol. **16**(4), 1373–1399 (2017)
11. Barrault, M., Maday, Y., Nguyen, N.C., Patera, A.T.: An 'empirical interpolation' method: application to efficient reduced-basis discretization of partial differential equations. Comptes Rendus Mathematique **339**(9), 667–672 (2004)
12. Benner, P., Ohlberger, M., Patera, A., Rozza, G., Urban, K. (eds.): Model Reduction of Parametrized Systems. MS&A – Modeling, Simulation and Applications, vol. 17. Springer, Berlin (2017)
13. Berkooz, G., Holmes, P., Lumley, J.: The proper orthogonal decomposition in the analysis of turbulent flows. Annu. Rev. Fluid Mech. **25**(1), 539–575 (1993)
14. Box, G.E., Draper, N.R.: Empirical Model-Building and Response Surfaces, vol. 424. Wiley, New York (1987)
15. Brown, S.A.: Building supermodels: emerging patient avatars for use in precision and systems medicine. Front. Physiol. **6**, 318 (2015)
16. Buhmann, M.D.: Radial Basis Functions: Theory and Implementations, vol. 12. Cambridge University Press, Cambridge (2003)
17. Caiazzo, A., Iliescu, T., John, V., Schyschlowa, S.: A numerical investigation of velocity-pressure reduced order models for incompressible flows. J. Comput. Phys. **259**, 598–616 (2014)
18. Carlberg, K., Farhat, C., Cortial, J., Amsallem, D.: The GNAT method for nonlinear model reduction: effective implementation and application to computational fluid dynamics and turbulent flows. J. Comput. Phys. **242**, 623–647 (2013)
19. Chen, P., Quarteroni, A.: Weighted reduced basis method for stochastic optimal control problems with elliptic PDE constraint. SIAM/ASA J. Uncertain. Quantif. **2**(1), 364–396 (2014)
20. Chen, P., Quarteroni, A., Rozza, G.: Comparison between reduced basis and stochastic collocation methods for elliptic problems. J. Sci. Comput. **59**(1), 187–216 (2014)
21. Chinesta, F., Keunings, R., Leygue, A.: The Proper Generalized Decomposition for Advanced Numerical Simulations: A Primer. Springer Science & Business Media, Berlin (2013)
22. Constantine, P.G.: Active Subspaces: Emerging Ideas for Dimension Reduction in Parameter Studies, vol. 2. SIAM, Philadelphia (2015)
23. Constantine, P., Gleich, D.: Computing active subspaces with Monte Carlo. arXiv preprint arXiv:1408.0545 (2015)
24. Constantine, P.G., Emory, M., Larsson, J., Iaccarino, G.: Exploiting active subspaces to quantify uncertainty in the numerical simulation of the HyShot II scramjet. J. Comput. Phys. **302**, 1–20 (2015)
25. Constantine, P.G., Eftekhari, A., Ward, R.: A near-stationary subspace for ridge approximation (2016). arXiv preprint arXiv:1606.01929
26. Constantine, P., Howard, R., Glaws, A., Grey, Z., Diaz, P., Fletcher, L.: Python active-subspaces utility library. J. Open Source Softw. **1**(5) (2016)
27. Cook, R.D.: Regression Graphics: Ideas for Studying Regressions Through Graphics, vol. 482. Wiley, New York (2009)

28. Cueto, E., Chinesta, F.: Real time simulation for computational surgery: a review. Adv. Model. Simul. Eng. Sci. **1**(1), 11:1–11:18 (2014)
29. Devore, J.L.: Probability and Statistics for Engineering and the Sciences. Cengage Learning, Boston (2015)
30. Doorly, D., Sherwin, S.: Geometry and flow. In: Formaggia, L., Quarteroni, A., Veneziani, A. (eds.) Cardiovascular Mathematics. MS&A – Modeling, Simulation and Applications, vol. 1. Springer Italia, Milano (2009)
31. Dryden, I., Mardia, K.: Statistical Analysis of Shape. Wiley, New York (1998)
32. Duchon, J.: Splines minimizing rotation-invariant semi-norms in Sobolev spaces. In: Constructive Theory of Functions of Several Variables, pp. 85–100. Springer, Berlin (1977)
33. Forti, D., Rozza, G.: Efficient geometrical parametrisation techniques of interfaces for reduced-order modelling: application to fluid–structure interaction coupling problems. Int. J. Comput. Fluid Dyn. **28**(3–4), 158–169 (2014)
34. Frey, P., George, P.: Mesh generation. Application to finite elements. Hermes Science Publishing, Paris, Oxford (2000)
35. González, D., Cueto, E., Chinesta, F.: Computational patient avatars for surgery planning. Ann. Biomed. Eng. **44**(1), 35–45 (2016)
36. Guibert, R., Mcleod, K., Caiazzo, A., Mansi, T., Fernández, M.A., Sermesant, M., Pennec, X., Vignon-Clementel, I.E., Boudjemline, Y., Gerbeau, J.F.: Group-wise construction of reduced models for understanding and characterization of pulmonary blood flows from medical images. Med. Image Anal. **18**(1), 63–82 (2014)
37. Gunzburger, M.D.: Perspectives in Flow Control and Optimization, vol. 5. SIAM, Philadelphia (2003)
38. Hesthaven, J.S., Rozza, G., Stamm, B.: Certified Reduced Basis Methods for Parametrized Partial Differential Equations. Springer Briefs in Mathematics. Springer, Berlin (2015)
39. Hokanson, J.M., Constantine, P.G.: Data-driven polynomial ridge approximation using variable projection (2017). arXiv preprint arXiv:1702.05859
40. Hu, X., Parks, G.T., Chen, X., Seshadri, P.: Discovering a one-dimensional active subspace to quantify multidisciplinary uncertainty in satellite system design. Adv. Space Res. **57**(5), 1268–1279 (2016)
41. INRIA 3D Meshes Research Database. Available at: https://www.rocq.inria.fr/gamma/gamma/download/download.php
42. Jefferson, J.L., Gilbert, J.M., Constantine, P.G., Maxwell, R.M.: Reprint of: Active subspaces for sensitivity analysis and dimension reduction of an integrated hydrologic model. Comput. Geosci. **90**, 78–89 (2016)
43. Kaipio, J., Somersalo, E.: Statistical and Computational Inverse Problems, vol. 160. Springer Science & Business Media, Berlin (2006)
44. Keiper, S.: Analysis of generalized ridge functions in high dimensions. In: 2015 International Conference on Sampling Theory and Applications (SampTA), pp. 259–263. IEEE, New York (2015)
45. Lassila, T., Rozza, G.: Parametric free-form shape design with PDE models and reduced basis method. Comput. Methods Appl. Mech. Eng. **199**(23–24), 1583–1592 (2010)
46. Logg, A., Mardal, K.A., Wells, G.N.: Automated Solution of Differential Equations by the Finite Element Method. Springer, Berlin (2012)
47. Lukaczyk, T.W., Constantine, P., Palacios, F., Alonso, J.J.: Active subspaces for shape optimization. In: 10th AIAA Multidisciplinary Design Optimization Conference, p. 1171 (2014)
48. Manzoni, A., Quarteroni, A., Rozza, G.: Model reduction techniques for fast blood flow simulation in parametrized geometries. Int. J. Numer. Methods Biomed. Eng. **28**(6–7), 604–625 (2012)
49. Marsden, A.L.: Optimization in cardiovascular modeling. Annu. Rev. Fluid Mech. **46**(1), 519–546 (2014)
50. McLeod, K., Caiazzo, A., Fernández, M., Mansi, T., Vignon-Clementel, I., Sermesant, M., Pennec, X., Boudjemline, Y., Gerbeau, J.F.: Atlas-based reduced models of blood flows for

fast patient-specific simulations. In: Camara, O., Pop, M., Rhode, K., Sermesant, M., Smith, N., Young, A. (eds.) Statistical Atlases and Computational Models of the Heart. Lecture Notes in Computer Science, vol. 6364, pp. 95–104. Springer, Berlin/Heidelberg (2010)
51. Metropolis, N., Ulam, S.: The monte carlo method. J. Am. Stat. Assoc. **44**(247), 335–341 (1949)
52. Morris, M.: Factorial sampling plans for preliminary computational experiments. Technometrics **33**(2), 161–174 (1991)
53. Morris, A., Allen, C., Rendall, T.: CFD-based optimization of aerofoils using radial basis functions for domain element parameterization and mesh deformation. Int. J. Numer. Methods Fluids **58**(8), 827–860 (2008)
54. Pinkus, A.: Ridge Functions, vol. 205. Cambridge University Press, Cambridge (2015)
55. PyGeM: Python Geometrical Morphing. Available at https://github.com/mathLab/PyGeM
56. Quarteroni, A., Rozza, G.: Reduced Order Methods for Modeling and Computational Reduction. MS&A – Modeling, Simulation and Applications, vol. 9. Springer, Berlin (2014)
57. Ravindran, S.: A reduced-order approach for optimal control of fluids using proper orthogonal decomposition. Int. J. Numer. Meth. Fluids **34**, 425–448 (2000)
58. Rozza, G., Veroy, K.: On the stability of the reduced basis method for Stokes equations in parametrized domains. Comput. Methods Appl. Mech. Eng. **196**(7), 1244–1260 (2007)
59. Sederberg, T., Parry, S.: Free-Form Deformation of solid geometric models. In: Proceedings of SIGGRAPH - Special Interest Group on GRAPHics and Interactive Techniques, pp. 151–159. SIGGRAPH (1986)
60. Shepard, D.: A two-dimensional interpolation function for irregularly-spaced data. In: Proceedings-1968 ACM National Conference, pp. 517–524. ACM, New York (1968)
61. Stabile, G., Rozza, G.: Finite volume POD-Galerkin stabilised reduced order methods for the parametrised incompressible Navier–Stokes equations. Comput. Fluids (2018). https://doi.org/10.1016/j.compfluid.2018.01.035
62. Tezzele, M., Salmoiraghi, F., Mola, A., Rozza, G.: Dimension reduction in heterogeneous parametric spaces with application to naval engineering shape design problems. Adv. Model. Simul. Eng. Sci. (2018, in press). Preprint, arXiv:1709.03298
63. Torlo, D., Ballarin, F., Rozza, G.: Stabilized reduced basis methods for advection dominated partial differential equations with random inputs (2017, submitted)
64. Venturi, L., Ballarin, F., Rozza, G.: Weighted POD–Galerkin methods for parametrized partial differential equations in uncertainty quantification problems (2017, submitted)
65. Wang, V.Y., Hoogendoorn, C., Frangi, A.F., Cowan, B.R., Hunter, P.J., Young, A.A., Nash, M.P.: Automated personalised human left ventricular FE models to investigate heart failure mechanics. In: Proceedings of the Third International Conference on Statistical Atlases and Computational Models of the Heart: Imaging and Modelling Challenges, STACOM'12, pp. 307–316. Springer, Berlin, Heidelberg (2013)
66. Witteveen, J., Bijl, H.: Explicit mesh deformation using Inverse Distance Weighting interpolation. In: 19th AIAA Computational Fluid Dynamics. AIAA, Washington (2009)
67. Zarins, C.K., Giddens, D.P., Bharadvaj, B., Sottiurai, V.S., Mabon, R.F., Glagov, S.: Carotid bifurcation atherosclerosis: quantitative correlation of plaque localization with flow velocity profiles and wall shear stress. Circ. Res. **53**(4), 502–514 (1983)

Marco Tezzele received a Bachelor of Science in Mathematics from the University of Pavia (Italy). Then he moved to Milan and earned a Master's degree in Applied Mathematics from the University of Milan in 2014. Moreover he holds a Master in High Performance Computing from SISSA, International School for Advanced Studies, and ICTP, International Centre for Theoretical Physics, in Trieste (2015).

Since 2015 he is a research assistant at SISSA mathLab group. His research focus is on geometrical parametrization, reduction of the parameter space for parametric PDEs, machine learning, and data driven model order reduction.

Francesco Ballarin is an assistant professor in numerical analysis at SISSA mathLab, Trieste, Italy, in the framework of the ERC CoG project AROMA-CFD (PI: Gianluigi Rozza). Francesco received his Bachelor's degree (2009) and Master's degree (2011) in Mathematical Engineering from Politecnico di Milano, Italy. He continued his study with a PhD in Mathematical Models and Methods in Engineering at Politecnico di Milano (2015), with a thesis on reduced order methods for patient-specific hemodynamics. He then joined SISSA mathLab in 2015, working as a post-doctoral research fellow and, since June 2018, as assistant professor. His research interests are related to numerical analysis, reduced order modelling, computational fluid dynamics, cardiovascular mathematics, inverse problems in optimal control and shape optimization.

Gianluigi Rozza is Full Professor of Numerical Analysis and Scientific Computing at SISSA, International School for Advanced Studies. He holds a degree in aerospace engineering from Politecnico di Milano (2002) and a PhD in applied mathematics from Ecole Polytechnique Federale de Lausanne (2005). He was a Postdoctoral Research Associate at the Massachusetts Institute of Technology (MIT) Center for Computational Engineering (2006–2008) and then a Researcher and Lecturer at EPFL (2008–2012). He is the author of over 100 scientific publications and recipient of the 2014 ECCOMAS young investigator Jacques Louis Lions Award in Computational Mathematics for researchers under the age of 40. Professor Rozza has been an Associate Editor of the *SIAM/ASA Journal of Uncertainty Quantification* since 2013, of the *SIAM Journal of Numerical Analysis* since 2015, and of *Computing and Visualization in Science* since 2016.

Extended Finite Elements Method for Fluid-Structure Interaction with an Immersed Thick Non-linear Structure

Christian Vergara and Stefano Zonca

Abstract We consider an Extended Finite Element method to solve fluid-structure interaction problems in the case of an immersed thick structure described by non-linear finite elasticity. This method, that belongs to the family of the Cut Finite Element methods, allows to consider unfitted meshes for the fluid and solid domains by maintaining the fluid mesh fixed in time as the solid moves. We review the state of the art about the numerical methods for fluid-structure interaction problems and we present an overview of the Cut Finite Element methods. We describe the numerical discretization proposed here to handle the case of a thick immersed structure with size comparable or smaller than the fluid mesh element size in the case of non-linear finite elasticity. Finally, we present some three-dimensional numerical results of the proposed method.

1 Introduction

The interaction between a fluid and an immersed structure may be significant in many applications, for example in aeronautic engineering to study the response of the air on the aircraft [10, 31, 32, 61], in civil engineering to understand the effect of wind on bridges [19, 75, 95], towers [54], and suspended cable [12, 82], in energy engineering to study the modeling of wind-turbines, heat exchangers and hydro-turbines [8, 81, 86], in sport engineering to investigate the impact of the waves over a rowing boat [33, 34] or the flow around a sailing yacht [79], in biomedical application, for instance in hemodynamics to study the stresses exerted by blood flow to the leaflets of an heart valve [68, 90], or to study the blood pressure exerted to the retinal vessels walls [2, 4].

In some cases, it may be interesting to consider a *full three-dimensional* (3D) model for the structure, even though the thickness of the structure is small with

C. Vergara · S. Zonca (✉)
MOX, Dipartimento di Matematica, Politecnico di Milano, Milano, Italy
e-mail: christian.vergara@polimi.it; stefano.zonca@polimi.it

D. Boffi et al. (eds.), *Mathematical and Numerical Modeling of the Cardiovascular System and Applications*, SEMA SIMAI Springer Series 16,
https://doi.org/10.1007/978-3-319-96649-6_9

respect to the characteristic size of the domain. For example, in the simulation of heart valves, one should consider the interaction between the blood and the valve leaflets. Often, for clinical purposes, there is the need to accurately evaluate the internal structural stresses, which can be computed only by means of a full 3D geometric model.

The numerical simulation of such a fluid-structure interaction problem is very challenging. First of all, the structure undergoes large displacements, thus its movement cannot be ignored from the geometric point of view. Second, the immersed structure is very thin, often smaller than the characteristic mesh size of the fluid problem, and this leads to numerical and computational difficulties due to the cut of some fluid mesh elements.

In this work, we propose an *Extended Finite Elements* discretization for FSI problem with an immersed 3D non-linear elastic structure in the regime of large displacements. In Sect. 2 we briefly review the most important numerical strategies introduced so far to handle this problem, whereas in Sect. 3 we specifically focus on the family of *Cut Finite Elements*, to which our method belongs. Then, in Sect. 4 we introduce the mathematical problem and the proposed numerical approximation, whereas in Sect. 5 we give some detail on the algorithm for the solution of the non-linear system arising after discretization. Finally, in Sect. 6 we present some numerical results.

2 State of the Art

Several numerical methods have been developed so far to solve the FSI problem with an immersed structure. We subdivide them depending on the treatment of the computational meshes. Accordingly, we arrange them into two main categories: *body-fitted* mesh methods and *fixed/unfitted* mesh methods.

In the first category, we place all the methods that use a conforming and fitted mesh at the fluid-structure interface. Among them, we cite the *Arbitrary Lagrangian Eulerian* approach introduced in [28, 55, 58]. In presence of very large displacements, this method may fail due to the high distortion of the fluid mesh, so that a remeshing procedure is required. Moreover, this procedure has the disadvantage that may introduce an artificial diffusivity due to the need of interpolating from one mesh to the new one. Nonetheless, the ALE method has been used by some authors to deal with immersed structures. For example, in [64] an ALE approach with remeshing is proposed to simulate heart valve closure on a 2D simplified geometry, in [76] a synthetic 3D model was employed to study the valve opening, in [63] a 3D model is used to study the influence of the sinus of Valsalva in the aortic valve, in [92] a 2D simulation of the aortic valve is performed on a plane of symmetry along the center of a leaflet for an entire cardiac cycle. A similar approach based on local adaptation is presented in [7]. This "Extended ALE" method allows the structure mesh to move independently of the fluid one that is kept fixed, resulting in a pair of meshes that are not fitted at the interface. Then, the fitting between the two meshes is obtained via local remeshing or local changes in the connectivity.

Another fitted method is given by *Space-Time Finite Elements* [29, 57, 59, 77]. The basic idea is to divide the time domain into slabs and then to use the Finite Element (FE) basis functions on each slab for both the spatial and temporal discretizations.

In the second family of numerical methods, we place the methods based on a fixed background mesh and on an overlapped unfitted mesh for the fluid and the solid, respectively. These approaches were developed to avoid the movement, or the remeshing, of the fluid mesh. In particular, they have been specifically designed for treating the case of large deformations.

The first fixed/unfitted mesh method proposed in the literature is the *Immersed Boundary* (IB) method, introduced in [80] in the context of Finite Differences for an immersed membrane and specifically realized for studying the fluid-dynamics in the heart. In this framework, the fluid is represented in Eulerian coordinates, while the structure in Lagrangian coordinates by means of a forcing term for the fluid problem that acts on the fluid-structure interface. The extension of the IB approach to the FE method is presented in [14, 15]. The FE formulation is extendable to the case of thick structures and allows to easily manage the forcing term given by the action of the structure on the fluid, see, e.g., [16, 91, 94]. As regards the applications, the Finite Difference IB method has been used in [47] to simulate the blood dynamics in a realistic domain of the heart and in [46, 48] to simulate the dynamics of the heart valves. The Finite Element IB method has been employed in [93] to study an immersed structure interacting with a viscous fluid and in [67] for several biological applications regarding valve dynamics, vessel stents, red blood cells interaction and cells migrations. See also [56] for an application to bioprosthetic heart valves. Another IB approach widely used in the context of heart valves is the *Curvilinear Immersed Boundary* (CURVIB) method [36] which is particularly suited for the 3D case. The CURVIB method was successfully employed for FSI problems for simulating the dynamics of prosthetic heart valves, see [17, 18, 36, 65].

A different approach in the category of the fixed/unfitted mesh methods is the *Fictitious Domain* (FD) method. This method was introduced in [41] for solving the viscous-plastic flow equations inside complex domains, then in [13, 42] it was used for solving the Navier-Stokes equations around immersed objects, and in [43] it was extended to treat the case of moving rigid bodies inside incompressible viscous flows with applications to particle flows, see also [44, 45]. Several works based on the FD method have been produced for the solution of FSI problems with immersed structures: in [87, 88] an application to FSI for heart valves including contact with rigid bodies is presented; in [6] a procedure for dealing with the interaction of an incompressible fluid and different structures is proposed, allowing the contact among the deformable bodies; in [84] an application for heart valves is compared with experimental data; in [62] an application to bioprosthetic heart valves is considered; in [89] a comparison of some FD approaches with the ALE one is presented for several FSI problems. A variant to the FD approach for dealing with valve dynamics is proposed in [25, 26]. In the latter works, the FD approach is combined with the ALE one in order to exploit their advantages: the ALE approach is used to describe the movement of the root of the valve, that undergoes a limited displacement, so that no remeshing is necessary, allowing to build a fitted mesh at the fluid-structure interface; the FD approach is instead used to describe the leaflets of the valve that, on the contrary, undergo large displacements.

3 Cut Finite Element Methods

The methods presented above feature two main limitations:

- the body-fitted mesh methods require the remeshing of the fluid domain due to the highly distorted fluid elements that appear when the displacement of the immersed structure is too large;
- the fixed/unfitted mesh methods require the implementation of ad-hoc strategies to sharply capture the interface. Indeed, in the case of "diffuse interface" methods (such as the IB method), the interface conditions are imposed through forcing terms which spread the effect of such conditions on a cluster of neighbouring cells. This results in a lack of sharpness in capturing the boundaries and the inability of enforcing boundary conditions for strongly fluctuating quantities, such as in turbulent flows, see, e.g., [73].

Here, we consider a specific class of methods belonging to the fixed/unfitted family that tries to overcome these limitations. The advantages of this class, referred to as *Cut Finite Element* (CFE) methods, are that they maintain the accuracy of classical FEM and can be developed by extending the features of FEM, see, e.g., [24] for a review of this class of methods.

Let us consider a finite set of domains $\Omega^i \in \mathbb{R}^d$, with $i = 1, \ldots, N$, and $d = 3$. We indicate with

- *background* domain, a domain Ω such that $\Omega \supseteq \bigcup_{i=1}^{N} \Omega^i$, i.e. a domain that covers all the domains Ω^i;
- *foreground* domain, each of the domains Ω^i, $i = 1, \ldots, N$, that overlaps the background domain Ω;
- *interface*, a curve Σ^i, $i = 1, \ldots, N$, of co-dimension one that separates the background domain Ω to the foreground domain Ω^i.

Moreover, we distinguish the foreground domains into three categories, depending on their thickness:

- *zero-thickness* domain, a foreground domain of co-dimension one (contained in the background domain Ω) that divides Ω into two parts (Σ in Fig. 1);
- *thin* domain, a foreground domain such that its thickness is smaller than the characteristic size of the background mesh (Ω^1 in Fig. 1);
- *thick* domain, a foreground domain Ω^i such that the thickness of the domain is comparable with its characteristic size (Ω^2 in Fig. 1).

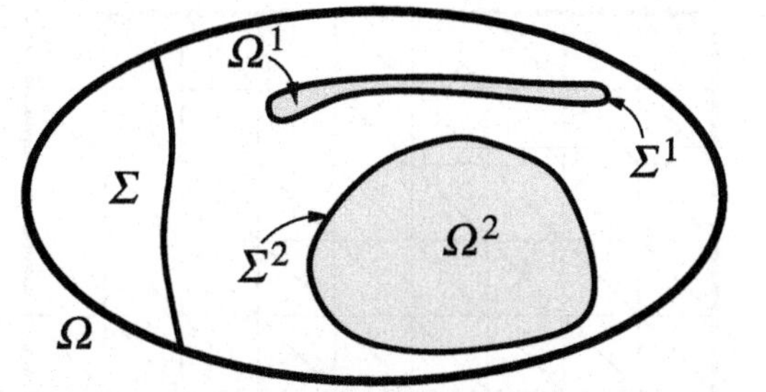

Fig. 1 The background domain Ω (white) contains the zero-thickness domain Σ, the thin foreground domain Ω^1 (grey) and the thick foreground domain Ω^2 (grey). The interfaces Σ^1 and Σ^2 delimit Ω^1 and Ω^2, respectively

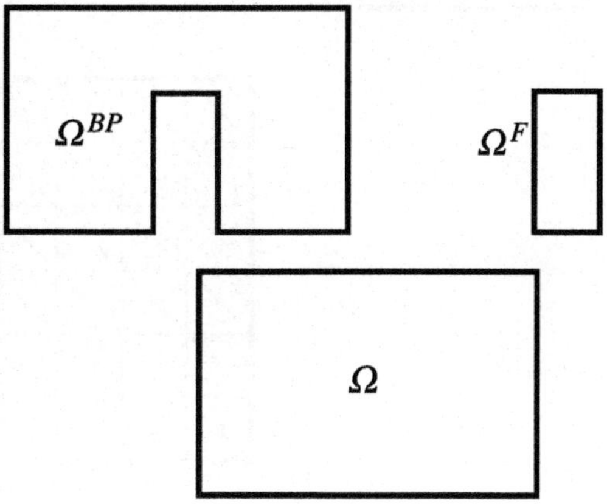

Fig. 2 Top, left: the background process domain Ω^{BP} associated with the process P^B. Top, right: the foreground domain Ω^F associated with the process P^F. Bottom: the background domain Ω is the union of the background process domain Ω^{BP} and the foreground domain Ω^F

With the final aim of describing how the physical processes are related to the domains and the meshes, in what follows we consider two processes, P^B and P^F, that represent the *background* and *foreground processes*, respectively. Process P^B occurs in the *background process domain* Ω^{BP}, while process P^F occurs in the *foreground domain* Ω^F, see Fig. 2 (top). The union of the domains Ω^{BP} and Ω^F generates the *background domain* Ω, see Fig. 2 (bottom).

For each of these three domains we generate the corresponding mesh. In particular, for the background domain Ω, we generate the *background mesh* $\mathscr{T}_h$, and for the foreground domain Ω^F, we generate the *foreground mesh* $\mathscr{T}_h^F$, see Fig. 3 (top). The *background process mesh* $\mathscr{T}_h^{BP}$ is instead generated by considering only the portions of the elements of the background mesh $\mathscr{T}_h$ that belong to the background process domain Ω^{BP}, see Fig. 3 (bottom), i.e.

$$\mathscr{T}_h^{BP} = \{K : K = K'_{|\Omega^{BP}}, \forall K' \in \mathscr{T}_h\}. \tag{1}$$

We have indicated by $h > 0$ the mesh size. Notice that the background process mesh contains elements of arbitrary shape, in particular polygons. We also define by $\widetilde{\mathscr{T}}_h^{BP}$ the smallest mesh contained in $\mathscr{T}_h$ that covers the background process

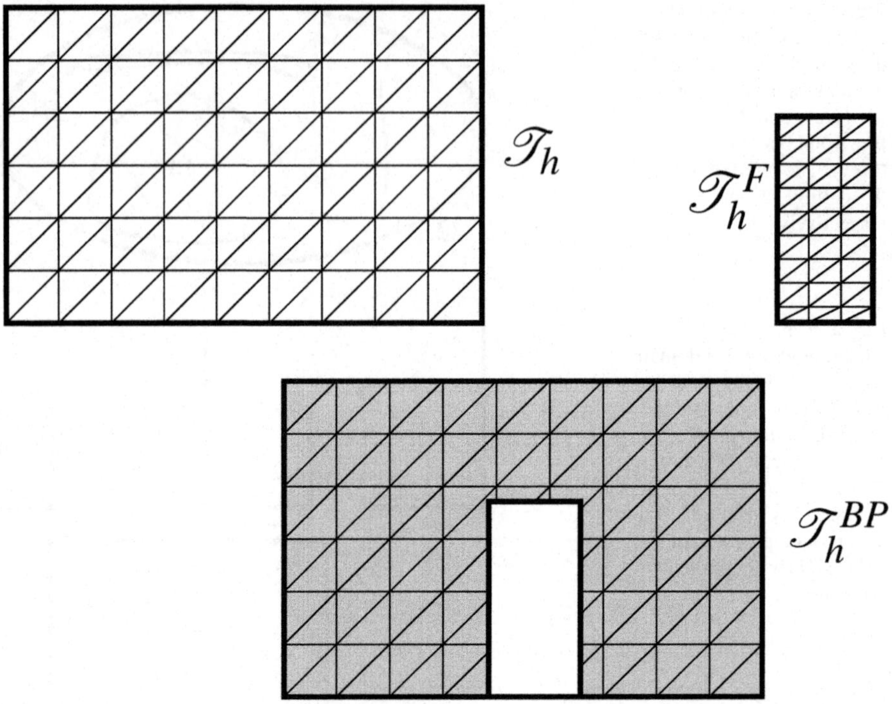

Fig. 3 Top, left: the background mesh $\mathscr{T}_h$ related to the background domain Ω. Top, right: the foreground mesh $\mathscr{T}_h^F$ related to the foreground domain Ω^F. Bottom: the background process mesh $\mathscr{T}_h^{BP}$ (grey) associated with the background process domain P^B

domain Ω^{BP} as

$$\widetilde{\mathscr{T}}_h^{BP} = \{K \in \mathscr{T}_h : K \cap \Omega^{BP} \neq \emptyset\}, \tag{2}$$

i.e. $\widetilde{\mathscr{T}}_h^{BP}$ is composed by the elements K that belong (also partially) to the domain Ω^{BP}, see Fig. 4 (left). Finally, we define by $\widetilde{\Omega}^{BP}$ the domain associated with the mesh $\widetilde{\mathscr{T}}_h^{BP}$, see Fig. 4 (right), i.e.

$$\widetilde{\Omega}^{BP} = int\left(\bigcup_{K \in \widetilde{\mathscr{T}}_h^{BP}} K\right). \tag{3}$$

With the above definitions, we can provide the following statement:
The common idea of Cut Finite Element methods is

(1) *to take a fixed background mesh overlapped by foreground meshes,*
(2) *to cut the elements of the background mesh with the zero-thickness mesh or with the interfaces of the thin and thick foreground meshes, generating elements of arbitrary shape (polytopes), and*

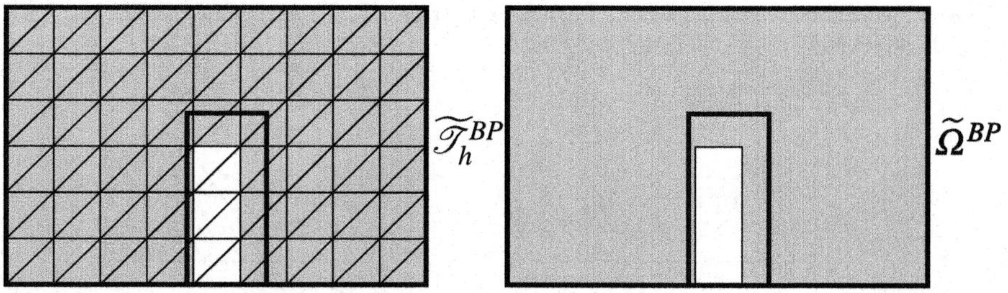

Fig. 4 Left: the mesh $\widetilde{\mathscr{T}}_h^{BP}$ (grey). Right: the domain $\widetilde{\Omega}^{BP}$ (grey)

(3) *to write a suitable weak formulation within the background process and foreground domains.*

We stress that the nature of the foreground domain may be either *geometric* or *physical*. In particular:

- for a geometric foreground domain, the zero-thickness domain describes a discontinuity in the properties of the background domain, but it is not subject to a process; instead, the thin and thick domains may represent (1) either a portion of the space with different physical properties with respect to the background one, however being described by the same process, (2) or a "hole" in the background geometry that is not subject to any process (*empty process*);
- for a physical foreground domain, the zero-thickness, thin and thick domains represent a different physical process with respect to the background one, represented by a different partial differential equation, as it happens in the FSI problem.

Different formulations have been proposed depending on the thickness and nature of the foreground domains, and each of these leads to a different CFE method. A graphical representation of all the possible combinations in the case of two processes is shown in Table 1. In the literature, the geometric cases are considered in the following works: the zero-thickness case is treated in [3, 11, 49, 74] for elliptic problems, in [9, 50, 78] for solid mechanics and in [52] for the Stokes problem; the thin case is treated in [83] for solving the Navier-Stokes equations; the thick case is considered in [51, 69] for elliptic problems and in [70] for the Stokes problem. The physical cases have been studied in the following works for FSI problems: the zero-thickness case in [1]; the thin case in [37–39] in a two-dimensional framework and in [96] for 3D problems; the thick case in [22] in two-dimensions, and in [37, 40, 71, 72] in three-dimensions. A particular physical case is considered in [53, 60] for solving PDEs only on the zero-thickness foreground domain, i.e. on immersed surfaces.

In particular, the thickness of the foreground domain has a strong effect on the approach employed to tackle the problem in the background process domain. In fact, when a foreground domain crosses the elements of the background mesh, three different configurations may appear, see Fig. 5.

Table 1 Schematic representation of possible scenarios when considering overlapping background (white) and foreground (grey) domains

	Geometric	Physical
Zero-thickness	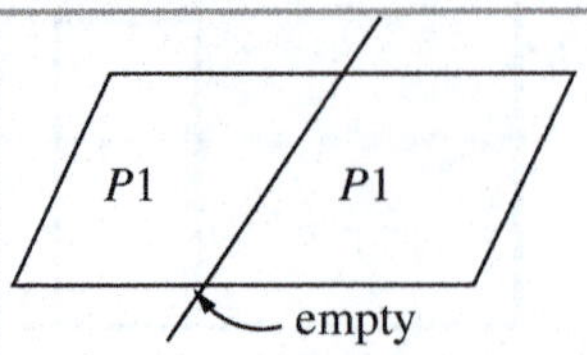	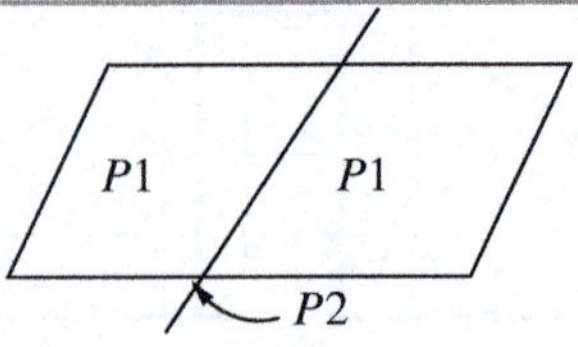
Thin	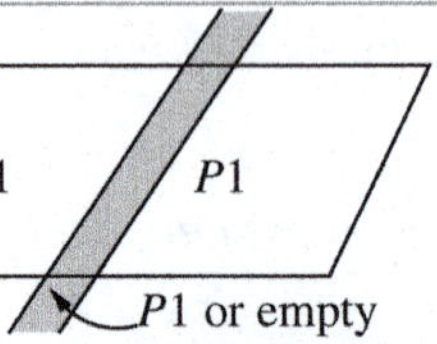	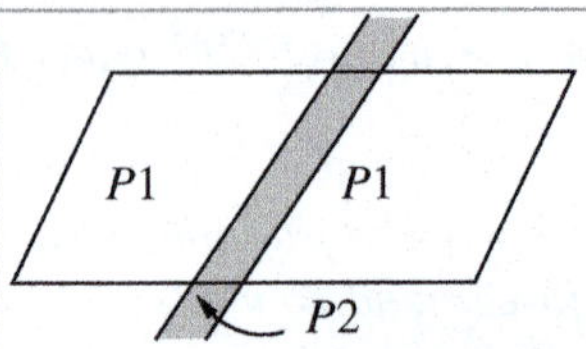
Thick	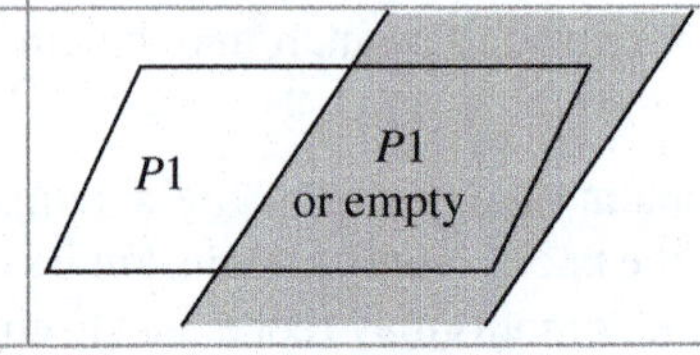	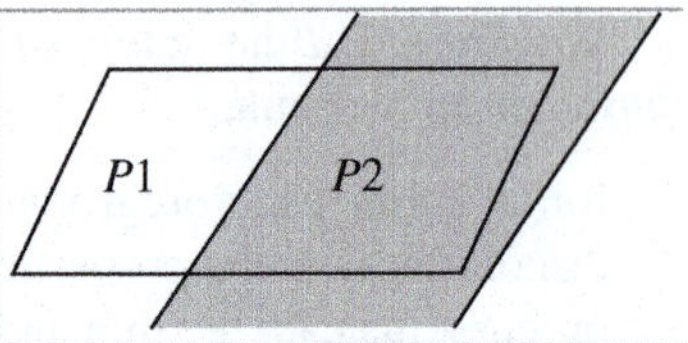

On each resulting domain, we have indicated the underlying process ($P1$, $P2$ or empty)

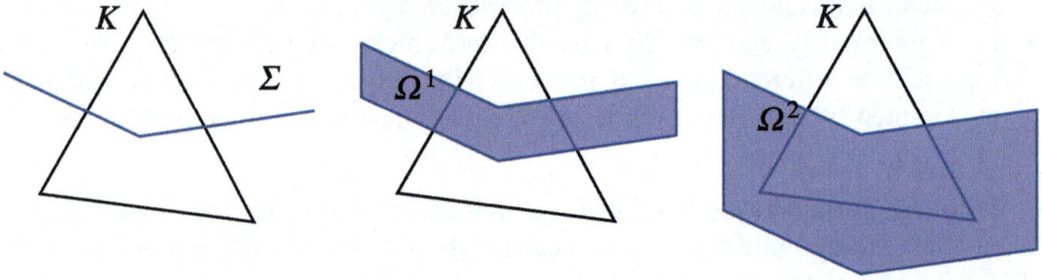

Fig. 5 Configurations of the background element K depending on the thickness of the foreground domain. Left: a foreground zero-thickness domain Σ intersects the background element K. Center: a thin foreground domain Ω^1 overlaps the background element K. Right: a thick foreground domain Ω^2 overlaps the background element K. In white, the background process element

In the case of a zero-thickness domain Σ, see Fig. 5 (left), the background element K is split into two parts and a numerical approximation is required on each portion of K. In the case of a thin domain Ω^1, see Fig. 5 (center), the background element K is partially overlapped and is divided into three parts: only the two parts that belong to the background process domain require a numerical approximation for the background process. Finally, in the case of thick domain Ω^2, see Fig. 5 (right), the background element K is divided into two portions and the numerical solution for the background process is required only on the background process mesh.

In any case, conforming grids at the interface between the background and foreground domains are difficult to generate, since the immersed foreground domain provides a severe constraint for the mesh generation. For this reason, unfitted overlapping meshes are considered in order to avoid the computational issue that may arise when generating fitted meshes.

The conditions to couple the problem at the interface are usually imposed via the *Discontinuous Galerkin* (DG) method (or, as some authors refer to, the *Nitsche*'s method). Lagrange multipliers are also considered, see, e.g., [37–40].

The majority of the methods presented above allow to consider a background domain overlapped by a thick foreground domain or crossed by a zero-thickness foreground domain. However, they do not deal with the case of a thin foreground domain immersed in the background one and, in fact, the extension of these methods to the thin case is not straightforward, since the thickness of the foreground domain is smaller than the characteristic size of the background mesh elements. Some works in this direction has been proposed in the literature: in [38–40], the authors consider two and three-dimensional approaches that are able to combine the feature of the *Extended Finite Element method* (XFEM) with overlapping meshes in the case of thin and thick foreground domains, by employing Lagrange multipliers for imposing the interface conditions; in [83] a method has been proposed to solve the Navier-Stokes equation solely with immersed fixed obstacles in a three-dimensional framework, where the interface conditions are imposed via the Nitsche's method. Recently, in [96], an XFEM method to handle the case of a thin foreground domain for 3D computations has been proposed for a linear structure, see also [35] for the case of ellipric problems.

The goal of this work is to describe a method for solving three-dimensional FSI problems with a non-linear immersed thin structure that combine the features of the approaches presented above, i.e. (1) the possibility of considering a numerical solution in each portion of the fluid background elements split by the interface and (2) the employment of composite grids to represent two domains, see Fig. 6.

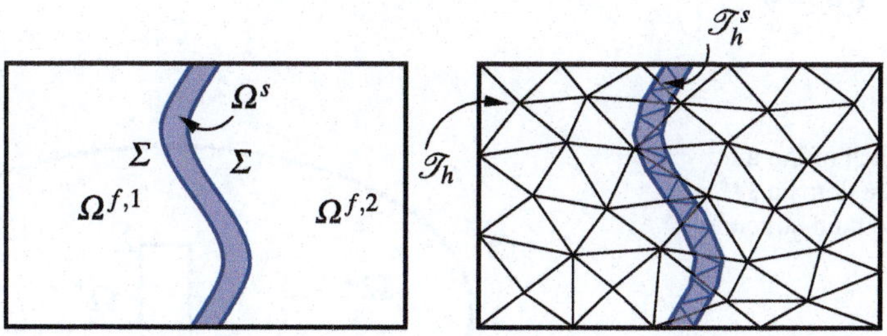

Fig. 6 Left: the background domain is overlapped by a thin foreground domain Ω^s. The interface Σ separates the fluid background process domains ($\Omega^{f,1}$ and $\Omega^{f,2}$) to the solid foreground domain. Right: The solid foreground mesh $\mathcal{T}_h^s$ overlaps the background mesh $\mathcal{T}_h$ that covers the entire domain. The thickness of $\mathcal{T}_h^s$ is smaller than the size of the background mesh elements

4 The Continuous Problem and the Extended Finite Elements Approximation

We consider a fluid background process domain $\Omega^{BP} = \Omega^f$ and a structure foreground domain $\Omega^F = \Omega^s$ such that $\Omega = \Omega^f \cup \Omega^s \subset \mathbb{R}^d$, $d = 2, 3$, is the background domain and $\Sigma = \overline{\Omega}^f \cap \overline{\Omega}^s$ is the fluid-structure interface, see Fig. 7. We denote by $\partial\Omega^f$ and $\partial\Omega^s$ the boundaries of the fluid and solid domains, respectively, and we define $\Gamma^f = \partial\Omega^f \setminus \Sigma$ and $\Gamma^s = \partial\Omega^s \setminus \Sigma$. Finally, we indicate by $\mathbf{n}^f$ and $\mathbf{n}^s$ the outward unit normal to the domain Ω^f and Ω^s, respectively. On the interface Σ we have $\mathbf{n}^f = -\mathbf{n}^s = \mathbf{n}$.

The continuous fluid-structure interaction problem reads as follows: Find for any $t \in (0, T]$, the fluid velocity $\mathbf{u}(t) : \Omega^f(t) \to \mathbb{R}^d$, the fluid pressure $p(t) : \Omega^f(t) \to \mathbb{R}$, the solid displacement $\widehat{\mathbf{d}}(t) : \widehat{\Omega}^s \to \mathbb{R}^d$, and the fluid domain displacement $\mathbf{d}^f(t) : \Sigma \to \mathbb{R}^d$, such that

- Fluid problem:

$$\begin{cases} \rho^f \partial_t \mathbf{u} + \rho^f \mathbf{u} \cdot \nabla \mathbf{u} - \nabla \cdot \mathbf{T}^f(\mathbf{u}, p) = \mathbf{0} & \text{in } \Omega^f(\mathbf{d}^f), \\ \nabla \cdot \mathbf{u} = 0 & \text{in } \Omega^f(\mathbf{d}^f), \\ \mathbf{u} = \mathbf{g} & \text{on } \Gamma^f_D(\mathbf{d}^f), \\ \mathbf{T}^f(\mathbf{u}, p)\mathbf{n}^f = \mathbf{h} & \text{on } \Gamma^f_N(\mathbf{d}^f); \end{cases} \tag{4}$$

- Solid problem:

$$\begin{cases} \rho^s \partial_{tt} \widehat{\mathbf{d}} - \nabla \cdot \widehat{\mathbf{T}}^s(\widehat{\mathbf{d}}) = \mathbf{0} & \text{in } \widehat{\Omega}^s, \\ \widehat{\mathbf{d}} = \mathbf{0} & \text{on } \widehat{\Gamma}^s; \end{cases} \tag{5}$$

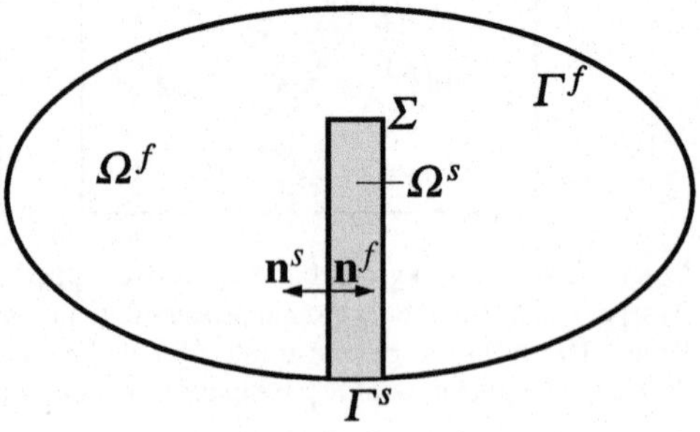

Fig. 7 Sketch of the fluid and structure domain Ω^f and Ω^s with the fluid-structure interface Σ

- Coupling conditions:

$$\begin{cases} \mathbf{u} = \partial_t \mathbf{d} & \text{on } \Sigma(\mathbf{d}^f), \\ \mathbf{T}^f(\mathbf{u}, p)\mathbf{n}^f = -\mathbf{T}^s(\mathbf{d})\mathbf{n}^s & \text{on } \Sigma(\mathbf{d}^f); \end{cases} \tag{6}$$

- Geometric condition:

$$\mathbf{d}^f = \mathbf{d} \qquad \text{on } \Sigma(\mathbf{d}^f), \tag{7}$$

where (4) are the Navier-Stokes equation, (5) are the equations of elastodynamics, (6) are the physical coupling conditions (kinematic and dynamic, respectively), and (7) is the condition that guarantees the geometric adherence between the fluid and solid domains. We have highlighted the dependence of the fluid domain and of its boundaries on the interface displacement $\mathbf{d}_f$, which in fact couples geometrically the fluid and the structure sub-problems. We point out that the fluid problem (4) is written in an Eulerian framework, i.e. in the deformed configuration, while the solid problem (5) is written in the Lagrangian framework, i.e. in the reference configuration. We have indicated with the superscript $\widehat{\cdot}$ the quantities evaluated in the reference configuration. The quantities without $\widehat{\cdot}$ are referred to the current instant t. We have $\Gamma^f = \Gamma_D^f \cup \Gamma_N^f$ and we have considered a Dirichlet boundary condition on Γ_D^f and a Neumann condition on Γ_N^f, with $\mathbf{g}$ and $\mathbf{h}$ suitable functions with the required regularity. . Moreover, ρ^f and ρ^s are the fluid and structure densities, $\mathbf{T}^f(\mathbf{u}, p) = -p\mathbf{I} + 2\mu^f\mathbf{D}(\mathbf{u})$ is the fluid Cauchy stress tensor, with μ^f the fluid viscosity and $\mathbf{D}(\mathbf{w}) = \dfrac{\nabla\mathbf{w} + \nabla^T\mathbf{w}}{2}$, $\widehat{\mathbf{T}}^s(\widehat{\mathbf{d}})$ is the first Piola-Kirchhoff solid stress tensor. Moreover, $\widehat{\mathbf{T}}^s = J\mathbf{T}^s\mathbf{F}^{-T}$ is the formula to pass from the solid Cauchy stress tensor $\mathbf{T}^s$ to the Piola-Kirchhoff tensor, with $\mathbf{F} = \nabla\mathbf{x}$ the deformation gradient, i.e. the gradient of the coordinates in the current position with respect to the reference space coordinates, and $J = \det(\mathbf{F})$ is its determinant. For the sake of simplicity we have considered homogeneous Dirichlet conditions on $\widehat{\Gamma}^s$. Moreover, the FSI problem given by Eqs. (4)–(7) has to be completed with initial conditions for the fluid and solid velocity and displacement.

The mechanical behaviour is described by a second order exponential model defined by the following the strain energy function:

$$\mathscr{W}(I_1) = \frac{\kappa}{4}\left((J-1)^2 + ln^2(J)\right) + \frac{\alpha}{2\gamma}\left(e^{\gamma(I_1-3)^2} - 1\right), \tag{8}$$

where $I_1 = tr(\mathbf{C})$, $\mathbf{C} = \mathbf{F}^T\mathbf{F}$ is the right Cauchy-Green tensor, α is the shear modulus that represents the mechanical stiffness of the material, κ is the bulk modulus and γ is a positive parameter that represents the level of non-linearity of the mechanical response of the body. The corresponding first Piola-Kirchhoff solid

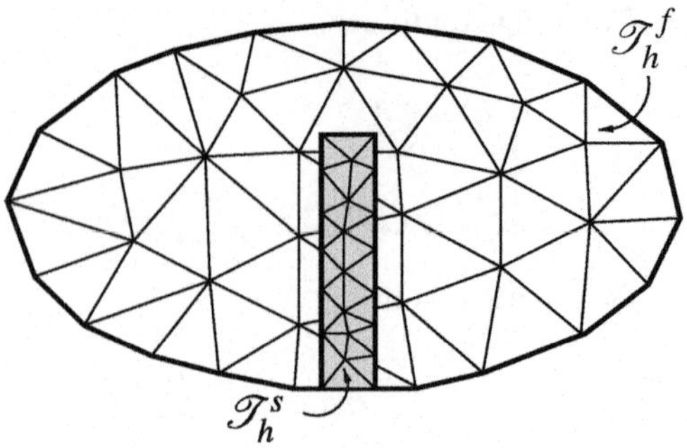

Fig. 8 Left: the structure foreground mesh $\mathcal{T}_h^s$ overlaps the background mesh $\mathcal{T}_h$. Right: representation of a background element split into three disconnected polyhedra (in blue/dark) by the solid foreground mesh (in grey/light)

stress tensor reads as follows:

$$\begin{aligned}\widehat{\mathbf{T}}^s(\mathbf{F}) =& \frac{\kappa}{2}\left(J^2 - J - ln\,(J)\right)\mathbf{F}^{-T} \\ &+ 2\alpha\left(J^{-2/3}I_1 - 3\right)e^{\gamma\left(J^{-2/3}I_1-3\right)^2}J^{-2/3}\left(\mathbf{F} - \frac{1}{3}I_1\mathbf{F}^{-T}\right).\end{aligned}$$

In what follows, with the aim of writing the Extended Finite Elements/Discontinuous Galerkin (XFEM/DG) discrete formulation, we follow [96] and we introduce the meshes and the numerical spaces. To ease the presentation, we assume that Ω^f, Ω^s and Σ are polyhedral. Referring to the notation of Sect. 4, we denote by $\mathcal{T}_h^F = \mathcal{T}_h^s$ the solid foreground mesh that covers the domain Ω^s and is fitted to $\partial\Omega^s$, and by $\mathcal{T}_h$ the background mesh that covers the whole domain Ω and is fitted to Γ^f, but in general not to Σ and Γ^s. We indicate by $h > 0$ the space discretization parameter, which is a function that may vary among the elements K of the meshes and between the background and foreground meshes. As a result, the solid foreground mesh $\mathcal{T}_h^s$ overlaps the background mesh $\mathcal{T}_h$, see Fig. 8 (left). Then, accordingly to definition (1), we denote by $\mathcal{T}_h^{BP} = \mathcal{T}_h^f$ the fluid background process mesh, i.e. the mesh generated by considering the restriction of $\mathcal{T}_h$ to Ω^f, defined as

$$\mathcal{T}_h^f = \{K : K = K'_{|\Omega^f}, \forall K' \in \mathcal{T}_h\}.$$

In the case of a thin foreground structure, the elements of the background mesh $\mathcal{T}_h$ could be cut by the foreground mesh and divided into several disconnected polyhedra, with a portion of the background elements overlapped by the foreground mesh, see Fig. 8. We refer to these elements as *split elements*.

We introduce the following mesh

$$\mathcal{G}_h = \{K : K \in \mathcal{T}_h,\ K \cap \Sigma \neq \emptyset,\ K \cap \Omega^f \text{ is a non}-\text{connected set}\},$$

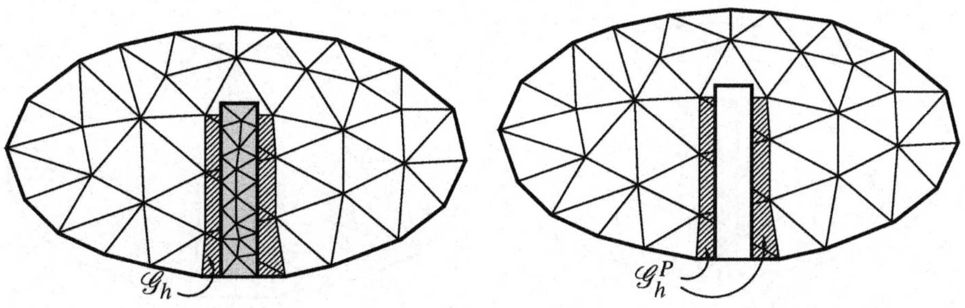

Fig. 9 Left: representation of the mesh $\mathscr{G}_h$. Notice that $\mathscr{G}_h$ contains also the portion of the elements overlapped by the structure. Right: representation of the non-connected mesh $\mathscr{G}_h^P$

that consists of all the elements K in $\mathscr{T}_h$ cut by the interface Σ which are split elements, see Fig. 9 (left). This means that each element $K \in \mathscr{G}_h$ is split into $N^K \geq 2$ fluid sub-parts, which in general are polyhedra. We denote by $P_i^K, i = 1, \ldots, N^K$, the polyhedra of a split element K. We define by $\mathscr{G}_h^P$ the union of all such polyhedra P_i^K, for $i = 1, \ldots, N^K$ and for each $K \in \mathscr{G}_h$, see Fig. 9 (right) where $N^K = 2$. More precisely

$$P \in \mathscr{G}_h^P \quad \longleftrightarrow \quad \exists K \in \mathscr{G}_h \text{ s.t. } P \subset K \cap \Omega^f \text{ is a connected set.}$$

The set $\mathscr{G}_h^P$ in now partitioned into its $N^f = \max_K N^K$ connected subsets $\Omega_h^{f,i}$. For example, by considering the same configuration in Fig. 10 (left), we have $N^f = 2$ connected subregions ($\Omega_h^{f,1}$ and $\Omega_h^{f,2}$).

Moving from these definitions, we set

$$\Omega_h^{f,0} = \Omega^f \setminus \bigcup_{K \subset \mathscr{G}_h} K.$$

Notice that $\Omega^f = \bigcup_{i=0,\ldots,N^f} \Omega_h^{f,i}$ and that $\Omega_h^{f,i} \cap \Omega_h^{f,j} = \emptyset, \forall i \neq j$. We denote by $\mathscr{T}_h^{f,0}$ the smallest mesh composed of the elements $K \in \mathscr{T}_h$ that covers the set $\Omega_h^{f,0}$, i.e.

$$K \in \mathscr{T}_h^{f,0} \quad \longleftrightarrow \quad K \cap \overline{\Omega}_h^{f,0} \neq \emptyset.$$

Finally, we denote by $\mathscr{T}_h^{f,i}$, for $i = 1, \ldots, N^f$, the smallest mesh that consists of all the elements of $\mathscr{G}_h$ that covers the set $\Omega_h^{f,i}$, i.e.

$$K \in \mathscr{T}_h^{f,i} \quad \longleftrightarrow \quad K \cap \overline{\Omega}_h^{f,i} \neq \emptyset, \quad i = 1, \ldots, N^f.$$

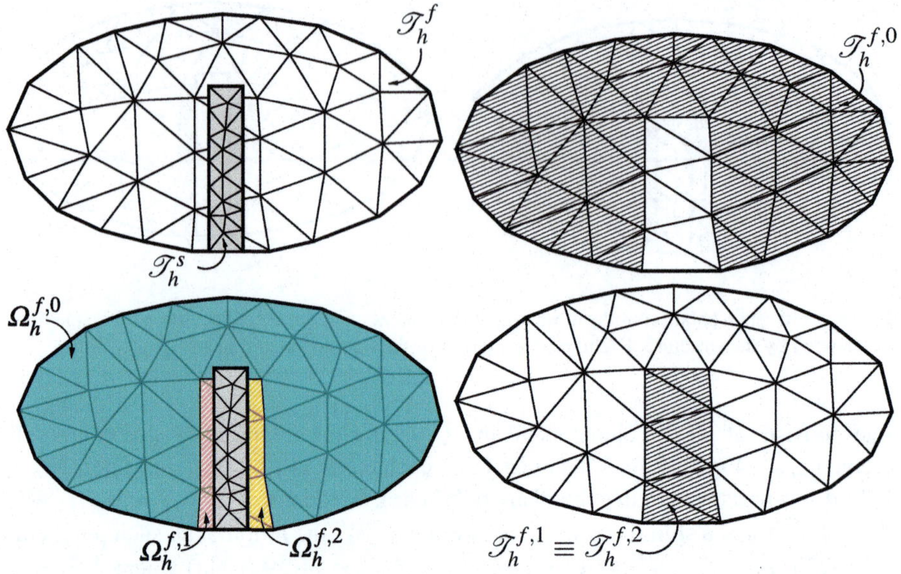

Fig. 10 Left: sketch of the background ($\mathscr{T}_h$) and foreground ($\mathscr{T}_h^s$) meshes (top) and the sets $\Omega_h^{f,0}$ (blue), $\Omega_h^{f,1}$ (pink) and $\Omega_h^{f,2}$ (yellow) (bottom). Right: the shaded regions represent the meshes $\mathscr{T}_h^{f,0}$ (top) and $\mathscr{T}_h^{f,i}$ (bottom)

Thus, each element $K \in \mathscr{G}_h$ belongs to N^K different meshes $\mathscr{T}_h^{f,i}$, and this will allow us to duplicate the dofs of K N^K times. The idea is to build the classical FEM approximation in $\mathscr{T}_h^{f,0}$, i.e. by using the classical dofs and shape functions, and to employ the XFEM strategy in $\mathscr{T}_h^{f,i}$, $i = 1, \ldots, N^f$, so that the dofs associated with the elements in $\mathscr{T}_h^{f,i}$, $i = 1, \ldots, N^f$, are duplicated: a set of dofs is used to compute the solution over each mesh $\mathscr{T}_h^{f,i}$.

The unfitted nature of the fluid and solid meshes requires a specific treatment of the coupling conditions between the corresponding fluid and solid problems at the interface Σ. A possibility, considered here, is to employ a Discontinuous Galerkin (DG) mortaring, see, e.g., [21, 22], in order to weakly impose the continuity of the fluid solution between the elements of the meshes $\mathscr{T}_h^{f,i}$, $i = 1, \ldots, N^f$, see below. On the contrary, in $\mathscr{T}_h^{f,0}$ it is possible to use either a non-conforming or a conforming discretization. For the sake of simplicity, we consider a conforming discretization, thus we impose a strong continuity in $\mathscr{T}_h^{f,0}$. We also notice that some operators of the discrete formulation will act on the domains $\Omega_h^{f,i} \subset \Omega^f$, while other operators, such as the stabilization terms, will act on the meshes $\mathscr{T}_h^{f,i}$, as we explain later on.

We observe that the set covered by $\mathscr{T}_h^{f,i}$ is larger than the one covered by the corresponding $\Omega_h^{f,i}$, see Fig. 10 (right) for the case $N^f = 2$. More complex configurations may happen for realistic three-dimensional domains.

Remark 1 We point out that the elements of the background mesh crossed by the interface Σ may be arbitrarily small due to the overlapping of the foreground domain. This may generate instabilities in the numerical solution and lead to an ill-conditioned matrix. For these reasons, to prevent instabilities and to maintain the robustness of the method, a possible strategy consists in the introduction of the *ghost-penalty* stabilization, see below and [20].

We classify the faces involved in the discrete formulation as follows:

- the faces belonging to the fluid-structure interface Σ, where we impose weakly the continuity of the velocity and stresses by means of the DG formulation, see, e.g., [21, 22];
- $\mathscr{F}_{h,p}^{f,i}$, the faces in $\mathscr{T}_h^{f,i}$, $i = 1, \ldots, N^f$, that belong to the background process domain Ω^f, where we impose weakly the continuity of the fluid velocity and stresses by means of the DG formulation, see below and, e.g., [5, 27];
- $\mathscr{F}_{h,\Sigma}^{f,i}$, the faces of $\mathscr{T}_h^{f,i}$, $i = 1, \ldots, N^f$, cut by the interface Σ, where the *ghost penalty* stabilization term (10) is applied, see below and [20].

For a representation of these faces, we refer to Fig. 11.

After a suitable time discretization of the FSI problem (4)–(7), we denote by $\Omega^{f,n}$ the approximation of Ω^f at time t^n. The discrete spaces for the fluid velocity and pressure read as follows:

$$\mathbf{V}_h^n = \{\mathbf{v}_h \in [X_h^{f,n}]^d : \mathbf{v}_{h|\Gamma^f} = \mathbf{0}\}, \quad Q_h^n = \{q_h \in X_h^{f,n}\},$$
$$\mathbf{W}_h = \{\mathbf{w}_h \in [X_h^s]^d : \mathbf{w}_{h|\Gamma^s} = \mathbf{0}\},$$

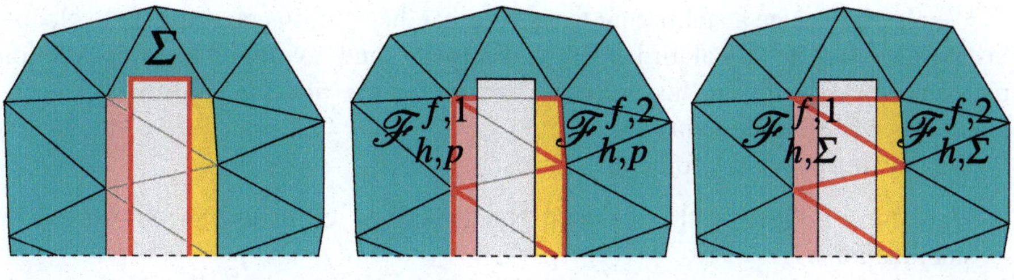

Fig. 11 Representation of the sets of faces involved in the discrete formulation (highlighted in red): (left) faces of the interface Σ; (center) faces $\mathscr{F}_{h,p}^{f,1}$ and $\mathscr{F}_{h,p}^{f,2}$ for the mortaring in the background process domain; (right) faces $\mathscr{F}_{h,\Sigma}^{f,1}$ and $\mathscr{F}_{h,\Sigma}^{f,2}$ for the ghost penalty stabilization

where

$$X_h^{f,n} = \{v_h \in L^2(\Omega^{f,n}) : v_h \in \mathscr{C}^0(\Omega_h^{f,0,n}),\, v_{h|K} \in \mathbb{P}_1(K),$$
$$\forall K \in \mathscr{T}_h^{f,i,n} \text{ for } i = 0, \ldots, N^f\},$$

and

$$X_h^s = \{v_h \in \mathscr{C}^0(\widehat{\Omega}^s) : v_{h|K} \in \mathbb{P}_1(K), \forall K \in \mathscr{T}_h^s\}.$$

To write the discrete formulation, we introduce the trace operators defined over an interface $\mathscr{I}$ that separates a generic domain $\Omega_{1,2}$ into Ω_1 and Ω_2. For a (scalar or vectorial) function q, we denote by $[\![\cdot]\!]_{\mathscr{I}}$ the jump and by $\{\!\{\cdot\}\!\}_{\mathscr{I},\epsilon}$ the ϵ-weighted mean across the interface $\mathscr{I}$, defined as

$$[\![q]\!]_{\mathscr{I}} = q_1 - q_2, \quad \{\!\{q\}\!\}_{\mathscr{I},\epsilon} = \epsilon q_1 + (1-\epsilon) q_2, \tag{9}$$

where q_1 and q_2 are the traces of q at the two sides of the interface and $\epsilon \in [0, 1]$. If the subscript ϵ is not indicated, we assume that $\epsilon = \frac{1}{2}$.

We consider a DG mortaring on Σ to impose the coupling conditions (6) and on the faces $\mathscr{F}_{h,p}^{f,i}$ to impose the continuity of the background fluid solution, by mimicking the (symmetric) interior penalty method, introduced for example in [5, 30] for the Poisson problem. Moreover, a ghost penalty term, see [20], is applied on $\mathscr{F}_{h,\Sigma}^{f,i}$ to guarantee robustness of the method with respect to the elements crossed by the interface Σ, defined as

$$g_h(\mathbf{u}_h, \mathbf{v}_h) = \gamma_g \sum_{i=1}^{N^f} \sum_{F \in \mathscr{F}_{h,\Sigma}^{f,i}} \mu^f h_F \int_F [\![\nabla \mathbf{u}_h]\!]_F \mathbf{n} \cdot [\![\nabla \mathbf{v}_h]\!]_F \mathbf{n}, \tag{10}$$

with $\gamma_g > 0$;

We also introduce a stabilizing term s_h to handle spurious pressure and velocity instabilities due to equal order Finite Elements and to dominating convection regimes, respectively. In this work we considered the *interior penalty* (IP) stabilization, see [23], as done in [83].

We now introduce the following forms.

– Fluid form collecting the classical Navier-Stokes terms and the ghost and IP stabilizations:

$$\mathscr{A}^f(\mathbf{z}, \mathbf{u}, p; \mathbf{v}, q)^r = \frac{\rho^f}{\Delta t}(\mathbf{u}, \mathbf{v})_{\Omega^{f,r}} + 2\mu^f(\mathbf{D}(\mathbf{u}), \mathbf{D}(\mathbf{v}))_{\Omega^{f,r}}$$
$$- (p, \nabla \cdot \mathbf{v})_{\Omega^{f,r}} + (q, \nabla \cdot \mathbf{u})_{\Omega^{f,r}} + \rho^f(\mathbf{z} \cdot \nabla \mathbf{u}, \mathbf{v})_{\Omega^{f,r}}$$
$$+ s_h(\mathbf{u}, p; \mathbf{v}, q)^r + g_h(\mathbf{u}, \mathbf{v})^r;$$

– Structure form:

$$\mathscr{A}^s\left(\widehat{\mathbf{d}};\widehat{\mathbf{w}}\right)=\frac{\rho^s}{\Delta t^2}(\widehat{\mathbf{d}},\widehat{\mathbf{w}})_{\widehat{\Omega}^s}+(\widehat{\mathbf{T}}^s(\widehat{\mathbf{d}}),\nabla\widehat{\mathbf{w}})_{\widehat{\Omega}^s};$$

– Form related to the DG terms involving only the fluid unknowns and test functions:

$$\begin{aligned}\mathscr{D}^{ff}(\mathbf{u},p;\mathbf{v},q)^r=&-\sum_{i=1}^{N^f}\sum_{F\in\mathscr{F}_{h,p}^{f,i,r}}\left(\left\{\left\{\mathbf{T}^f(\mathbf{u},p)\right\}\right\}_F\mathbf{n},[\![\mathbf{v}]\!]_F\right)_F\\&-\sum_{i=1}^{N^f}\sum_{F\in\mathscr{F}_{h,p}^{f,i,r}}\left([\![\mathbf{u}]\!]_F,\left\{\left\{\mathbf{T}^f(\mathbf{v},-q)\right\}\right\}_F\mathbf{n}\right)_F\\&+\sum_{i=1}^{N^f}\sum_{F\in\mathscr{F}_{h,p}^{f,i,r}}\frac{\gamma_p\mu^f}{h_F}([\![\mathbf{u}]\!]_F,[\![\mathbf{v}]\!]_F)_F,\\&-\left(\epsilon\mathbf{T}^f(\mathbf{u},p)\mathbf{n},\mathbf{v}\right)_{\Sigma^r}-\left(\mathbf{u},\epsilon\mathbf{T}^f(\mathbf{v},-q)\mathbf{n}\right)_{\Sigma^r}+\frac{\gamma_\Sigma\mu^f}{h}(\mathbf{u},\mathbf{v})_{\Sigma^r};\end{aligned}\tag{11}$$

– Form related to the DG terms involving only the structure unknowns and test functions:

$$\begin{aligned}\mathscr{D}^{ss}\left(\widehat{\mathbf{d}};\widehat{\mathbf{w}}\right)=&-\left((1-\epsilon)\widehat{\mathbf{T}}^s(\widehat{\mathbf{d}})\mathbf{n},-\widehat{\mathbf{w}}\right)_{\widehat{\Sigma}}\\&-\left(-\frac{\widehat{\mathbf{d}}}{\Delta t},(1-\epsilon)\widehat{\mathbf{T}}^s(\widehat{\mathbf{w}})\mathbf{n}\right)_{\widehat{\Sigma}}\\&+\frac{\gamma_\Sigma\mu^f}{h}\left(-\frac{\widehat{\mathbf{d}}}{\Delta t},-\widehat{\mathbf{w}}\right)_{\widehat{\Sigma}};\end{aligned}\tag{12}$$

– Form related to the DG terms involving mixed (fluid and structure) unknowns and test functions:

$$\begin{aligned}\mathscr{D}^{fs}(\mathbf{u},p,\mathbf{d};\mathbf{v},q,\mathbf{w})^r=&-\left(\epsilon\mathbf{T}^f(\mathbf{u},p)\mathbf{n},-\mathbf{w}\right)_{\Sigma^r}-((1-\epsilon)\mathbf{T}^s(\mathbf{d})\mathbf{n},\mathbf{v})_{\Sigma^r}\\&-\left(\mathbf{u},(1-\epsilon)\mathbf{T}^s(\mathbf{w})\mathbf{n}\right)_{\Sigma^r}-\left(-\frac{\mathbf{d}}{\Delta t},\epsilon\mathbf{T}^f(\mathbf{v},-q)\mathbf{n}\right)_{\Sigma^r}\\&+\frac{\gamma_\Sigma\mu^f}{h}(\mathbf{u},-\mathbf{w})_{\Sigma^r}+\frac{\gamma_\Sigma\mu^f}{h}\left(-\frac{\mathbf{d}}{\Delta t},\mathbf{v}\right)_{\Sigma^r};\end{aligned}\tag{13}$$

– Right hand side given by terms coming from time discretization and forcing terms:

$$\mathscr{F}\left(\mathbf{u}^m, p^m, \mathbf{d}^m; \mathbf{v}, q, \mathbf{w}\right)^r = \frac{\rho^f}{\Delta t}(\mathbf{u}^m, \mathbf{v})_{\Omega^{f,r}} + \frac{2\rho^s}{\Delta t^2}(\widehat{\mathbf{d}}^m, \widehat{\mathbf{w}})_{\widehat{\Omega}^s} - \frac{\rho^s}{\Delta t^2}(\widehat{\mathbf{d}}^{m-1}, \widehat{\mathbf{w}})_{\widehat{\Omega}^s}$$
$$+ \left(\frac{\mathbf{d}^m}{\Delta t}, \epsilon \mathbf{T}^f(\mathbf{v}, -q)\mathbf{n} + (1-\epsilon)\mathbf{T}^s(\mathbf{w})\mathbf{n}\right)_{\Sigma^r}$$
$$- \frac{\gamma_\Sigma \mu^f}{h}\left(\frac{\mathbf{d}^m}{\Delta t}, \mathbf{v} - \mathbf{w}\right)_{\Sigma^r} + (\mathbf{f}^{f,m+1}, \mathbf{v})_{\Omega^{f,r}} + (\widehat{\mathbf{f}}^{s,m+1}, \widehat{\mathbf{w}})_{\widehat{\Omega}^s}.$$

Notice that the DG mortaring terms introduce two penalty parameters, $\gamma_p > 0$ and $\gamma_\Sigma > 0$. The first parameter appears in the term $\mathscr{D}^{ff}$ and it is related to the mortaring on the faces in $\mathscr{F}^{f,i,r}_{h,p}$, while the latter appears in terms $\mathscr{D}^{ff}$, $\mathscr{D}^{fs}$, $\mathscr{D}^{ss}$ and it is related to the mortaring on the fluid-structure interface Σ.

Thus, the XFEM/DG approximation of the monolithic FSI problem (4)–(7) reads: For each n, find $(\mathbf{u}_h^{n+1}, p_h^{n+1}, \widehat{\mathbf{d}}_h^{n+1}) \in \mathbf{V}_h^{n+1} \times Q_h^{n+1} \times \mathbf{W}_h$ such that

$$\begin{aligned}
&\mathscr{A}^f\left(\mathbf{u}_h^{n+1}, \mathbf{u}_h^{n+1}, p_h^{n+1}; \mathbf{v}_h, q_h\right)^{n+1} + \mathscr{A}^s\left(\widehat{\mathbf{d}}_h^{n+1}; \widehat{\mathbf{w}}_h\right) \\
&+\mathscr{D}^{ff}\left(\mathbf{u}_h^{n+1}, p_h^{n+1}; \mathbf{v}_h, q_h\right)^{n+1} + \mathscr{D}^{ss}\left(\widehat{\mathbf{d}}_h^{n+1}; \widehat{\mathbf{w}}_h\right) + \mathscr{D}^{fs}\left(\mathbf{u}_h^{n+1}, p_h^{n+1}, \mathbf{d}_h^{n+1}; \mathbf{v}_h, q_h, \mathbf{w}_h\right)^{n+1} \\
&= \mathscr{F}\left(\mathbf{u}_h^n, p_h^n, \mathbf{d}_h^n; \mathbf{v}_h, q_h, \widehat{\mathbf{w}}_h\right)^n
\end{aligned} \tag{14}$$

for all $(\mathbf{v}_h, q_h, \widehat{\mathbf{w}}_h) \in \mathbf{V}_h^{n+1} \times Q_h^{n+1} \times \mathbf{W}_h$. In compact form we write

$$\mathscr{H}\left(\mathbf{u}_h^{n+1}, p_h^{n+1}, \widehat{\mathbf{d}}_h^{n+1}; \mathbf{v}_h, q_h, \widehat{\mathbf{w}}_h\right)^{n+1} = 0$$

for all $(\mathbf{v}_h, q_h, \widehat{\mathbf{w}}_h) \in \mathbf{V}_h^{n+1} \times Q_h^{n+1} \times \mathbf{W}_h$

Remark 2 Notice that in the previous formulation we could consider also a correction in the trilinear form to maintain the condition that the latter vanishes for $\mathbf{z} = \mathbf{u}$ at the discrete level [27, 85] and a term to maintain the consistency of the formulation [96]. This is what we did in the numerical experiments. However, to simplify the notation and focus on the XFEM/DG discretization, we omitted these terms in (14).

5 An Inexact-Newton Method for the Solution of the FSI Problem

For the solution of the FSI problem (14), we introduce in what follows an inexact Newton-Krylov method, used in combination with a block Gauss-Seidel preconditioner.

To this aim, we indicate by $\delta y_{(k)} = y_{(k)} - y_{(k-1)}$ the increment of a quantity y and we consider the following linearized forms:

–

$$\widetilde{\mathscr{A}}^s\left(\delta\widehat{\mathbf{d}}_{(k)}; \widehat{\mathbf{w}}\right) = \frac{\rho^s}{\Delta t^2}\left(\delta\widehat{\mathbf{d}}_{(k)}, \widehat{\mathbf{w}}\right)_{\widehat{\Omega}^s} + \left(D_F\widehat{\mathbf{T}}^s(\widehat{\mathbf{d}}_{(k-1)}) : \delta\widehat{\mathbf{d}}_{(k)}, \nabla\widehat{\mathbf{w}}\right)_{\widehat{\Omega}^s},$$

where D_F indicated the Gateaux derivative with respect to $\mathbf{F}$;

–

$$\begin{aligned}\widetilde{\mathscr{D}}^{ss}\left(\delta\widehat{\mathbf{d}}_{(k)}; \widehat{\mathbf{w}}\right) = &- \left(\left((1-\epsilon)D_F\widehat{\mathbf{T}}^s(\widehat{\mathbf{d}}_{(k-1)}) : \delta\widehat{\mathbf{d}}_{(k)}\right)\mathbf{n}, -\widehat{\mathbf{w}}\right)_{\widehat{\Sigma}}\\
&- \left(-\frac{\delta\widehat{\mathbf{d}}_{(k)}}{\Delta t}, \left((1-\epsilon)D_F\widehat{\mathbf{T}}^s(\widehat{\mathbf{d}}_{(k-1)}) : \nabla\widehat{\mathbf{w}}\right)\mathbf{n}\right)_{\widehat{\Sigma}}\\
&+ \frac{\gamma_\Sigma\mu^f}{h}\left(-\frac{\delta\widehat{\mathbf{d}}_{(k)}}{\Delta t}, -\widehat{\mathbf{w}}\right)_{\widehat{\Sigma}};\end{aligned}$$

–

$$\begin{aligned}\widetilde{\mathscr{D}}^{fs}\left(\mathbf{u}_{(k)}, p_{(k)}, \delta\mathbf{d}_{(k)}; \mathbf{v}, q, \mathbf{w}\right)^r = &- \left(\epsilon\mathbf{T}^f(\mathbf{u}_{(k)}, p_{(k)})\mathbf{n}, -\mathbf{w}\right)_{\Sigma^r}\\
&- \left(\left((1-\epsilon)D_F\mathbf{T}^s(\mathbf{d}_{(k-1)}) : \nabla\delta\mathbf{d}_{(k)}\right)\mathbf{n}, \mathbf{v}\right)_{\Sigma^r}\\
&- \left(\mathbf{u}_{(k)}, \left((1-\epsilon)D_F\mathbf{T}^s(\mathbf{d}^m_{(k-1)}) : \nabla\mathbf{w}\right)\mathbf{n}\right)_{\Sigma^r}\\
&- \left(-\frac{\delta\mathbf{d}_{(k)}}{\Delta t}, \epsilon\mathbf{T}^f(\mathbf{v}, -q)\mathbf{n}\right)_{\Sigma^r}\\
&+ \frac{\gamma_\Sigma\mu^f}{h}\left(\mathbf{u}_{(k)}, -\mathbf{w}\right)_{\Sigma^r} + \frac{\gamma_\Sigma\mu^f}{h}\left(-\frac{\delta\mathbf{d}_{(k)}}{\Delta t}, \mathbf{v}\right)_{\Sigma^r}.\end{aligned}$$

Moreover, we consider the approximated form $\widehat{\mathscr{D}}^{fs}$ instead of $\widetilde{\mathscr{D}}^{fs}$ obtained by considering the following approximation

$$\frac{\partial\mathbf{T}^s}{\partial\mathbf{F}} = \frac{\partial}{\partial\mathbf{F}}(J^{-1}\widehat{\mathbf{T}}^s\mathbf{F}^T) \approx J^{-1}\frac{\partial\widehat{\mathbf{T}}^s}{\partial\mathbf{F}}\mathbf{F}^T.$$

This, together with the fixed point iteration strategy (instead of the full Newton method) used for the fluid problem (see (15)) leads to an inexact Newton method.

Finally, we point out that at each iteration of the inexact Newton method, we have to update the fluid mesh obtained by the intersections generated by the moving structure mesh onto the background fixed one (see Fig. 4), and, accordingly, the velocity and pressure spaces. In particular, we introduce the compact notation *updatedDomainsAndSpaces()* that at time t^{n+1}, iteration k, performs the following steps:

1. given the displacement at the previous iteration $\widehat{\mathbf{d}}^{n+1}_{h,(k-1)}$, computation of the new position of the solid mesh $(\mathscr{T}_h^s)^{n+1}_{(k-1)}$;
2. computation of the new fluid mesh $(\mathscr{T}_h^f)^{n+1}_{(k-1)}$. This is done by intersecting the background mesh $\mathscr{T}_h$ and the solid mesh $(\mathscr{T}_h^s)^{n+1}_{(k-1)}$;
3. definition of the new discrete spaces $\mathbf{V}^{n+1}_{h,(k-1)}$ and $Q^{n+1}_{h,(k-1)}$.

The FSI problem (14) is solved by means of the following algorithm:

Remark 3 Notice that in this case, due to the fixed nature of the background mesh, we do not have any geometric problem, thus no shape derivatives appear in the exact Jacobian.

6 Numerical Results

In this section, we present some numerical results for the FSI problem given by Eqs. (4)–(7). We consider Algorithm 1 for its numerical solution. We present the following test cases:

- *Blocked channel*: A time-dependent FSI problem with an immersed non-linear elastic structure that completely blocks a channel;
- *Non-linear elastic slab*: A time-dependent FSI problem with an immersed non-linear elastic slab with a high Reynolds number;

Algorithm 1 Inexact Newton method for the FSI problem (14)

At time t^{n+1}, given an initial solution $\mathbf{u}^{n+1}_{h,(0)}, p^{n+1}_{h,(0)}, \widehat{\mathbf{d}}^{n+1}_{h,(0)}$:
for $k = 1 : k_{max}$ **do**
1. *updateDomainsAndSpaces()*;
2. Find $(\mathbf{u}^{n+1}_{h,(k)}, p^{n+1}_{h,(k)}, \delta\widehat{\mathbf{d}}^{n+1}_{h,(k)}) \in \mathbf{V}^{n+1}_{h,(k-1)} \times Q^{n+1}_{h,(k-1)} \times \mathbf{W}_h$ such that

$$
\begin{aligned}
&\mathscr{A}^f\left(\mathbf{u}^{n+1}_{h,(k-1)}, \mathbf{u}^{n+1}_{h,(k)}, p^{n+1}_{h,(k)}; \mathbf{v}_h, q_h\right)^{n+1}_{(k-1)} + \widetilde{\mathscr{A}^s}\left(\delta\widehat{\mathbf{d}}^{n+1}_{h,(k)}; \widehat{\mathbf{w}}_h\right)\\
&+\mathscr{D}^{ff}\left(\mathbf{u}^{n+1}_{h,(k)}, p^{n+1}_{h,(k)}; \mathbf{v}_h, q_h\right)^{n+1}_{(k-1)} + \widetilde{\mathscr{D}}^{ss}\left(\delta\widehat{\mathbf{d}}^{n+1}_{h,(k)}; \widehat{\mathbf{w}}_h\right)\\
&+\widehat{\mathscr{D}}^{fs}\left(\mathbf{u}^{n+1}_{h,(k)}, p^{n+1}_{h,(k)}, \delta\mathbf{d}^{n+1}_{h,(k)}; \mathbf{v}_h, q_h, \mathbf{w}_h\right)^{n+1}_{(k-1)}\\
&= -\mathscr{H}\left(\mathbf{u}^{n+1}_{h,(k-1)}, p^{n+1}_{h,(k-1)}, \mathbf{d}^{n+1}_{h,(k-1)}; \mathbf{v}_h, q_h, \widehat{\mathbf{w}}_h\right)^{n+1}_{(k-1)},
\end{aligned}
\tag{15}
$$

for all $(\mathbf{v}_h, q_h, \widehat{\mathbf{w}}_h) \in \mathbf{V}^{n+1}_{h,(k-1)} \times Q^{n+1}_{h,(k-1)} \times \mathbf{W}_h$;
3. $\widehat{\mathbf{d}}^{n+1}_{h,(k)} = \widehat{\mathbf{d}}^{n+1}_{h,(k-1)} + \delta\widehat{\mathbf{d}}^{n+1}_{h,(k)}$.
end for

– *Ideal aortic valve*: A time-dependent FSI problem in the case of three immersed linear elastic leaflets.

The proposed examples are simulated in a three-dimensional (3D) framework, and for the structure we use the non-linear strain energy function (8) for the blocked channel and non-linear elastic slab tests, whereas the Hooke law for the ideal aortic valve. The mortaring parameter ϵ for the fluid-solid coupling in the forms $\mathscr{D}^{ff}$, $\mathscr{D}^{fs}$, $\mathscr{D}^{ss}$ and in functional $\mathscr{F}$ is set equal to 1, see [22]. Moreover, we point out that at iteration k of Algorithm 1, the fluid velocity $\mathbf{u}_h^n$ at the previous time step appearing in the term coming from time discretization (which is defined in $\Omega^{f,n}$) and the fluid velocity $\mathbf{u}_{h,(k-1)}^{n+1}$ at the previous iteration used in the convective term (which is defined in $\Omega_{(k-2)}^{f,n+1}$) are not defined in the current domain $\Omega_{(k-1)}^{f,n+1}$ (remember that at iteration k the fluid problem is solved in $\Omega_{(k-1)}^{f,n+1}$). Thus, these terms should be properly defined in the new computational domain $\Omega_{(k-1)}^{f,n+1}$ in order to be used in the discrete formulation. In particular, issues may occur when the uncovered portion of a fluid element change between time n and $n+1$ and/or between iteration $k-1$ and k. For the numerical treatment of these cases, we employ the procedure proposed in [96].

The simulations have been performed with the Finite Element library LifeV [66].

6.1 Blocked Channel

In this experiment we consider a thick membrane placed in the middle of a channel so that the structure completely blocks the flow in the channel, see Fig. 12. The aim of this example is to assess the validity of the proposed method. We consider

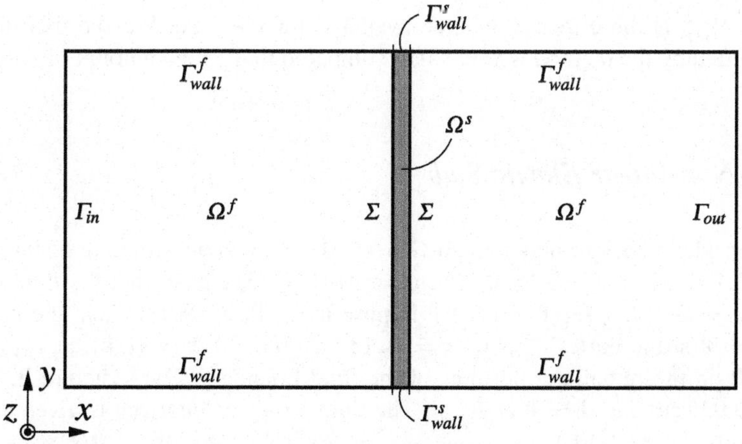

Fig. 12 Top view of the fluid Ω^f and structure Ω^s domains. Blocking channel test

a background domain $\Omega = 0.4\,\text{cm} \times 0.2\,\text{cm} \times 0.00625\,\text{cm}$ and a structure domain $\Omega^s = 0.01\,\text{cm} \times 0.2\,\text{cm} \times 0.00625\,\text{cm}$. The resulting fluid domain is $\Omega^f = \Omega \setminus \Omega^s$. Notice that, to ease the computational cost, we reduce the size of the domain along the z-axis. We impose $\mathbf{T}^f\mathbf{n} = (-1000, 0, 0)\,\text{dyne/cm}^2$ at the inlet Γ_{in}, $\mathbf{T}^f\mathbf{n} = \mathbf{0}$ at the outlet Γ_{out}, $\mathbf{u} = \mathbf{0}$ on Γ^f_{wall}, and $\mathbf{u} \cdot \widehat{\mathbf{k}} = 0$, $(\mathbf{T}^f\mathbf{n}) \cdot \widehat{\mathbf{l}} = 0$, $\mathbf{l} = \{\mathbf{i}, \mathbf{j}\}$, on the remaining portions of the fluid boundary, i.e. for $z = \{0, 0.00625\}$. Notice that the latter choice allows the fluid to move in the xy-plane at the extreme surfaces $z = \{0, 0.00625\}$. The solid is fixed at Γ^s_{wall}, i.e. $\mathbf{d} = \mathbf{0}$, and, like the fluid, is allowed to move in the xy-plane on the remaining portions of the boundary, i.e. $\mathbf{d} \cdot \widehat{\mathbf{k}} = 0$, $(\mathbf{T}^s\mathbf{n}) \cdot \widehat{\mathbf{l}} = 0$, $\mathbf{l} = \{\mathbf{i}, \mathbf{j}\}$, for $z = \{0, 0.00625\}$. As initial conditions, we set $\mathbf{u}(\mathbf{x}, 0) = \mathbf{d}(\mathbf{x}, 0) = \dot{\mathbf{d}}(\mathbf{x}, 0) = \mathbf{0}$. We also use the following values for the parameters: $\rho^f = 1\,\text{g/cm}^3$, $\rho^s = 1.2\,\text{g/cm}^3$, $\mu^f = 0.035\,\text{poise}$, $\alpha = 1 \cdot 10^8\,\text{dyne/cm}^2$, $\kappa = 1.034 \cdot 10^7\,\text{dyne/cm}^2$, $\gamma = 1$, and $T = 0.02\,\text{s}$.

We employ a background mesh $\mathscr{T}_h$ composed of $16 \cdot 10^3$ tetrahedra ($h_{ave} = 0.00571\,\text{cm}$) and a solid mesh $\mathscr{T}^s_h$ composed of $1.4 \cdot 10^3$ tetrahedra ($h_{ave} = 0.00377\,\text{cm}$). Notice that, the thickness of the structure domain (0.01 cm) is higher than the average size of the fluid elements, so that we are in the thick case. The time step Δt is $2 \cdot 10^{-4}\,\text{s}$. We choose $\gamma_\Sigma = 10^4$ (see Eqs. (11)–(13)), $\gamma_p = 10^3$ (see Eq. (11)) and $\gamma_g = 1$ (see Eq. (10)).

In Figs. 13 and 14, we show the numerical solution at different time steps. In particular, we plot the velocity and the pressure fields in the fluid domain and the structure displacement in the solid domain. A quantitative plot of the displacement of the center of mass of the structure is reported in Fig. 15.

Finally, in Fig. 16 (left), we plot the behaviour in time of the total amount of fluid (in cm^3) that goes through Γ_{in} (indicated by V_{in}) and Γ_{out} (indicated by V_{out}), the variation of the structure volume (in cm^3) with respect to the initial time (indicated by ΔV^s), and the sum of these three quantities (indicated by $V_{balance}$) which represents the error with respect to the balance of volume. In Fig. 16 (right), we plot relative error of the balance of volume, i.e. the ratio $r = V_{balance}/V^f_{eff}$, where V^f_{eff} is the effective volume available for the fluid. We see that the error committed by the method is very small compared to the total amount of volume.

6.2 Non-linear Elastic Slab

We consider a background domain $\Omega = (0, 0.5)^3\,\text{cm}$ and a structure domain $\Omega^s = (0.025, 0.425)\,\text{cm} \times (0.15, 0.35)\,\text{cm} \times (0.10, 0.13)\,\text{cm}$, so that the fluid domain is $\Omega^f = \Omega \setminus \overline{\Omega^s}$, see Fig. 17. We impose $\mathbf{u} = (0, 0, 100)\,\text{cm/s}$ at the inlet Γ_{in}, $\mathbf{T}^f\mathbf{n} = \mathbf{0}$ at the outlet Γ_{out}, $\mathbf{u} \cdot \widehat{\mathbf{i}} = 0$, $(\mathbf{T}^f\mathbf{n}) \cdot \widehat{\mathbf{l}} = 0$, $\mathbf{l} = \{\mathbf{j}, \mathbf{k}\}$, at Γ_{symm}, and $\mathbf{u} = \mathbf{0}$ on the remaining portions of the fluid boundary. The structure is fixed at $x = 0.025\,\text{cm}$, i.e. $\mathbf{d} = \mathbf{0}$ at Γ^s_{wall}. The fluid-structure interface is given by $\Sigma = \partial\Omega^s \setminus \Gamma^s_{wall}$. As initial conditions, we set $\mathbf{u}(\mathbf{x}, 0) = \mathbf{d}(\mathbf{x}, 0) = \dot{\mathbf{d}}(\mathbf{x}, 0) = \mathbf{0}$. We also use the following values for the parameters: $\rho^f = 1\,\text{g/cm}^3$, $\rho^s = 1.2\,\text{g/cm}^3$,

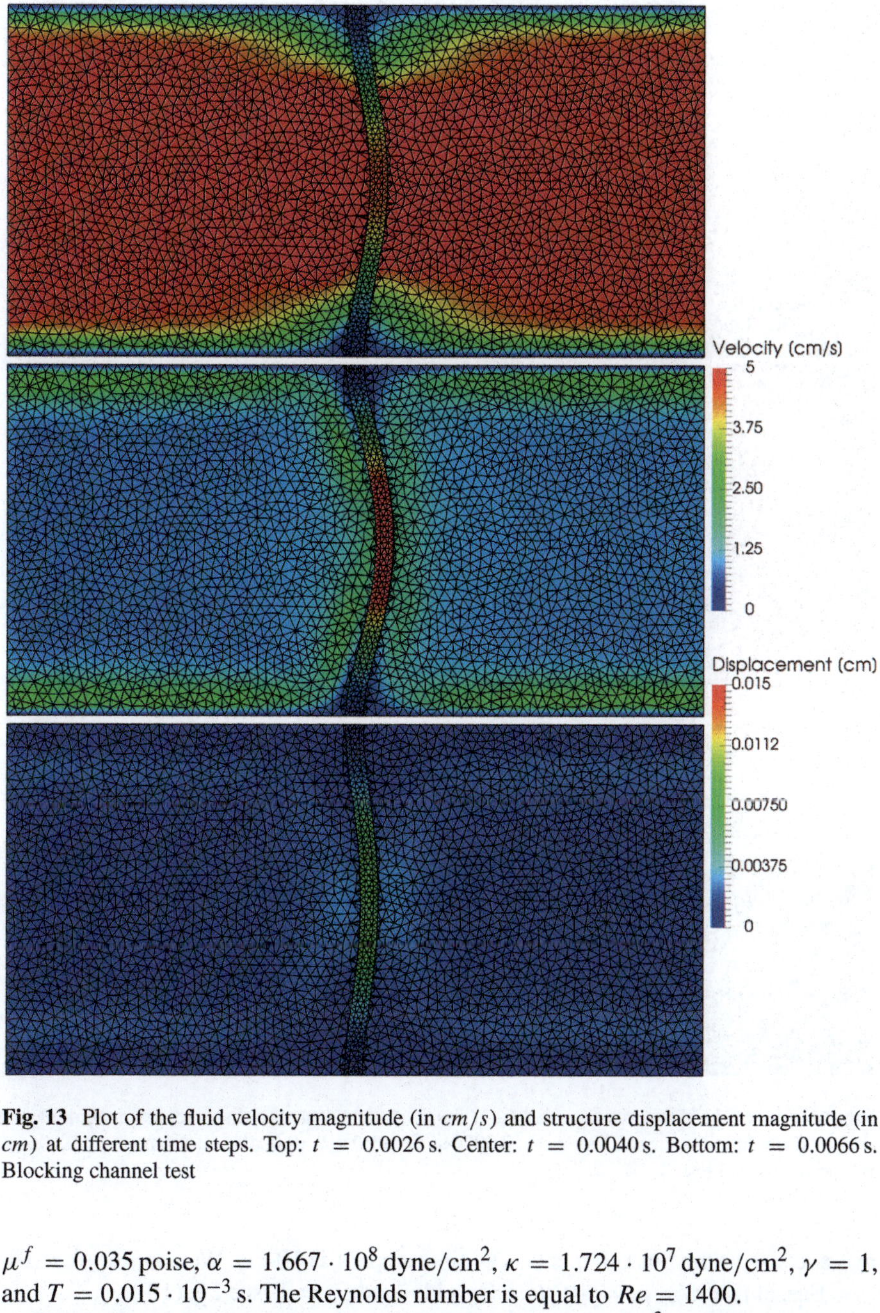

Fig. 13 Plot of the fluid velocity magnitude (in cm/s) and structure displacement magnitude (in cm) at different time steps. Top: $t = 0.0026$ s. Center: $t = 0.0040$ s. Bottom: $t = 0.0066$ s. Blocking channel test

$\mu^f = 0.035$ poise, $\alpha = 1.667 \cdot 10^8$ dyne/cm^2, $\kappa = 1.724 \cdot 10^7$ dyne/cm^2, $\gamma = 1$, and $T = 0.015 \cdot 10^{-3}$ s. The Reynolds number is equal to $Re = 1400$.

We employ a background mesh $\mathscr{T}_h$ composed of $56 \cdot 10^3$ tetrahedra ($h_{ave} = 0.025$ cm) and a solid mesh $\mathscr{T}_h^s$ composed of $30 \cdot 10^3$ tetrahedra ($h_{ave} = 0.0078$ cm). Notice that, the thickness of the structure domain (0.03 cm) is comparable to the

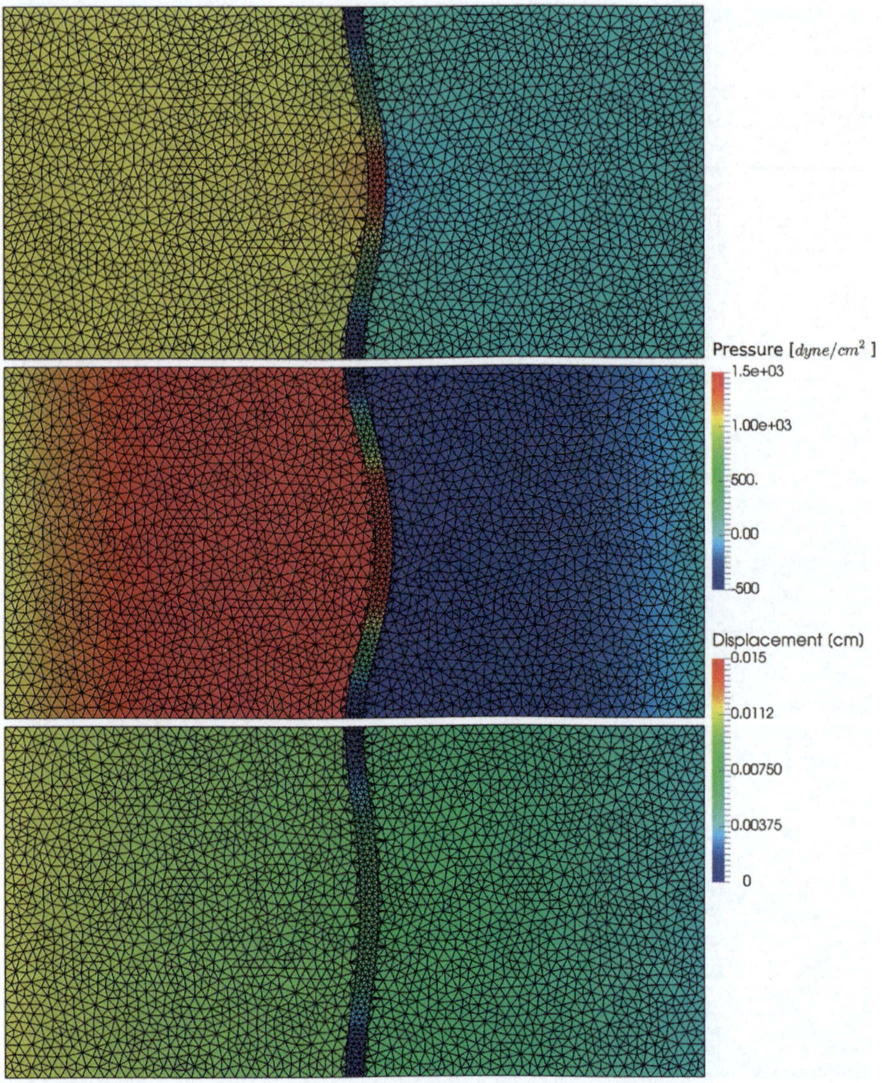

Fig. 14 Plot of the fluid pressure filed (in $dyne/cm^2$) and structure displacement magnitude (in cm) at different time steps. Top: $t = 0.0026$ s. Center: $t = 0.0040$ s. Bottom: $t = 0.0066$ s. Blocking channel test

average size of the fluid elements. The time step Δt is 10^{-3} s. We choose $\gamma_\Sigma = 10^2$ (see Eqs. (11)–(13)), $\gamma_p = 10^3$ (see Eq. (11)) and $\gamma_g = 1$ (see Eq. (10)).

In Fig. 18, we show the fluid velocity (in cm/s) and the structure displacement (in cm) at four different time steps. The maximum velocity is about 160 cm/s and the maximum displacement reached by the structure is 0.35 cm. We see that the method is able to deal with high Reynolds number and large displacement.

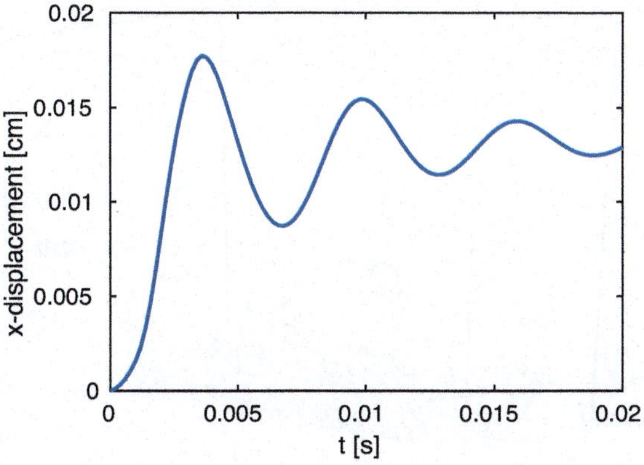

Fig. 15 Plot of the x-displacement (in cm) at the center of mass of the structure. Blocking channel test

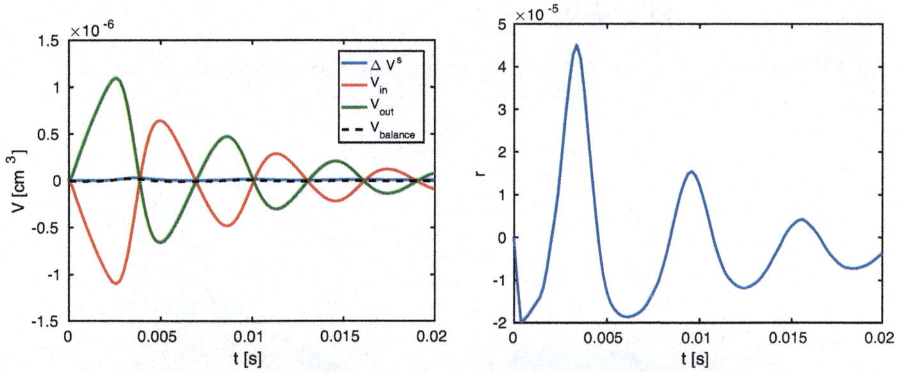

Fig. 16 Left: Plot of volumes (in cm^3) ΔV^s, V_{in}, V_{out} and $V_{balance}$ in time. Right: Plot of the relative error of the volume in time. Blocking channel test

In Fig. 19, we show the z-displacement (in cm) of the tip of the structure, i.e. at $\mathbf{x}_{tip} = (0.425, 0.25, 0.115)$ cm, in time.

In Fig. 20, we plot the velocity field in the fluid domain and we represent the moving structure accordingly to the computed displacement at different time-steps. We see that the fluid elements crossed by the structure may change in time. We point out that the refinement appearing near the structure is made only for a visualization purpose, in fact the background fluid mesh never changes.

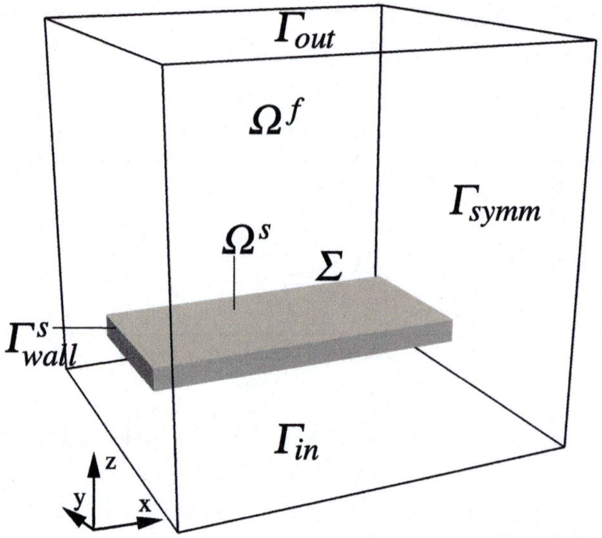

Fig. 17 Sketch of the fluid Ω^f and structure Ω^s domains. Non-linear elastic slab test

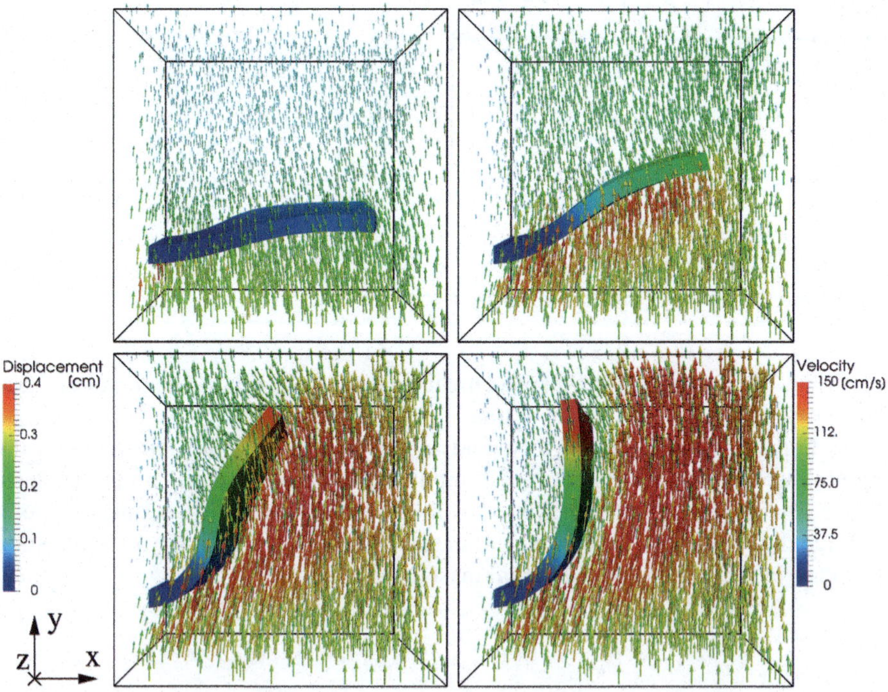

Fig. 18 Solution at two different time steps. We plot the fluid velocity (in cm/s) and the solid displacement (in cm). Top, left: $t = 1 \cdot 10^{-3}$. Top, right: $t = 2 \cdot 10^{-3}$. Bottom, left: $t = 4 \cdot 10^{-3}$. Bottom, right: $t = 6 \cdot 10^{-3}$. Non-linear elastic slab test

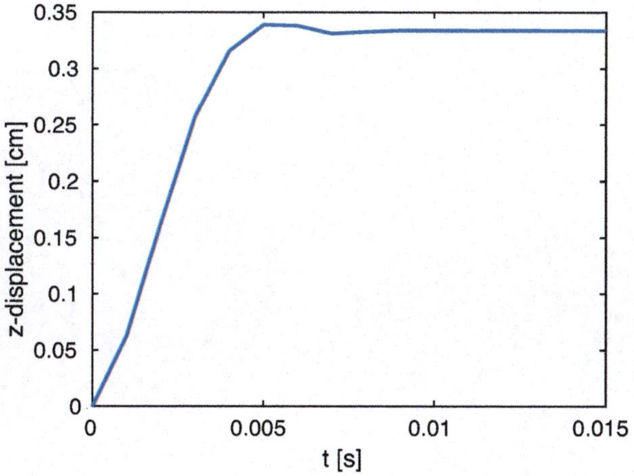

Fig. 19 Plot of the z-displacement (in cm) at the tip of the structure in time. Non-linear elastic slab test

Fig. 20 Velocity magnitude on the slice $y = 0.25\,\text{cm}$ at time $t = 1 \cdot 10^{-3}$ s (top) and at time $t = 2 \cdot 10^{-3}$ s (bottom). The element highlighted in red at time $t = 1 \cdot 10^{-3}$ s is partially overlapped by the interface, while at time $t = 2 \cdot 10^{-3}$ s is not crossed by the structure. Non-linear elastic slab test

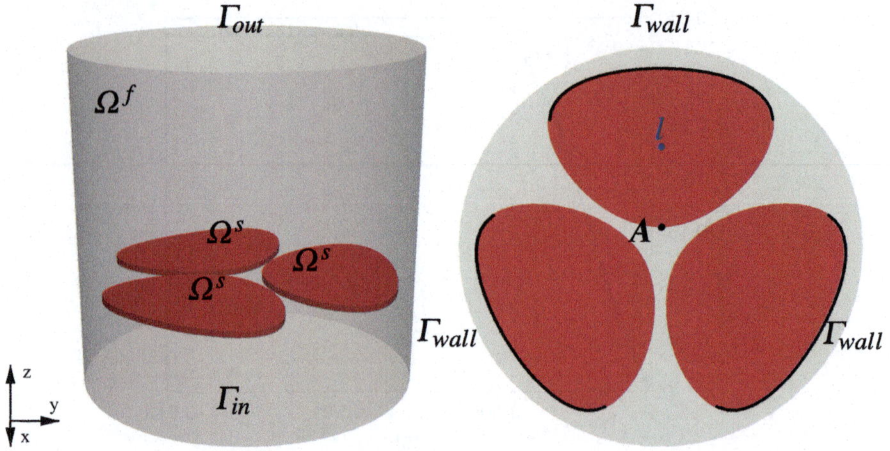

Fig. 21 Sketch of the fluid background domain Ω^f and the three foreground domains Ω^s. To the right, we report the top view of the domains. Notice in black the region Γ_{wall} where the leaflets are clamped. Ideal aortic valve test

6.3 Ideal Aortic Valve

In this example, we consider the domain Ω defined by a cylinder of radius 0.5 cm and height 1 cm and three linear immersed structures that are an ideal representation of the leaflets of an aortic valve, see Fig. 21. The thickness of the leaflets is 0.02 cm. We impose the velocity profile $\mathbf{u} = (0, 0, 50\sin(\frac{\pi}{8}t))$ cm/s at the inlet Γ_{in}, we set $\mathbf{T}^f\mathbf{n} = \mathbf{0}$ at the outlet Γ_{out}, and $\mathbf{u} = \mathbf{0}$ on the remaining portions of the fluid boundary. At the fluid-structure interface, we impose the kinematic and dynamic coupling conditions, except on Γ_{wall} where the leaflets are fixed, i.e. $\mathbf{d} = \mathbf{0}$. As initial conditions, we set $\mathbf{u}(\mathbf{x}, 0) = \mathbf{d}(\mathbf{x}, 0) = \dot{\mathbf{d}}(\mathbf{x}, 0) = \mathbf{0}$. We use the following values for the physical parameters: $\rho^f = 1\,\text{g/cm}^3$, $\rho^s = 1.2\,\text{g/cm}^3$, $\mu^f = 0.035$ poise, $\alpha = 1.667 \cdot 10^8\,\text{dyne/cm}^2$, $\kappa = 1.724 \cdot 10^7\,\text{dyne/cm}^2$ and $\gamma = 1$. We simulate only the initial phase of the movement of the leaflets, i.e. $T = 0.45$ s. The Reynolds number is equal to $Re = 1430$.

For the numerical simulation, we employ a background mesh $\mathscr{T}_h$ of $110 \cdot 10^3$ elements with the average size of the mesh elements $h_{ave} = 0.035$ cm, while each structure mesh is composed of $16 \cdot 10^3$ elements with $h_{ave} = 0.011$ cm. We set $\Delta t = 0.05$ s. We choose $\gamma_\Sigma = 10^2$ (see Eqs. (11)–(13)), $\gamma_p = 10^3$ (see Eq. (11)) and $\gamma_g = 1$ (see Eq. (10)).

In what follows we report preliminary results for this test. At the instant where the fluid flow reverses, numerical instabilities occurs. For this reason, we have reported the numerical results until the solution features a stable behaviour. The study of such oscillations is under investigation. A qualitative representation of the solution at time $t = 0.4$ s is shown in Fig. 22. More specifically, in Fig. 23 (left), we plot the z-displacement at the tip (point A in Fig. 21) of the three leaflets in time. We observe that the three leaflets behave very similarly during time. In Fig. 23 (right),

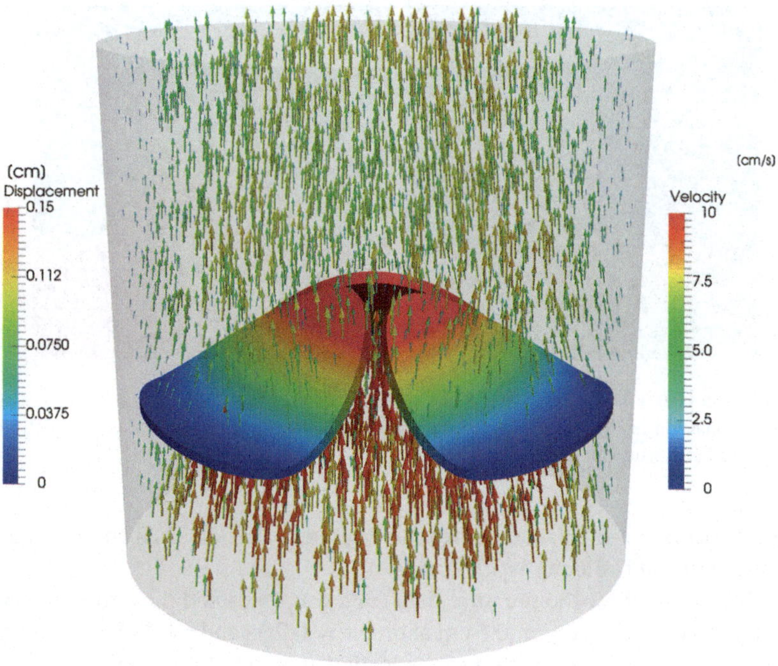

Fig. 22 Solution at time $t = 0.4$ s. We plot the fluid velocity (in cm/s) and the structure displacement (in cm). Ideal aortic valve test

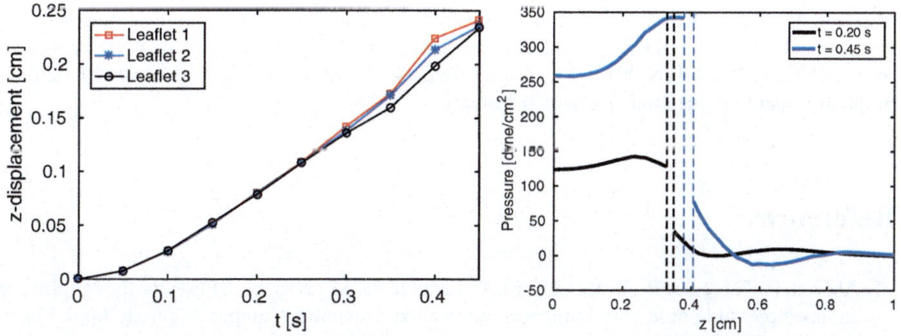

Fig. 23 Left: evolution of the displacement (in cm) at the tip of the three leaflets. Right: fluid pressure (in $dyne/cm^2$) along the line l at two different time steps. The position of the leaflet is denoted by the dashed lines. Ideal aortic valve test

we plot the fluid pressure along line $l : x = 0\,\text{cm},\ y = 0.25\,\text{cm},\ 0 \le z \le 1\,\text{cm}$ at two different time steps, namely, $t = 0.20$ s and $t = 0.45$ s. From this result, we see the different value of the fluid pressure upstream and downstream the leaflet. Notice that, the position of the leaflet (dashed lines) has changed in time.

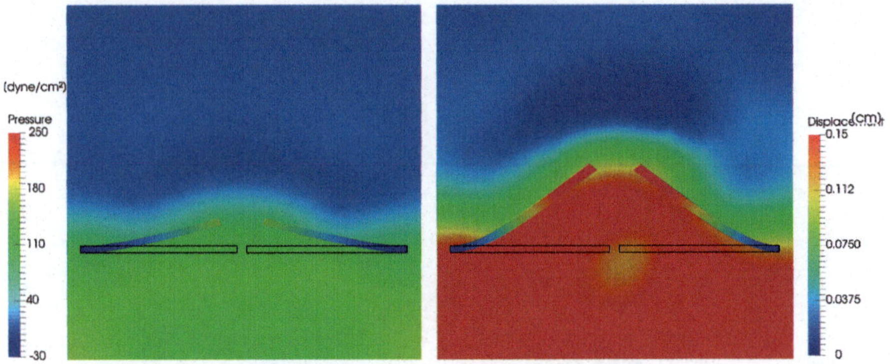

Fig. 24 Fluid pressure (in $dyne/cm^2$) and structure displacement (in cm) at slice $y = 0.5$ cm. The initial position of the leaflets is denoted by the black lines. Left: time $t = 0.2$ s. Right: time $t = 0.45$ s. Ideal aortic valve test

The maximum z-displacement reached by the leaflets is 0.24 cm and the maximum value of fluid velocity is 17.5 cm/s.

In Fig. 24, we show the pressure field (in $dyne/cm^2$) and the structure displacement (in cm) on the slice $y = 0.5$ cm at time $t = 0.20$ s and $t = 0.45$ s. In particular, we observe the different position of the leaflets with respect to their initial position outlined in black. Again, it is possible to see the different values of the fluid pressure upstream and downstream the leaflets.

Acknowledgements The authors gratefully acknowledge the financial support of the Italian MIUR by the grant PRIN12, number 201289A4LX, "Mathematical and numerical models of the cardiovascular system, and their clinical applications". C. Vergara has been partially supported by the H2020-MSCA-ITN-2017, EU project 765374 "ROMSOC—Reduced Order Modelling, Simulation and Optimization of Coupled systems".

References

1. Alauzet, F., Fabrèges, B., Fernández, M.A., Landajuela, M.: Nitsche-XFEM for the coupling of an incompressible fluid with immersed thin-walled structures. Comput. Methods Appl. Mech. Eng. **301**, 300–335 (2016)
2. Aletti, M., Gerbeau, J.-F., Lombardi, D.: Modeling autoregulation in three-dimensional simulations of retinal hemodynamics. J. Model. Ophthalmol. **1**, 88–115 (2015)
3. Annavarapu, C., Hautefeuille, M., Dolbow, J.E.: A robust Nitsche's formulation for interface problems. Comput. Methods Appl. Mech. Eng. **225–228**, 44–54 (2012)
4. Arciero, J., Harris, A., Siesky, B., Amireskandari, A., Gershuny, V., Pickrell, A., Guidoboni, G.: Theoretical analysis of vascular regulatory mechanisms contributing to retinal blood flow autoregulation mechanisms contributing to retinal autoregulation. Invest. Ophthalmol. Vis. Sci. **54**(8), 5584–5593 (2013)
5. Arnold, D.N., Brezzi, F., Cockburn, B., Marini, L.D.: Unified analysis of discontinuous Galerkin methods for elliptic problems. SIAM J. Numer. Anal. **39**(5), 1749–1779 (2001)

6. Astorino, M., Gerbeau, J.-F., Pantz, O., Traoré, K.-F.: Fluid-structure interaction and multi-body contact: application to the aortic valves. Comput. Methods Appl. Mech. Eng. **198**, 3603–3612 (2009)
7. Basting, S., Quaini, A., Čanić, S., Glowinski, R.: Extended ALE Method for fluid–structure interaction problems with large structural displacements. J. Comput. Phys. **331**, 312–336 (2017)
8. Bazilevs, Y., Hsu, M.-C., Kiendl, J., Wüchner, R., Bletzinger, K.-U.: 3D simulation of wind turbine rotors at full scale. Part II: fluid–structure interaction modeling with composite blades. Int. J. Numer. Methods Fluids **65**(1–3), 236–253 (2011)
9. Becker, R., Burman, E., Hansbo, A.: A Nitsche extended finite element method for incompressible elasticity with discontinuous modulus of elasticity. Comput. Methods Appl. Mech. Eng. **198**(41–44), 3352–3360 (2009)
10. Beckert, A., Wendland, H.: Multivariate interpolation for fluid-structure-interaction problems using radial basis functions. Aerosol Sci. Technol. **5**(2), 125–134 (2001)
11. Belytschko, T., Moës, N., Usui, S., Parimi, C.: Arbitrary discontinuities in finite elements. Int. J. Numer. Methods Eng. **50**(4), 993–1013 (2001)
12. Benedettini, F., Rega, G., Alaggio, R.: Non-linear oscillations of a four-degree-of-freedom model of a suspended cable under multiple internal resonance conditions. J. Sound Vib. **182**(5), 775–798 (1995)
13. Bertrand, F., Tanguy, P.A., Thibault, F.: A three-dimensional fictitious domain method for incompressible fluid flow problems. Int. J. Numer. Methods Fluids **25**(6), 719–736 (1997)
14. Boffi, D., Gastaldi, L: A finite element approach for the immersed boundary method. Comput. Struct. **81**(8–11), 491–501 (2003). K.J Bathe 60th Anniversary Issue
15. Boffi, D., Gastaldi, L., Heltai, L.: Numerical stability of the finite element immersed boundary method. Math. Models Methods Appl. Sci. **17**(10), 1479–1505 (2007)
16. Boffi, D., Gastaldi, L., Heltai, L., Peskin, C.: On the hyper-elastic formulation of the immersed boundary method. Comput. Methods Appl. Mech. Eng. **197**(25–28), 2210–2231 (2008)
17. Borazjani, I., Ge, L., Sotiropoulos, F.: Curvilinear immersed boundary method for simulating fluid structure interaction with complex 3D rigid bodies. J. Comput. Phys. **227**(16), 7587–7620 (2008)
18. Borazjani, I., Ge, L., Sotiropoulos, F.: High-resolution fluid–structure interaction simulations of flow through a bi-leaflet mechanical heart valve in an anatomic aorta. Ann. Biomed. Eng. **38**(2), 326–344 (2010)
19. Braun, A.L., Awruch, A.M.: Finite element simulation of the wind action over bridge sectional models: Application to the Guamá river bridge (Pará State, Brazil). Finite Elem. Anal. Des. **44**(3), 105–122 (2008)
20. Burman, E.: Ghost penalty. C. R. Math. Acad. Sci. Paris **348**(21–22), 1217–1220 (2010)
21. Burman, E., Fernández, M.A.: Stabilized explicit coupling for fluid-structure interaction using Nitsche's method. C. R. Acad. Sci. Paris Sér. I Math. **345**, 467–472 (2007)
22. Burman, E., Fernández, M.A.: An unfitted Nitsche method for incompressible fluid-structure interaction using overlapping meshes. Comput. Methods Appl. Mech. Eng. **279**, 497–514 (2014)
23. Burman, E., Fernández, M.A., Hansbo, P.: Continuous interior penalty finite element method for Oseen's equations. SIAM J. Numer. Anal. **44**(3), 1248–1274 (2006)
24. Burman, E., Claus, S., Hansbo, P., Larson, M.G., Massing, A.: CutFEM: discretizing geometry and partial differential equations. Int. J. Numer. Methods Eng. **104**(7), 472–501 (2015)
25. De Hart, J., Baaijens, F.P.T., Peters, G.W.M., Schreurs, P.J.G.: A computational fluid-structure interaction analysis of a fiber-reinforced stentless aortic valve. J. Biomech. **36**(5), 699–712 (2003). Cardiovascular Biomechanics
26. De Hart, J., Peters, G.W.M., Schreurs, P.J.G., Baaijens, F.P.T.: A three-dimensional computational analysis of fluid–structure interaction in the aortic valve. J. Biomech. **36**(1), 103–112 (2003)
27. Di Pietro, D.A., Ern, A.: Mathematical Aspects of Discontinuous Galerkin Methods. Mathématiques et Applications, vol. 69. Springer, Berlin (2012)

28. Donea, J.: An arbitrary Lagrangian-Eulerian finite element method for transient dynamic fluid-structure interaction. Comput. Methods Appl. Mech. Eng. **33**, 689–723 (1982)
29. Donea, J., Huerta, A.: Finite Element Methods for Flow Problems. Wiley, New York (2003)
30. Douglas, J., Dupont, T.: Interior Penalty Procedures for Elliptic and Parabolic Galerkin Methods, pp. 207–216. Springer, Berlin (1976)
31. Dowell, E.H., Hall, K.C.: Modeling of fluid-structure interaction. Ann. Rev. Fluid Mech. **33**(1), 445–490 (2001)
32. Farhat, C., Lesoinne, M., Le Tallec, P.: Load and motion transfer algorithms for fluid/structure interaction problems with non-matching discrete interfaces: momentum and energy conservation, optimal discretization and application to aeroelasticity. Comput. Methods Appl. Mech. Eng. **157**(1–2), 95–114 (1998)
33. Formaggia, L., Miglio, E., Mola, A., Parolini, N.: Fluid–structure interaction problems in free surface flows: application to boat dynamics. Int. J. Numer. Methods Fluids **56**(8), 965–978 (2008)
34. Formaggia, L., Miglio, E., Mola, A., Montano, A.: A model for the dynamics of rowing boats. Int. J. Numer. Methods Fluids **61**(2), 119–143 (2009)
35. Formaggia, L., Vergara, C., Zonca, S.: Unfitted extended finite elements for composite grids. Comput. Math. Appl. **76**(4), 893–904 (2018)
36. Ge, L., Sotiropoulos, F.: A numerical method for solving the 3D unsteady incompressible Navier–Stokes equations in curvilinear domains with complex immersed boundaries. J. Comput. Phys. **225**(2), 1782–1809 (2007)
37. Gerstenberger, A.: An XFEM based fixed-grid approach to fluid-structure interaction. PhD thesis, Technical University of Munich (2010)
38. Gerstenberger, A., Wall, W.A.: An extended finite element method based approach for large deformation fluid-structure interaction. In: Wesseling, P., Onate, E., Periaux, J. (eds.) Proceedings of the European Conference on Computational Fluid Dynamics (2006)
39. Gerstenberger, A., Wall, W.A.: An extended finite element method/Lagrange multiplier based approach for fluid–structure interaction. Comput. Methods Appl. Mech. Eng. **197**(19), 1699–1714 (2008)
40. Gerstenberger, A., Wall, W.A.: An embedded Dirichlet formulation for 3D continua. Int. J. Numer. Methods Eng. **82**(5), 537–563 (2010)
41. Glowinski, R., Pan, T.-W., Periaux, J.: A fictitious domain method for Dirichlet problem and applications. Comput. Methods Appl. Mech. Eng. **111**(3–4), 283–303 (1994)
42. Glowinski, R., Pan, T.-W., Periaux, J.: A fictitious domain method for external incompressible viscous flow modeled by Navier-Stokes equations. Comput. Methods Appl. Mech. Eng. **112**(1), 133–148 (1994)
43. Glowinski, R., Pan, T.-W., Periaux, J.: A Lagrange multiplier/fictitious domain method for the numerical simulation of incompressible viscous flow around moving rigid bodies: (I) case where the rigid body motions are known a priori. C. R. Acad. Sci. Ser. I-Math. **324**(3), 361–369 (1997)
44. Glowinski, R., Pan, T.-W., Hesla, T.I., Joseph, D.D.: A distributed lagrange multiplier/fictitious domain method for particulate flows. Int. J. Multiphase Flow **25**(5), 755–794 (1999)
45. Glowinski, R., Pan, T.-W., Hesla, T.I., Joseph, D.D., Periaux, J.: A fictitious domain approach to the direct numerical simulation of incompressible viscous flow past moving rigid bodies: application to particulate flow. J. Comput. Phys. **169**(2), 363–426 (2001)
46. Griffith, B.E.: Immersed boundary model of aortic heart valve dynamics with physiological driving and loading conditions. Int. J. Numer. Methods Biomed. Eng. **28**(3), 317–345 (2012)
47. Griffith, B.E., Hornung, R.D., McQueen, D.M., Peskin, C.S.: An adaptive, formally second order accurate version of the immersed boundary method. J. Comput. Phys. **223**(1), 10–49 (2007)
48. Griffith, B.E., Luo, X., McQueen, D.M., Peskin, C.S.: Simulating the fluid dynamics of natural and prosthetic heart valves using the immersed boundary method. Int. J. Appl. Mech. **1**, 137–176 (2009)

49. Hansbo, A., Hansbo, P.: An unfitted finite element method, based on Nitsche's method, for elliptic interface problems. Comput. Methods Appl. Mech. Eng. **191**(47–48), 5537–5552 (2002)
50. Hansbo, A., Hansbo, P.: A finite element method for the simulation of strong and weak discontinuities in solid mechanics. Comput. Methods Appl. Mech. Eng. **193**(33–35), 3523–3540 (2004)
51. Hansbo, A., Hansbo, P., Larson, M.G.: A finite element method on composite grids based on Nitsche's method. ESAIM: Math. Model. Numer. Anal. **37**(3), 495–514 (2003)
52. Hansbo, P., Larson, M.G., Zahedi, S.: A cut finite element method for a Stokes interface problem. Appl. Numer. Math. **85**, 90–114 (2014)
53. Hansbo, P., Larson, M.G., Zahedi, S.: Characteristic cut finite element methods for convection–diffusion problems on time dependent surfaces. Comput. Methods Appl. Mech. Eng. **293**, 431–461 (2015)
54. Harun, Z., Reda, E., Abdullah, S.: Large eddy simulation of the wind flow over skyscrapers. Recent Adv. Mech. Mech. Eng. **15**, 72–79 (2015)
55. Hirt, C.W., Amsden, A.A., Cook, J.L.: An arbitrary lagrangian-eulerian computing method for all flow speeds. J. Comput. Phys. **14**(3), 227–253 (1974)
56. Hsu, M.-C., Kamensky, D., Bazilevs, Y., Sacks, M.S., Hughes, T.J.R.: Fluid–structure interaction analysis of bioprosthetic heart valves: significance of arterial wall deformation. Comput. Mech. **54**(4), 1055–1071 (2014)
57. Hughes, T.J.R., Hulbert, G.M.: Space-time finite element methods for elastodynamics: formulations and error estimates. Comput. Methods Appl. Mech. Eng. **66**(3), 339–363 (1988)
58. Hughes, T.J.R., Liu, W.K., Zimmermann, T.K.: Lagrangian-Eulerian finite element formulation for incompressible viscous flows. Comput. Methods Appl. Mech. Eng. **29**(3), 329–349 (1981)
59. Johnson, C., Nävert, U., Pitkäranta, J.: Finite element methods for linear hyperbolic problems. Comput. Methods Appl. Mech. Eng. **45**(1–3), 285–312 (1984)
60. Jonsson, T., Larson, M.G., Larsson, K.: Cut finite element methods for elliptic problems on multipatch parametric surfaces. Comput. Methods Appl. Mech. Eng. **324**, 366–394 (2017)
61. Kamakoti, R., Shyy, W.: Fluid–structure interaction for aeroelastic applications. Prog. Aerosp. Sci. **40**(8), 535–558 (2004)
62. Kamensky, D., Hsu, M.-C., Schillinger, D., Evans, J.A., Aggarwal, A., Bazilevs, A., Sacks, M.S., Hughes, T.J.R.: An immersogeometric variational framework for fluidstructure interaction: application to bioprosthetic heart valves. Comput. Methods Appl. Mech. Eng. **284**, 1005–1053 (2015)
63. Katayama, S., Umetani, N., Sugiura, S., Hisada, T.: The sinus of Valsalva relieves abnormal stress on aortic valve leaflets by facilitating smooth closure. J. Thorac. Cardiovasc. Surg. **136**(6), 1528–1535 (2008)
64. Lai, Y.G., Chandran, K.B., Lemmon, J.: A numerical simulation of mechanical heart valve closure fluid dynamics. J. Biomech. **35**(7), 881–892 (2002)
65. Le, T.B., Sotiropoulos, F.: Fluid–structure interaction of an aortic heart valve prosthesis driven by an animated anatomic left ventricle. J. Comput. Phys. **244**, 41–62 (2013)
66. LifeV.: The parallel finite element library for the solution of PDEs (2018). http://www.lifev.org
67. Liu, W.K., Liu, Y., Farrell, D., Zhang, L.T., Wang, X.S., Fukui, Y., Patankar, N., Zhang, Y., Bajaj, C., Lee, J., Hong, J., Chen, X., Hsu, H.: Immersed finite element method and its applications to biological systems. Comput. Methods Appl. Mech. Eng. **195**(13), 1722–1749 (2006)
68. Marom, G.: Numerical methods for fluid–structure interaction models of aortic valves. Arch. Comput. Meth. Eng. **22**(4), 595–620 (2015)
69. Massing, A., Larson, M.G., Logg, A.: Efficient implementation of finite element methods on nonmatching and overlapping meshes in three dimensions. SIAM J. Sci. Comput. **35**(1), C23–C47 (2013)
70. Massing, A., Larson, M.G., Logg, A., Rognes, M.E.: A stabilized Nitsche overlapping mesh method for the Stokes problem. Numer. Math. **128**(1), 73–101 (2014)

Printed by Printforce, the Netherlands